Ceramics: Properties, Processing and Applications

Ceramics: Properties, Processing and Applications

Editor: Carl Burt

MURPHY & MOORE
www.murphy-moorepublishing.com

Published by Murphy & Moore Publishing
1 Rockefeller Plaza,
New York City, NY 10020, USA
www.murphy-moorepublishing.com

Ceramics: Properties, Processing and Applications
Edited by Carl Burt

International Standard Book Number: 978-1-63987-099-8 (Hardback)

Cataloging-in-Publication Data

Ceramics : properties, processing and applications / edited by Carl Burt.
p. cm.
Includes bibliographical references and index.
ISBN 978-1-63987-099-8
1. Ceramics. 2. Ceramic materials. 3. Ceramic engineering. I. Burt, Carl.
TP810.5 .C47 2022
666--dc23

Contents

Preface

Over the recent decade, advancements and applications have progressed exponentially. This has led to the increased interest in this field and projects are being conducted to enhance knowledge. The main objective of this book is to present some of the critical challenges and provide insights into possible solutions. This book will answer the varied questions that arise in the field and also provide an increased scope for furthering studies.

Ceramics are corrosion resistant and heat resistant materials which are formed by treating non-metallic minerals such as clay at high temperatures. Some examples of ceramics are porcelain, earthenware and brick. They are hard and brittle in nature and generally exhibit a semi-crystalline structure. Depending upon the treatment, they can be either vitrified or non-vitrified. Some general properties of ceramics are poor electrical conductivity, high melting temperature, chemical resistance and high moduli of elasticity. They find extensive application across various fields such as dentistry, aerospace and automobile industry. This book covers in detail some existent theories and innovative concepts revolving around properties, processing and applications of ceramics. It is a valuable compilation of topics, ranging from the basic to the most complex advancements in the field of ceramics. This book will serve as a valuable source of reference for graduate and post graduate students.

I hope that this book, with its visionary approach, will be a valuable addition and will promote interest among readers. Each of the authors has provided their extraordinary competence in their specific fields by providing different perspectives as they come from diverse nations and regions. I thank them for their contributions.

Editor

1

Non-Contact Measurements of the Apparent Density of Green Ceramics with Complex Shape

G.M. Revel, E.P. Tomasini, G. Pandarese and A. Cavuto
Università Politecnica delle Marche, Dipartimento di Meccanica, Ancona Italia

1. Introduction

The apparent density (or, as often reported in the technical literature, the bulk density) of green ceramic tiles is a fundamental parameter for the quality of final products. In fact, the apparent density determines the entity of the dimensional shrinkage of the ceramic body during firing and is proportional to the mechanical resistance.

In industrial production an irregular shrinkage can be responsible for example for non-planar tiles or residual strengths. Measuring the apparent density allows to control this parameter to limit the production waste and increase the quality of the final product. The apparent density, in the ceramic industries, can be currently measured off-line [Standard UNI EN 1097 - 3, (1999)], and the most used method is generally the hydrostatic weighting in a mercury bath on a small sample [Pei at al, (1999)]. This technique has a limit on the control sample and a continuous mercury use is dangerous for the human health. This, together with ISO 14000 Standard application, poses severe limitations to the use of mercury in Europe, in the United States and other countries. So new methods have been developed to measure the bulk density of ceramic tiles. Some are based on the gas laws utilizing air pressure to determine test specimen volume in a measuring chamber [Dietrich, (1999)], whilst more recent ones are based on X-ray absorption [Amorós at al, (2010)]. The first is a destructive and discrete method, so it is not suitable for on-line application. The second is accurate and non-destructive, but can be expensive and X-ray devices still have some acceptance problems in industry.

In this chapter an innovative method for non-contact measurement of the apparent density of green ceramics is presented, which is also suitable to be used during production. This method, previously developed by the authors, is based on ultrasonic wave propagation within the material. The waves are generated and received by dedicated air-coupled probes. The time of flight is measured in transmission mode on the tile together with the thickness. The thickness can be derived by the ultrasonic signals or, if higher accuracy is required on complex shapes, by a laser triangulation sensor. The velocity of the ultrasound wave can be calculated by these measurements and it is proportional to the apparent density. The conversion factor between velocity and apparent density is evaluated by a calibration procedure with a reference method of known uncertainty (e.g. mercury bath).

The chapter not only summarizes the research steps performed in this field in the last 10 years for measurements both at the laboratory and industrial level, but also presents a new challenging application to pieces with complex shapes.

2. Bulk density and ultrasonic propagation

The ultrasound propagation is the physical phenomenon on which the theory for measurement of the apparent density in green ceramics is founded. A probe generates a ultrasound wave that propagates in the medium with a characteristic velocity so the time of flight, given a certain length of the propagation path, can be measured.
In a solid material longitudinal and transversal waves can be propagated. In an isotropic material, the relationships between longitudinal (oscillation along the propagation direction) waves velocity v and transversal (oscillation orthogonal to the propagation direction) waves velocity v_t with the elasticity properties of the material are given by [Krautkramer at al, (1983)]:

$$v = \sqrt{\frac{E}{\rho} \cdot \frac{(1-\mu)}{(1+\mu)\cdot(1-2\mu)}} \tag{1}$$

$$v_t = \sqrt{\frac{E}{\rho} \cdot \frac{1}{2(1+\mu)}} = \sqrt{\frac{G}{\rho}} \tag{2}$$

where E is the Young's modulus, G is the elasticity tangential modulus, μ is the Poisson coefficient and ρ is the bulk density of the material. For a porous body, like a green tile (isotropic in first approximation, as obtained by powder compaction), there are many theoretical models linking the elastic properties with porosity and consequently with the apparent density [Wachtman, (1996)]. Commonly for ceramic bodies, the empiric Spriggs model is used [Spriggs, (1961)], which states that Young's and elasticity tangential modules exponentially decrease with porosity p. Porosity can be defined as:

$$p = \frac{\rho_s - \rho}{\rho_s} = \frac{V_a - V_s}{V_a} \tag{3}$$

where V_a is the apparent (bulk) volume of the body gross open and closed pores, considered as the total volume in the macroscopic external surface of the body, i.e. in its envelope and V_s is the effective volume of the body net of the closed and open pores; ρ_s is the average actual density. So the Spriggs model can be written as:

$$E = E_0 e^{-bp} \tag{4}$$

$$G = G_0 e^{-bp} \tag{5}$$

where E_0 and G_0 are the values at null porosity (p=0, $\rho=\rho_s$), and b is a parameter to be experimentally determined, which can differ by some percentage between equations (4) and (5), but it is normally b= 4,0 – 4,5 when p= 0 – 40%.
In this approach only the longitudinal waves velocity is considered, because the v_t is not easily measurable in the industrial application. This is supported by the assumption that the Poisson coefficient is invariant with the apparent density [Revel, (2007)]. The longitudinal waves velocity, like those propagated by a non contact ultrasound probe, are measured

orthogonally to the sample surface and the relationship linking the velocity v and the apparent density ρ is obtained by *(Eq. 1), (Eq. 3)* and *(Eq. 4):*

$$v = \sqrt{\frac{E_0 e^{-b\cdot\left(1-\frac{\rho}{\rho_s}\right)}}{\rho} \cdot \frac{(1-\mu)}{(1+\mu)\cdot(1-2\mu)}} \tag{6}$$

In Fig. 1 (a) the function of (Eq. 6) is plotted for typical values (e.g. [Wachtman, (1996)]) for ceramic green tiles (E_0 = 6 GPa, b = 4, μ= 0,2, ρ_s = 2,600 g/cm³) obtained by traditional mix of raw materials (clays, feldspars, etc.) and with a fixed constant humidity content.

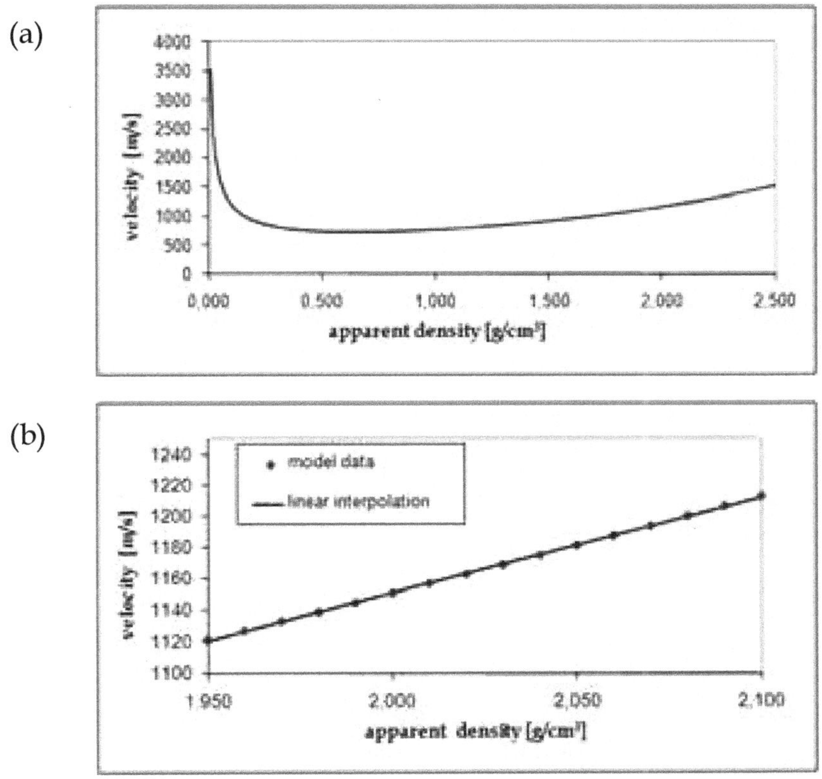

Fig. 1. (a) Plot of the longitudinal waves velocity v in function of the apparent density (E_0 = 6 GPa, b = 4, μ = 0,2, ρ_s = 2,600 g/cm³); (b) Zoom of the diagram in a typical range of industrial interest with relative interpolating straight line

The graphic shows that, increasing the apparent density up to porosity values lower than 60 % (i.e. about ρ > 1,000 g/cm³), the longitudinal waves velocity tends to increase with an almost linear trend. If a typical range of apparent density of industrial interest for green tiles is considered (Fig. 1 (b)), with a good approximation the model can be considered as linear. In fact the discrepancies between the model's data and their linear interpolation is always lower than 2 m/s (i.e. lower than 0,2 %).

In [Revel, (2007)] the results of a parametric analysis of *(Eq. 6)* with the variation of E_0, b and μ was presented. This analysis clearly showed that the uncertainty on the model parameters can generate significant uncertainty on the results of the correlation between velocity and apparent density. On the contrary the discrepancies caused by the linearization seem to be always very low.

Ultrasound measurements are usually performed with contact probes using film of gel, oil or water as coupling media. This set-up is capable of guaranteeing high signal-to-noise ratio, thanks to the reduced attenuation at the interfaces.
In the case of non contact inspection (air-coupled probes), the propagation path is more complex and the attenuation significantly higher. The energy of the signal that passes through the material in this case is very limited, due to the high difference between the acoustic impedance of the air and of the solid material. In fact, if an ultrasonic wave is propagating from a medium with acoustic impedance Z_1 to another medium with acoustic impedance Z_2, the reflected signal energy will be proportional to the difference between Z_1 and Z_2. This is quantified by the coefficients of reflection R and transmission T of the sound pressure, defined as:

$$R = \frac{P_r}{P_i} = \left(\frac{Z_2 - Z_1}{Z_2 + Z_1}\right) \tag{7}$$

$$T = \frac{P_t}{P_i} = \left(\frac{2Z_2}{Z_2 + Z_1}\right) \tag{8}$$

where P_t, P_r and P_i are sound pressures of the reflected, transmitted and incident waves respectively (Fig. 2).

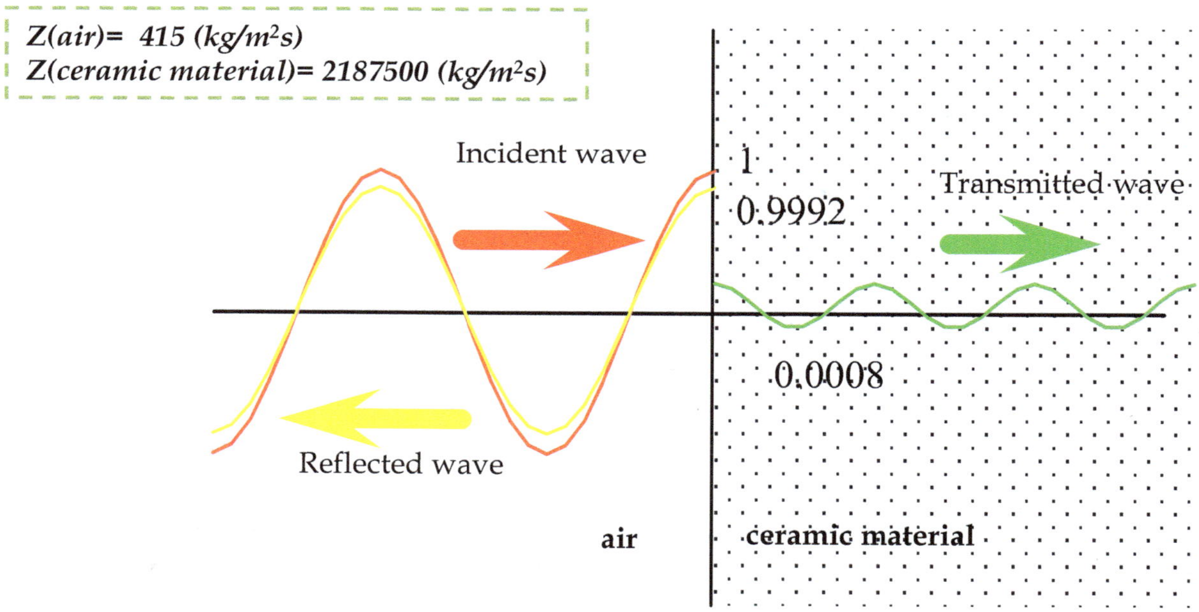

Fig. 2. Scheme of Transmission and Reflection energy: air-ceramic material

In terms of energy, these coefficients become:

$$\bar{R} = \left(\frac{Z_2 - Z_1}{Z_2 + Z_1}\right)^2 \tag{9}$$

$$\bar{T} = \frac{4Z_2 Z_1}{(Z_2 + Z_1)^2} \tag{10}$$

As depicted in Fig. 2, if we consider the energy transmitted at a single interface between air and ceramic, this is in the order of about 0,1 %. If multiple reflections are used for time of flight measurement, the wave goes through the interfaces several times, thus creating a dramatic signal attenuation of several order of magnitudes.

Therefore air-coupled ultrasonic techniques require proper signal conditioning (filtering, averaging, signal enhancement, etc.) and measurement set-up to reach an acceptable signal-to-noise ratio (> 20 dB).

On the other hand, this technique offers several advantages:

- possibility of measurement on moving objects or with moving probes;
- easy set-up of a scanning system;
- no contamination of the samples with coupling medium;
- the total production can be inspected with reduced testing time.

3. The measurement method - laboratory tests

According to velocity theory the propagation velocity is proportional to the apparent density ρ. Once the speed is measured, *(Eq. 6)* could allow to determine ρ, nevertheless it would be also necessary to know E_0, b, μ and ρ_s. Considering that these parameters change by tile typology (body, raw materials, etc.), that their measure is very complex and uncertain and that this uncertainty can reduce the accuracy in the apparent density determination, it is not here suggested to use *(Eq. 6)* to measure ρ, whilst a more practical approach is proposed.

In the proposed measurement method the time of flight of ultrasound waves is firstly measured with non contact probes in transmission configuration through the tile. The scheme of this measurement method is showed in Fig. 3.

The times of flight are measured with cross-correlation algorithms between the excitation signal of the transmission probe and the received signal.

As shown in Fig. 3 (b), t_c is the time of flight of the wave from probe to probe through the tile, while t_m is the propagation time inside the material. A typical cross-correlation signal measured on a green tile is showed in Fig. 3 (c), t_c is represented by the first peak, while the difference in the time axis between the first and second peak represents $2t_m$.

From the time of flight measured in air t_a, if the distance D between probes is known, the propagation velocity can be achieved.

In this first approach the distance among the probes membranes D (= 100 mm in this case) has to be measured using a calliper. It follows that the ultrasound measurement method consists in two steps: (a) measure of the time of flight t_a of the ultrasound wave in air in the environmental condition of use; (b) measure of the time of flight of the ultrasound wave through the tile body t_c (where $t_c < t_a$) where, through the measurement of the arrival time of the second echo, it is possible to know also the time of propagation t_m. These steps are repeated every time a tile passes through the probes during the production.

From the measure in air the propagation velocity of sound in air in the application environmental condition can be derived:

$$v_a = \frac{D}{t_a} \tag{11}$$

It is necessary to often repeat this measure because, as well known, the sound velocity v_a changes in environmental condition mainly in function of temperature and humidity. For example with a temperature of 20 °C and with relative humidity of 50 % there is a speed of 344 m/s.

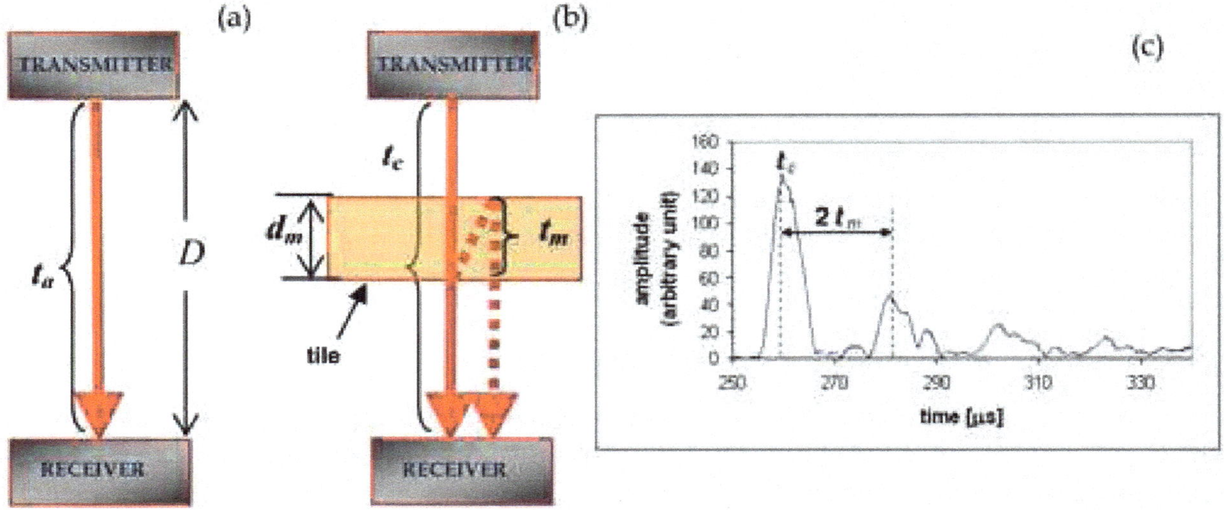

Fig. 3. Scheme of the measurement method: (a) Measurement in air; (b) Measurement on the tile; (c) Example of cross-correlation signal measured on a tile (amplitude in arbitrary unit)

In literature there are many formulas for the sound velocity estimation, for example one can be approximated from [Cramer, (1993)]:

$$v_a(T,\theta) = (331,5 + 0,59 \times T[°C])\left(1 + 0,004\frac{\theta[\%]}{100}\right) \tag{12}$$

where T is the air temperature and θ the relative humidity.
Nevertheless in order to avoid the measurement of temperature and humidity, it is better to compensate this effect by directly measuring v_a, especially in the production line where it is not possible to take under control the environmental conditions.
By the measurement of t_c and t_m the average thickness of the tile in the area of the ultrasonic beam (about 1 cm^2) can be obtained

$$d_m = D - v_a(t_c - t_m) \tag{13}$$

and in the end the propagation velocity of longitudinal waves as:

$$v = \frac{d_m}{t_m} \tag{14}$$

Considering that in the range of interest *Eq. 6* can be considered as linear with an error lower than 0,2 %, it is proposed to determine a linear experimental correlation between velocity and apparent density. This relationship can be obtained by an experimental calibration procedure to be carried out with a reference method of known uncertainty, such as the one based on mercury.

The Fig. 4 provides an example of correlation diagram, which can be later used to determine the apparent density of unknown samples, once measured the propagation velocity.

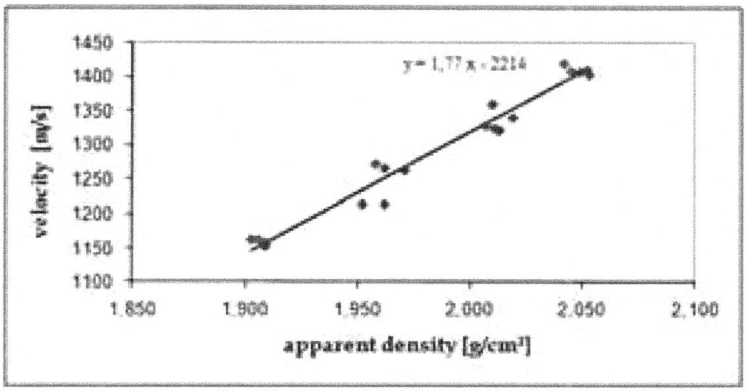

Fig. 4. Correlation diagram (with relative straight fitting line) between longitudinal wave velocity and apparent density measured on green ceramic tiles

In reality also the humidity content has an influence on the propagation velocity, but this effect is considered negligible when measurements are performed on dried tiles or on tiles with constant humidity level. A preliminary attempt to deal with the humidity influence on such ultrasonic measurements has been presented in [Cantavella at al, (2006)]. The problem of moisture influence in on –line measurements has been tackled by the authors in [Revel at al, (2009)] and [Rivola at al, (2007)] and will be discussed in the next chapter.

The experimental apparatus used in the first investigations was based on piezoelectric probes [Bhardwaj, (2002)] with a membrane diameter of 12,5 mm working at around 1 MHz. Air coupled piezoelectric probes have usually a dedicated layer of low impedance material (porous plastic, pressed fibres, polymers, etc. depending on the specific patent) placed between the piezoelectric crystal and the air, in order to reduce the impedance difference between emitting element and air.

Green ceramic tiles having flat faces on both sides, produced in laboratory, have to be used as calibration samples. The planarity of the sides allows to have a higher *SNR* because the grid in the rear side of the tile represents a modifying input for the ultrasound measurement, caused by the effects of diffraction and thickness variations. In addition, the samples are dried in order to avoid any effect due to moisture. For the calibration of Fig. 4 , 20 samples of known density (5 samples for 4 different pressing levels) were used. The propagation velocity was measured in a point of each sample with 200 averages on the signal. It is worth to underline that the determined parameters are valid only for ceramic green tiles with body and raw materials used for the calibration samples. In industrial field, it is necessary to repeat the calibration with every change of body. Examples of correlation diagrams achieved for 3 different kinds of clay and body are reported in Fig. 5, showing the possibility of generalizing the method.

Therefore measurement of the apparent density with the proposed method consists of 3 main steps (once fixed and measured the distance *D* between the probes):

Step 1. times of flight measurement.

Step 2. estimation of the longitudinal waves velocity through.
Step 3. experimental correlation between velocity and apparent density through calibration.

In each step there are uncertainty components which could be generated and propagated up to the final results. In order to optimize the measurement process, it is necessary to determine the weight of each uncertainty source on the final results. This has been addressed in detail in [Revel, (2007)] where it is shown that the method guarantees an uncertainty below ± 0,009 g/cm^3.

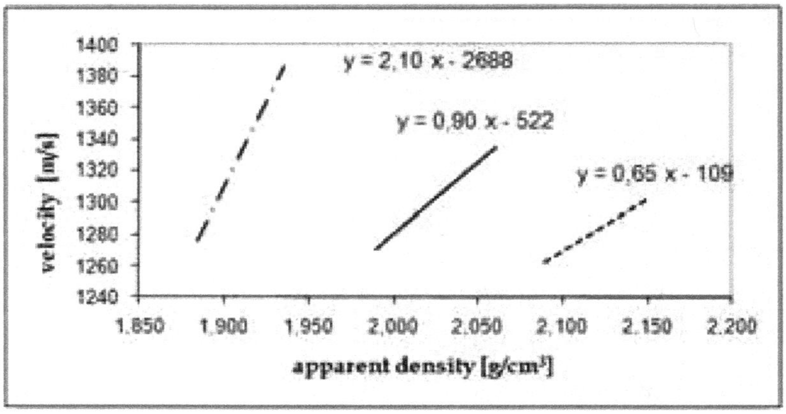

Fig. 5. Examples of correlation diagrams achieved on green tiles with different raw materials and body

4. Industrial application

After some years of research in the EU funded project SENSOCER techniques based on non-contact ultrasonic probes have been developed to properly measure the bulk density of ceramic tiles directly on the line after pressing. These systems are based on air-coupled electro-capacitive transducers and laser triangulation sensors for the detection of the thickness. They have recently demonstrated enough accuracy for industrial applications (repeatability below 0,01 g/cm^3) [Revel at al, (2009)] and are being now exploited in an industrialisation phase, after application of an international patent [Rivola at al, (2007)]. The measurement apparatus is conceived to monitor both the average production value and the spatial distribution of the bulk density and takes into account the effect of moisture content on the measurement accuracy.

The on-line system is a development of the basic one described in the previous Par. 3 for laboratory tests [Revel, (2007)], upgraded for further improvement of accuracy and flexibility. It is constituted by electro-capacitive transducers and a couple of laser triangulation sensors. The electro-capacitive transducers, specifically developed for the application, are driven by a Metalscan ultrasonic board and have an active area diameter of 20 mm and a working frequency of 190 kHz. They have been chosen for this application because of their high sensitivity, in particular in the low frequency range. They are set in a through-transmission configuration for measuring the time of flight of the ultrasonic signal through the tile, while the laser triangulation systems, M7L/10-B24 produced by MEL, measure the tile thickness. Thanks to these devices, it is possible to evaluate the ultrasound velocity v in the green ceramic tiles, using equation 14.

The main difference with respect to the method in Par. 3 is that the thickness is independently measured (by the laser probes) and it is not derived using the same ultrasonic signals. From one side, this makes the system more expensive and complicated, but on the other hand the accuracy is improved. Therefore, if the basic system was already able to guarantee an uncertainty in the order of ± 0,009-0,01 g/cm^3, here the performances are better (uncertainty in the order of ± 0,006-0,007 g/cm^3, having an accuracy on the measurement of the thickness of ± 0,02 mm). In addition, since the thickness has not to be measured by ultrasounds, a lower frequency can be used (190 kHz), allowing lower attenuation in air and thus higher *signal-to-noise* ration (30 – 34 dB in this work with 200 averages for each acquisition). This solution showed to be the best one also for the application on complex shapes as will be reported in next Par. 5.

In the production line there is very few time for the measurement and for a good calibration it is important to measure the real produced tiles. So the linear correlation between apparent density and velocity, at the production moisture level, has been estimated using samples taken directly from the line at 3 different density levels (achieved with 3 different pressure levels varying about 70 bar above and below the normal press operating conditions, of 444 bar).

Concerning the effect of the moisture on the ultrasound measurements, it is known that moisture decreases the propagation velocity [Cantavella at al, (2006)], so also the influence of this parameter was evaluated to guarantee the required uncertainty (max ± 0,009-0,01 g/cm^3).

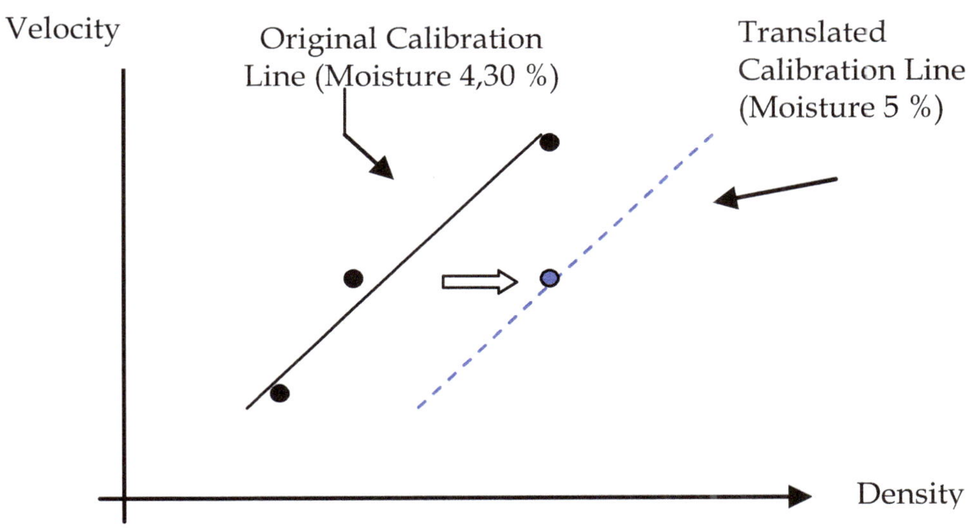

Fig. 6. Adjustment of the calibration line to different moisture values

An extensive sensitivity analysis performed during the SENSOCER project demonstrated that, on average, humidity variations higher than 0,3 % can cause a deviation in apparent density measurements in the order of 0,005 g/cm^3. To deal with this problem, two methods have been implemented and tested. In the first one, the results of continuous humidity measurements using an IR probe serve as a correction factor when evaluating the density readings. Long on-line observations showed, however, that in many cases the variations are lower than 0,3 % for many hours. Thus a second method, simpler and cheaper but less accurate, has been introduced. Here, the moisture content is manually measured by an operator on the line (e.g. every hour, as standard practice already is in industry) and, when the moisture level varies more than 0,3 %, a new calibration point has to be measured by the mercury reference method and used to rigidly translate the calibration line. In fact, it has been demonstrated that, for moisture variations in the order of 0,5–1 %, the slope of the calibration line does not significantly change (effects in the order of 0,001 g/cm^3).

A last important point of the on-line measurement system is that the probes have been mounted on an automatic scanning system. This system is capable of translating together the ultrasonic and the laser probes along the direction perpendicular to the tile motion. This system, coupled with an additional control device suitable to stop the tiles in different positions along the line, allows to scan the apparent density values over different areas of the produced tiles. An average trend of the density spatial distribution can be thus finally achieved and used to predict and correct the differential shrinkage in the different zones.

4.1 Results of on-line test

A schematic presentation of the installation in the on-line ceramic factory (for details see [Revel at al, (2009)] is shown in Fig. 7 (a). The system is located between the press and the dryer, taking up less than 1 m of the production line. Both the signals coming from the ultrasonic device and the laser triangulation systems are acquired and processed with an industrial PC and visualized in real-time together with the current density value.

Fig. 8 (a) shows the moving head, holding both the non-contact ultrasonic transducers and the laser triangulation sensors. A general view of the scanning system with remote PLC control can be seen in Fig. 8 (b). This positioning system, perfectly integrated with the production line process, has been specially developed as a reference system for the test: the direction of the product flow is labeled as Y axis, while the X axis is the perpendicular one.

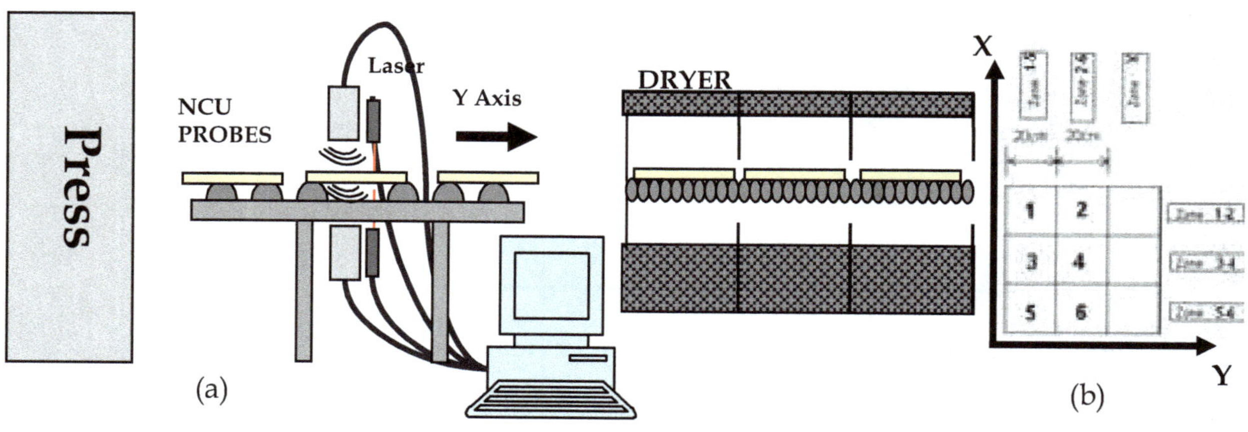

Fig. 7. (a) Schematic drawing of the on-line installation; (b) Reference system for the inspected tile zones

The measurement procedure is as follows: one measurement is performed per tile, each time in a different position, until all the defined areas are covered. Then a new scanning cycle is started. In practice, if 9 areas have to be measured, 9 tiles are required to pass along the line. In this way, a large data set is obtained from which it is possible to derive: a) information on the average density of the products; b) information on the average spatial density distribution.

(a) (b)

Fig. 8. (a) Head with the sensor holder; (b) Support for on-line positioning

Averaging (both the mean value and spatial distribution) ensures that corrective action on the press is taken only in the event of systematic problems (e.g. wear of a die, non-uniformity in powder loading, significant variations in powder moisture, etc.) and not due to local or instantaneous deviations of the measured values.

The results obtained from measurements on a 60 cm × 60 cm format with a regular rectangular grid on the back are shown. The tile surface is divided into 9 square areas of 20 cm × 20 cm. For the sake of better clarity, the combinations of the areas along the X or Y axis are defined as zones and marked with the number of the first and last included areas (Fig. 7 (b)). For these tests, only 6 areas have been monitored, neglecting the last zone of the tile (zone X), thus reducing the necessary production controller modifications to a minimum.

In order to simulate a production variation, the pressure was increased by 60 bar.

The interval where the pressure has been varied (from tile 24 to tile 40) was detected by the system in its whole duration, with an increase in the average apparent density of about 0,016 g/cm^3. Some few samples have been taken from the line and measured with the mercury-based method, which revealed a maximum deviation of about 0,005 g/cm^3. In order to confirm the possibility of predicting variations in the tile dimensions (calibre) after firing, i.e. dimensional shrinkage, each single measured tile was marked and its dimensions after firing were measured.

Fig. 9 (b) illustrates the calibre trend measured along Zone 5-6 for the same tiles, whose apparent density measurements on the Area 5 are plotted in Fig. 9 (a). The trend clearly confirms the behaviour expected from the ultrasonic measurements: the increase in pressure generates higher density and, consequently, larger tile dimensions. In addition, it is possible to note that in both the apparent density and the calibre measurements there is a correlated drift: in fact, their average values at the beginning and at the end of the interval considered seem to increase, probably due to a variation in the moisture content.

The Fig. 9 (a) shows also the relative moving average used to highlight the trends on the single investigated area (with a delay of some samples, due to the algorithm itself). This drift is thus well predicted by the proposed method.

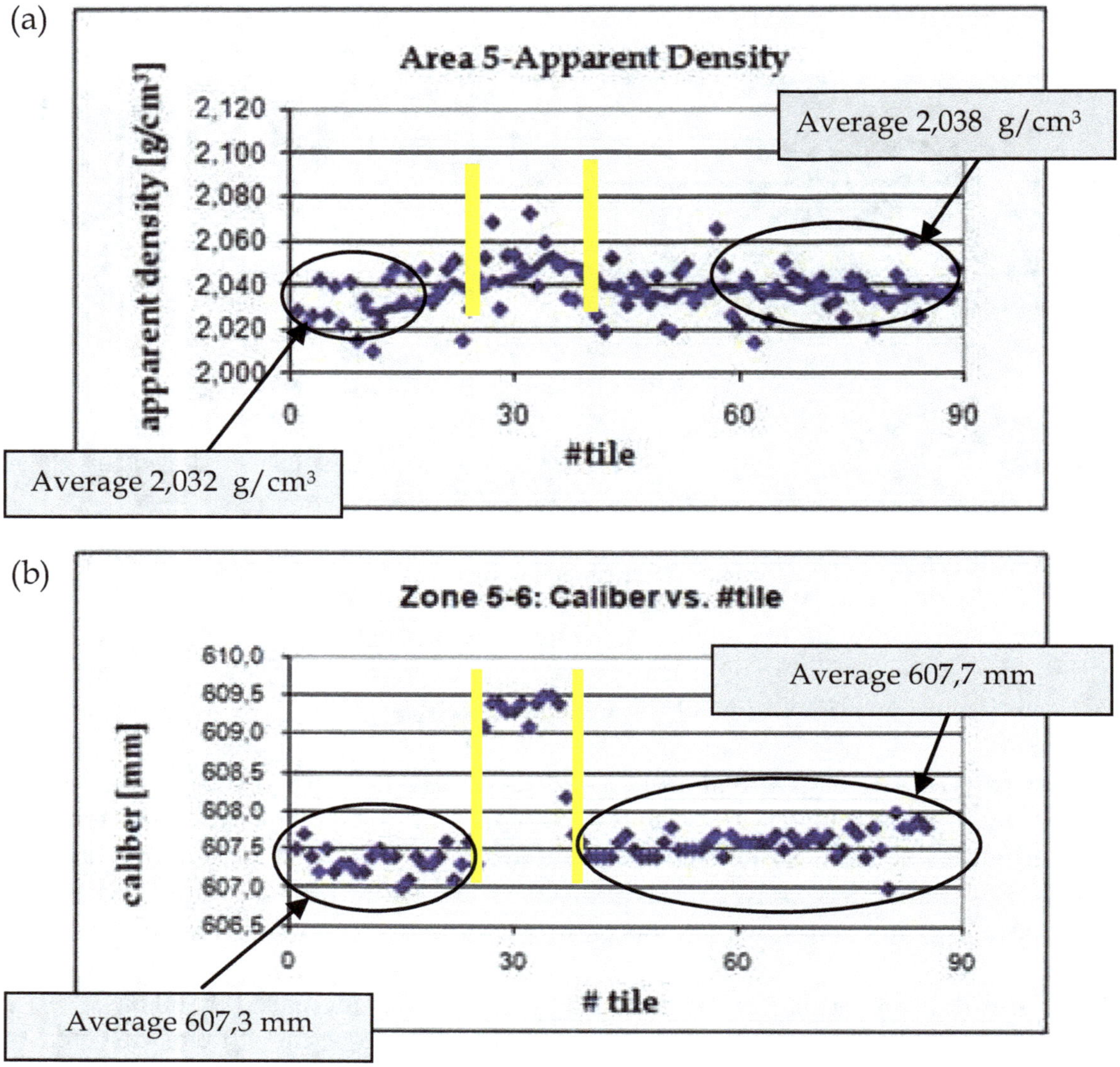

Fig. 9. (a) On-line apparent density measurement for Area 5 in consecutive tiles: the continuous lines represent the interpolation curve obtained with a moving average evaluated on 10 samples; (b) Tile dimension (caliber) trend along Zone 5–6 (measured after firing)

5. Advanced application on complex shapes

An additional new goal for the developed non-contact measurement techniques is to inspect also ceramic bodies with complex shape. In tile industry this is becoming of wide interest, due to potential irregular compaction achieved by new production processes.

Here therefore a feasibility study of the method applicability on tiles with high thickness variation and irregular superficial profile (see Fig. 10) is approached. The test has been focused on the central part of the tile, which presents relevant thickness variations higher than a normal tile. The monitoring of the central part is an important achievement, as this is the part with lower strength section: a strong density gradient here is reflected in dimensional and angular problems also in the other 3D parts.

This study was performed in laboratory using the low-frequency electro-capacitive ultrasonic probes (working frequency 190 kHz) interfaced with a dedicated acquisition board, plugged into a PCI slot of a personal computer.

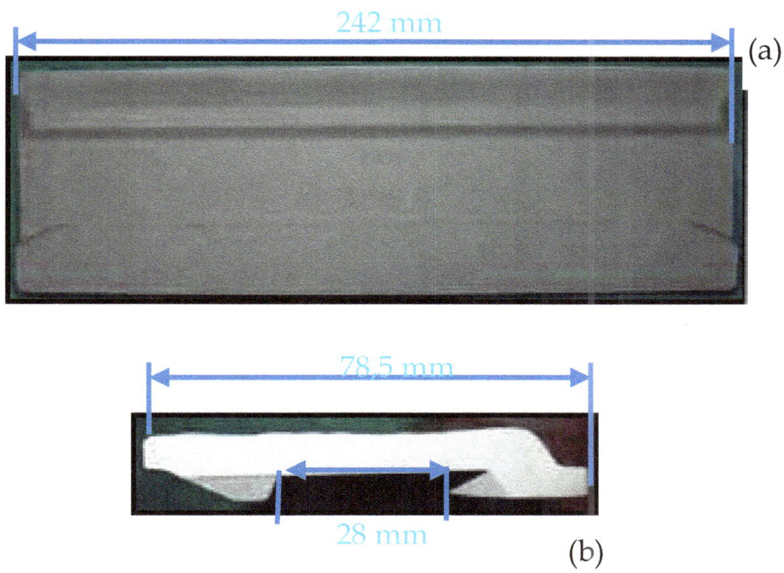

Fig. 10. Tile: (a) Bottom view; (b) Lateral view

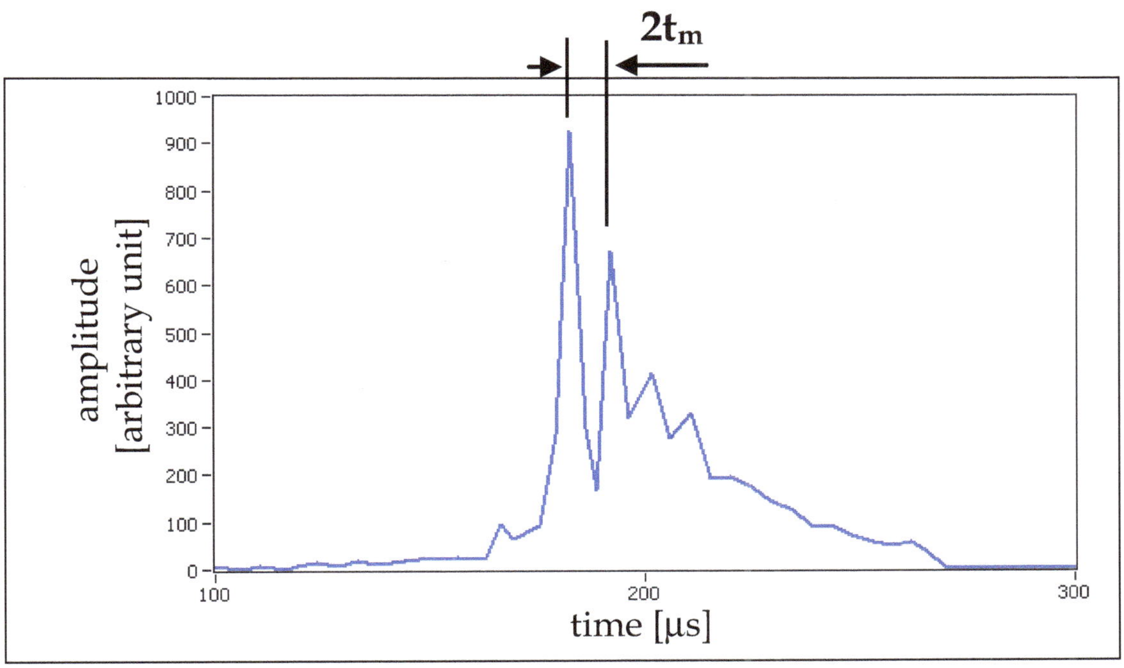

Fig. 11. Waveform of cross-correlation between signal in air and signal through the material

Following the developed procedure, firstly the time of flight t_a of the ultrasound wave in air in the environmental condition of use is measured by acquisition with no samples between the probes. Then the time of flight in the material t_m is measured with cross-correlation algorithms between the signal in air without sample and the signal with the sample between the probes (300 averages for each acquisition). An example of cross-correlation signal is shown in Fig. 11, demonstrating the good quality of the measurements. The thickness was independently measured with two laser triangulation sensors (Laser Mel M5L, resolution 6 µm and laser Keyence LC-2100, resolution 0,5 µm).

The system was calibrated with four laboratory planar samples at know density (referred to mercury bath measurements performed by the tile manufacturer). The samples had different compaction and were manufactured using same batch composition of the production tiles. For each calibration sample, time of flight and thickness were measured. Fig. 12 shows the achieved calibration line, i.e. the relationship between velocity and bulk density.

After calibration, the system was applied on the real tiles firstly in laboratory conditions and then in simulated on-line tests.

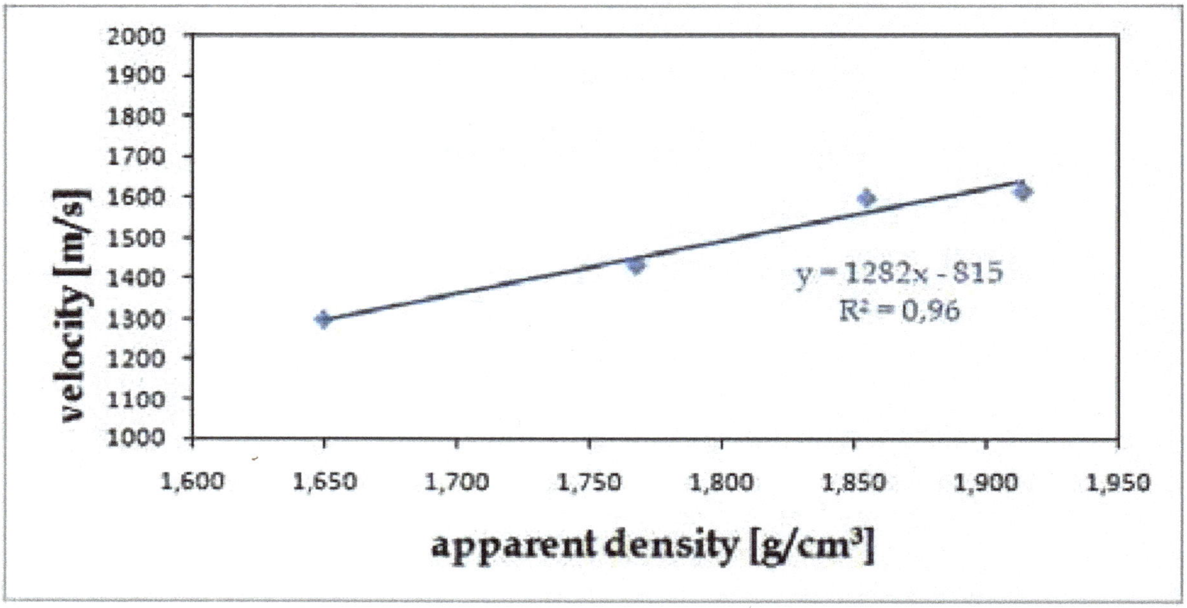

Fig. 12. Calibration line

5.1 Test in laboratory conditions

To verify the calibration line and the measurement system on the real tile with complex shape, tests were performed on the central portion of the tile using a manual planar scanning system for transducers. The analyzed portion was divided in 3 zones (Fig. 13 (a)), each having dimensions of 24x32 mm, thus comparable with the diameter of the ultrasonic transducer. In each area the propagation velocity was measured using an average thickness for each of them. The average thickness was evaluated by measuring 5 points taken randomly within each area.

Using the calibration line shown in Fig. 12, once velocity is measured, a density value can be estimated for each zone. In order to have a reference for the achieved values, the 3 areas have been cut and for each of them bulk density was measured with hydrostatic buoyancy method with the samples sealed by wax. This method has a lower accuracy (about ± 0,01 g/cm³) with respect to the mercury bath (about ± 0,002 g/cm³), but it was the only one available in our lab. Comparison results are reported in Table 1: it is shown that the growing density trend is well detected, even if the thickness variations are relevant, with an average discrepancy of about 0,024 g/cm³. This discrepancy is slightly higher than expected, but this can be also due to the lower accuracy of the hydrostatic buoyancy method used as reference.

(a) (b)

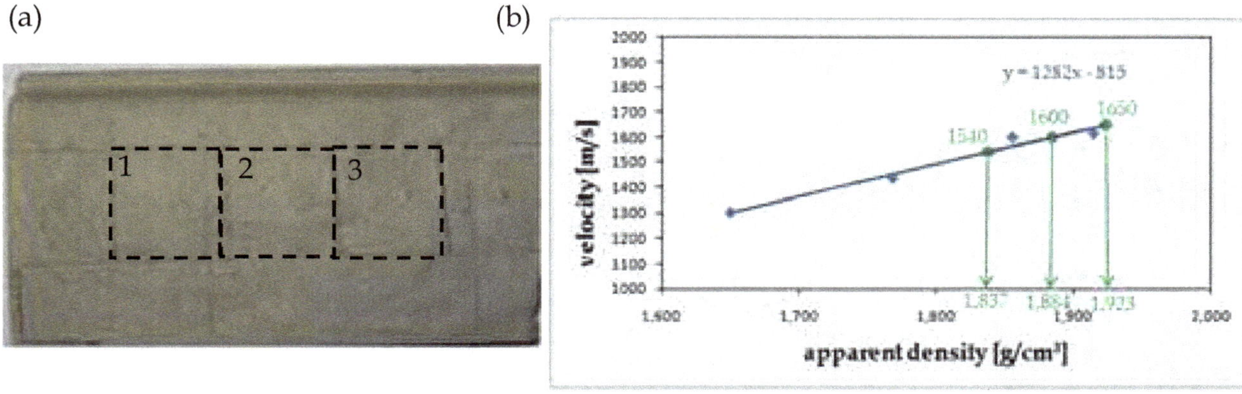

Fig. 13. (a) The 3 investigated areas; (b) Bulk density measured by calibration line (green values)

Zone	Average Thickness (mm)	Velocity (m/s)	Apparent density Ultrasonic method/buoyancy method (g/cm³)	Δ (g/cm³)	Δ (%)
1	8,74	1540	1,837/1,807	0,030	1,7
2	8,52	1600	1,884/1,856	0,028	1,5
3	7,85	1650	1,923/1,908	0,015	0,8

Table 1. Measurement results on the 3 considered zones

5.2 Tests in simulated on-line conditions

In order to simulate on-line measurements, an automatic planar scanning system was used to move the ultrasonic and laser triangulation transducers with a 1 mm step while the tile was at rest (Fig. 14). So in this case the probes are moved, but in a real on-line installation it could be possible to use the tile movement to acquire thickness and density profiles.

Whilst the density measurements in Par. 5.1 are related to an area of 24x32 mm, here the density profile is measured along a longitudinal section of the tile using a now approach. Fig. 15 shows the measured thickness profile and morphological reconstruction of the tile.

The tile thickness profile seems to be quite irregular with maximum deviations up to 3 mm. This probably can also lead to expect bulk density and compaction variations along the profile. There construction of the propagation velocity profile was made along the same longitudinal section in the middle of the tile, where the time of flight was measured. The edges with variable thickness were not considered, as it was assessed that here ultrasound beam propagation is complex and depends on the size of the probes, incidence angle and scattering phenomena.

Fig. 14. The scanning system with ultrasonic and laser triangulation transducers

It is worth noting that, whilst thickness measurements by triangulation laser systems are punctual, the times of flight are measured in an area as large as the beam size (about 20 mm). So, even if also the ultrasonic probes are moved by the scanning system at steps of 1 mm, there is the need to define which thickness value has to be associated to the times of flight measured at each different step. It is clear that it has no sense to associate to the measurements performed at one position the single thickness value measured by the laser on the same ultrasonic probe axis position.

As a consequence, it has been decided that, in order to take into account an average thickness within the beam area, a moving average algorithm was adopted: the idea is that, from one position to the next one, the average thickness value is updated discarding the first value (now outside the beam area) and considering the new last value (just entered in the beam area).

Fig. 16 shows respectively: the raw data of the thickness profile along the longitudinal section indicated in Fig. 15 (a) and the relative moving average; the times of flight along the same section and, finally, the velocity/density along the longitudinal section calculated from the "moving average" thickness. As an example of the adopted procedure the first time of flight value is measured with the ultrasonic probe axis placed in x= 10 mm. At this value, the moving average thickness value computed with the first 20 sample (i.e. x=20 mm in Fig. 16 (a)) is associated to estimate velocity and density. The use of the moving average allows to achieve a sort of continuous bulk density profile (with 1 mm spatial resolution), which can be useful to evaluate undesired trends. It is clear that, from one point to the other, many information are overlapping. So, it could also be useful to estimate average values for larger areas (e.g. 20 mm as the beam size): this can be easily done from the measured data and results are reported in the same graph of Fig. 16 (c). On these values also error bars with amplitude proportional to the uncertainty estimated in previous Paragraphs are reported.

Measured density distribution seems to be coherent with the sample values measured in Par. 5.1. In addition it can be noted that thickness and density profiles tends to have inversely proportional trends, which is coherent with the compaction process, thus demonstrating the feasibility of the proposed measurements.

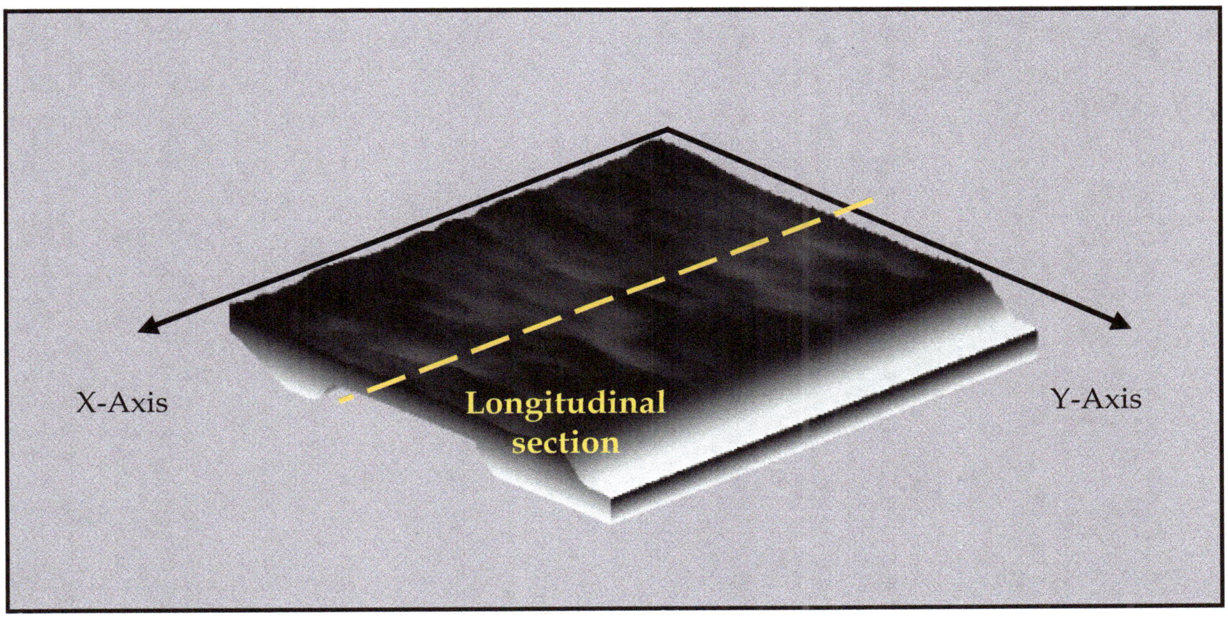

(a)

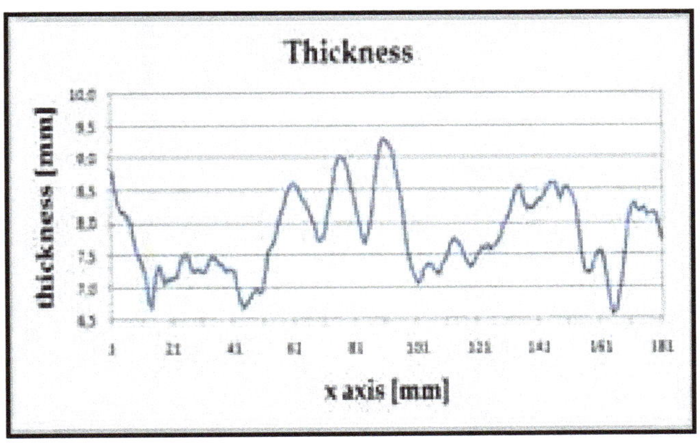

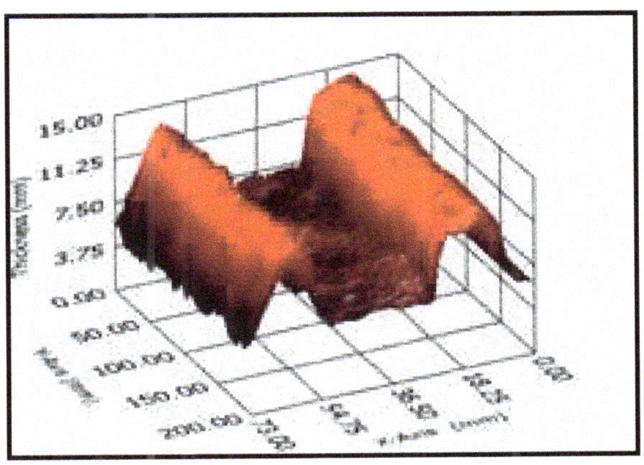

(b) (c)

Fig. 15. (a) Morphological reconstruction of the tile; (b) Thickness profile along the longitudinal section; (c) Tile thickness distribution

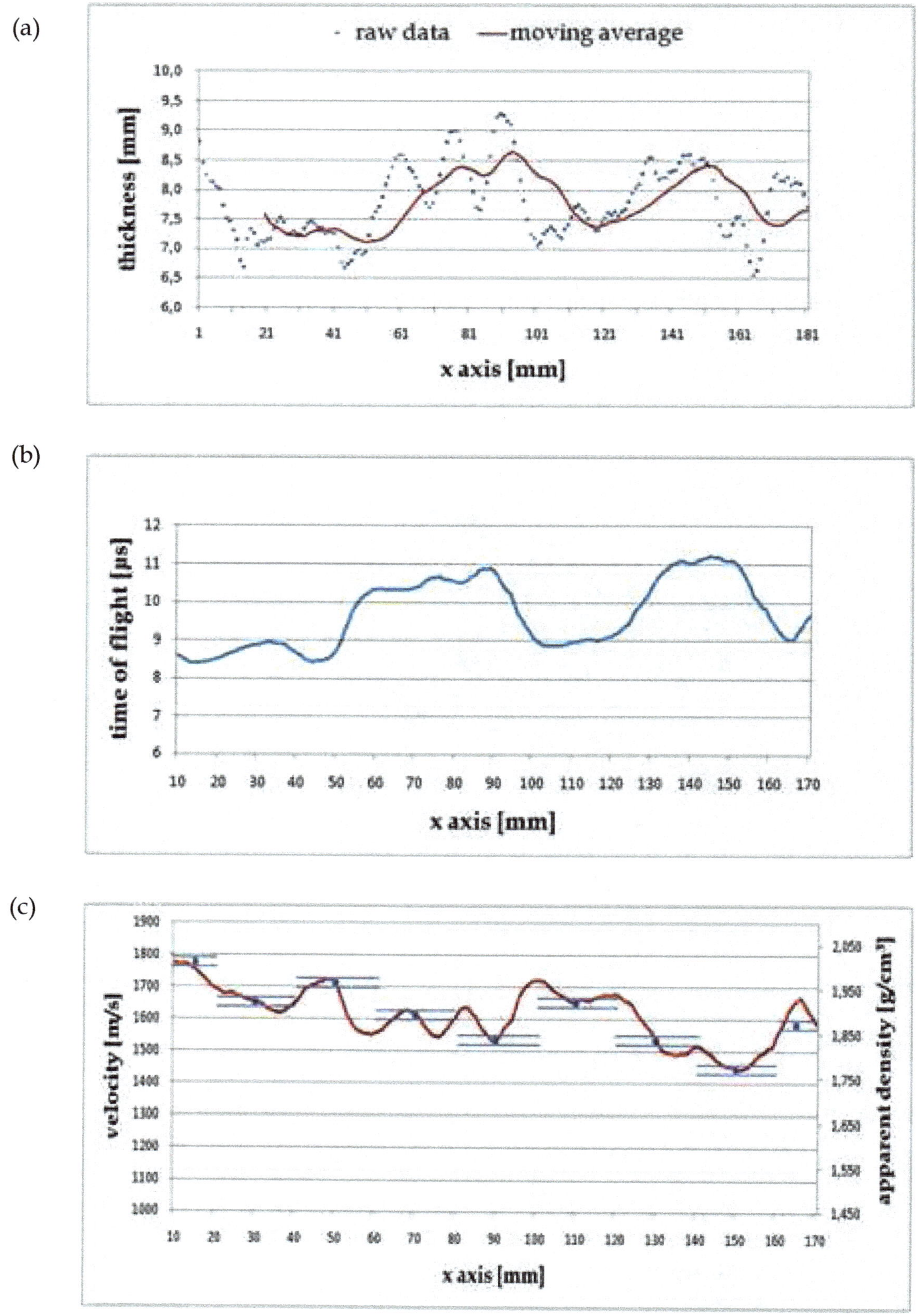

Fig. 16. Diagrams of: (a) Thickness (raw data and moving average); (b) times of flight and (c) velocity/density estimated with "moving average" thickness, with discrete average values computed on 20 mm lengths superimposed

6. Conclusion

This Chapter has shown the complete development of the non-contact ultrasonic method for the laboratory and on-line measurement of apparent density on green ceramic tiles. The research was developed in several years and the outcomes have been protected by an international patent. Both the basics of the method and experimental set-ups for laboratory and on-line applications are described, showing that the system is capable of tracking average production values and spatial distribution of apparent density with the industrially required accuracy (max ± 0,01 g/cm^3).

An original application on green ceramics with complex shape has been also developed and presented. The results showed the feasibility of measuring on parts with irregular thickness variations up to 3 mm. However the method can be further developed to determine the density values even at the edges of the tile where the profile is curve. The main objective will be to implement dedicated post-processing algorithms to identify and isolate the portions of the measurement signals related to the different paths of the ultrasound wave in the tile.

7. Acknowledgments

The research has been partly funded by the EU project SENSOCER. The authors would like to thank the company SACMI for the cooperation in the on-line implementation of the prototype. In addition they would like to thank INAX Corporation for the samples with complex shape.

8. References

Amorós, J.L.; Boix, J.; Llorens, D.; Mallol, G.; Fuentes, I. & Feliu, C. (2010). Non-destructive Measurement Of Bulk Density Distribution In Large-Sized Ceramic Tiles. *Journal of the European Ceramic Society, 30, pp. 2927–2936,* Available online at www.sciencedirect.com, 12 March 2010

Bhardwaj, M.C. (2002). High Transduction Piezoelectric Transducers And Introduction Of Non-Contact Analysis, In*: Chapter of Encyclopedia of Smart Materials*, Wiley, J.A. Harvey, (Ed), New York, ISBN 0-471-17780-6

Cantavella, V.; Llorens, D.; Mezquita, A.; Moltò, C.; Bhardwaj, M. C.; Vilanova, P.; Ferrando, J. & Maldonado-Zagal, S. (2006). Use of ultrasound techniques to measure green tile bulk density and optimise the pressing process, *Proceedings of QUALICER 2006*, Vol. 2, P.BC - 161, ISBN 84-95931-21-4

Cramer, O. (1993). The variation Of The Specific Heat Ratio And The Speed Of Sound In Air With Temperature, Pressure, Humidity, And CO_2 Concentration, *J. Acoust. Soc. Am.*, 93(5) pp. 2510-2616, ISSN 0001-4966

Dietrich, E. (1999) *New Measurement Principle for Determining Bulk Density*, EP 0936451.

Krautkramer, J. & Krautkramer, H. (1983). *Ultrasonic Testing of Materials*, Springer-Verlag, ISBN 3-540-11733-4

Pei, P.; Minor, D. & Onoda, G. (1999). Laboratory Techniques For Bulk Density Measurement, *In: Advances Process Measurements For The Ceramic Industry*, A. Jillavenkatesa and G. Onoda, (Ed). 293-306, Westerville, OH, ISBN 1574980866.

Revel, G.M. (2007). Measurement Of The Apparent Volumic Mass Of Green Ceramic Tiles By A Non-Contact Ultrasonic Method, *Experimental Mechanics* 47 pp. 637–648

Revel, G.M.; Pietroni, P.; Tomasini, E.P.; Pandarese, G. & Cocquio, A. (2009). On-Line Measurement of Green Tiles Apparent Density: Industrial Implementation and Test of an Ultrasonic-Based System, *Tile&Brik International Manual 2009* pp. 42-46. ISSN 0938-9806

Rivola, P.; Cocquio, A.; Tomasini, E.P.; Revel, G.M.; Pietroni, P. & Pandarese, G. (2007). *A Process For Manufacturing Ceramic Tiles,* WO2007093481AL - 2007

Spriggs, R.M. (1961). Expression for Effect of Porosity on Elastic Modolus of Polycrystalline Refractory Materials, particularly Aluminioum Oxide, *J. Am. Ceram. Soc.*, Vol. 45, 94, , ISSN 0002-7820

Standard UNI EN 1097 - 3: (1999), *Determinazione della massa volumica in mucchio e dei vuoti intergranulari,* ICS 91.100.15

Wachtman, J.B. (1996). Mechanical Properties of Ceramics, *John Wiley & Sons Inc.*, ISBN: 0-471-13316-7

2

Strength of a New All-Ceramic Restorative Material "Turkom-Cera" Compared to Two Other Alumina-Based All-Ceramic Systems

Bandar M. A. AL-Makramani[1], Abdul A. A. Razak[2] and Mohamed I. Abu-Hassan[3]

[1]Department of Conservative Dentistry, Faculty of Dentistry, Sana'a University,
[2]Department of Conservative Dentistry, Faculty of Dentistry,
University of Malaya, Kuala Lumpur,
[3]Department of Restorative Dentistry, Faculty of Dentistry,
University Teknology MARA, Selangor,
[1]Yemen
[2,3]Malaysia

1. Introduction

Early types of metal-free ceramics did not enjoy success in dentistry, especially in the posterior region (Shimada et al., 2002). The high crystalline content ceramic systems have been developed in an attempt to improve the strength of metal-free restorations as well as deliver more esthetic results than conventional metal-fused-to-ceramic restorations (Ozcan et al., 2001; Valandro et al., 2006). Glass infiltrated alumina ceramic (eg. In-Ceram Alumina, Vita Zahnfabrik, Bad Sackingen, Germany), densely sintered aluminum oxide ceramic (eg, Procera AllCeram, Nobel Biocare AB, Gothenburg, Sweden) and zirconium oxide ceramic (eg, Lava 3M ESPE, St. Paul, MN, USA) are popular oxide-based high-strength ceramic materials that offer favorable esthetic characteristics, mechanical properties and biocompatibility (Blatz et al., 2004; Della Bona et al., 2007).

It is obvious from the different studies in relation to the occlusal fracture resistance of all-ceramic systems that the values reported are highly variable. This is because the testing of the occlusal fracture resistance of crowns is not a standard procedure like a bending test for a geometrically well-defined bar. As reported by some researchers (Webber et al., 2003; AL-Makramani et al., 2008a; Di Iorio et al., 2008), the results of the fracture load of all-ceramic crowns may be influenced by different factors. These include the microstructure of the ceramic material, preparation design, shape and thickness of the restoration, size and distribution of surface flaws, the magnitude, direction and location of the applied load, luting methods, the elastic modulus of the restoration components and storage conditions before loading to fracture.

Regarding the effect of finish line on the occlusal fracture resistance of all-ceramic crowns, it was shown that preparation with a 1.2 mm shoulder finish line and sharp axiogingival line angle produced the strongest Dicor crowns, while crowns prepared with a chamfer finish line produced the weakest restoration when cemented to metal dies (Doyle et al., 1990a; Doyle et al., 1990b). A similar in-vitro study on Procera AllCeram crowns found that the shoulder finish line showed significantly higher fracture resistance than chamfer finish line (Di Iorio et al., 2008). In contrast, some authors reported that the fracture resistance of Dicor crowns luted with a resin luting cement was unaffected by the type of finish line used (Bernal et al., 1993; Malament and Socransky, 1999).

The present study investigated the effect of three variables on the occlusal fracture resistance of all-ceramic crowns. Attempts were made to standardize the other variables that may have an influence on the results of the fracture load. The studies concerning the effect of ceramic material and margin design on the occlusal fracture resistance of Turkom-Cera restorations are limited. Therefore, the objectives of this study were to:-

1. Determine the occlusal fracture resistance of Turkom-Cera copings compared to In-Ceram and Procera All-Ceram copings cemented to extracted teeth.
2. Examine the effect of finish line design on the occlusal fracture resistance of Turkom-Cera copings.

Null hypotheses

1. There is no difference in the occlusal fracture resistance of Turkom-Cera, In-Ceram and Procera AllCeram copings when cemented to natural teeth.
2. There is no effect of finish line design on the occlusal fracture resistance of Turkom-Cera copings.

2. Materials and methods

2.1 Materials used

40 sound and crack-free maxillary premolar teeth, extracted for orthodontic reasons (the patient's ages ranged from 15-20 years), were used for this study. In addition, three types of all-ceramic systems were used for coping production namely, Turkom-Cera™ (Turkom-Ceramic (M), Puchong, Malaysia), In-Ceram (Vita Zahnfabrik, Bad Sackingen, Germany) and Procera AllCeram (Nobel Biocare, Goteborg, Sweden), and one type of resin luting cement (Panavia-F, Kuraray Medical Inc., Okayama, Japan) with its silane coupling agent (Clearfil Silane Kit, Kuraray), were used in this study.

2.2 Methods

2.2.1 Specimen collection and storage

Based on criteria, the selected teeth were free of cracks and fractures, had no evidence of caries or restorations and had no previous endodontic treatment. The average bucco-lingul, mesio-distal crowns width and teeth length were 9.1 mm, 7.3 mm and 22.3 mm, respectively. The teeth were obtained directly after extraction and stored in 0.5% Chloramine-T trihydrate solution for one week (ISO/TS 11405/2003). Both calculus deposits and residual periodontal tissues were removed by Ultrasonic Scaler (Peizon® Master 400, EMS, Nyon, Switzerland). All teeth were examined under stereo microscope (Olympus SZ61, Olympus Corp., Tokyo, Japan) at 30x to detect cracks before including them in the study. Throughout this study, the teeth were kept hydrated in distilled water as this storage solution does not seem to alter dentine permeability (Goodis et al., 1993). The storage solution was changed every one week and the teeth were stored at 4 degrees Celsius (ISO/TS 11405/2003).

2.2.2 Preparation of teeth

The teeth were embedded in epoxy resin 2.0 mm below the cemento-enamel junction using a plastic mould with 30 mm diameter and 30 mm height (Fig. 1). Two layers of nail polish were applied to the external surface of the entire roots. A dental surveyor was used to position the long axis of the teeth vertically.

Fig. 1. The tooth embedded in the epoxy resin

The preparation of the teeth was carried out using high-speed handpiece attached to a paralleling apparatus (Fig. 2), which allowed standardized preparations. The apparatus consists of a specimen fixture as well as vertical and horizontal arms. The specimen fixture holds the specimen and designed in a way that the fixture can rotate the specimen against a diamond bur (Fig. 3a & b). The vertical arm of the apparatus which holds the handpiece permits vertical as well as rotational movement around the tooth. The high-speed handpiece was attached to the vertical arm of the paralleling apparatus using a custom made jig (Fig. 4). This jig secures the handpiece to the vertical arm in such a manner that the attached bur can be fixed at a set angle to that of the tooth during preparation.

Fig. 2. The paralleling apparatus used

Fig. 3a & b. Two views of the specimen fixture

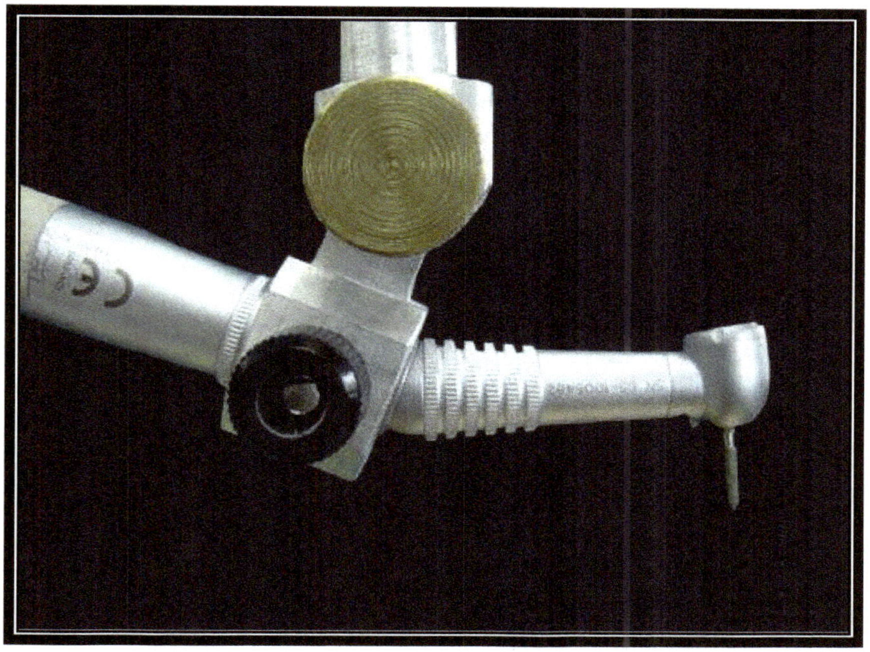

Fig. 4. The jig used to fix the handpiece to the vertical arm of the apparatus

The axial taper angle used in the present study was 6°. According to Shillingburg et al., (1997) a tapered bur will impart an inclination of 2-3 degrees to any surface it cuts if the shank of the instrument is held parallel to the intended path of insertion of the preparation. A specially designed jig which consists of a semi-circular transparent Perspex block with a protractor fixed to its side was used to set the degree of taper (Fig. 5). A hole was drilled along the outer side of the perspex block corresponding to a taper angle of 3°.

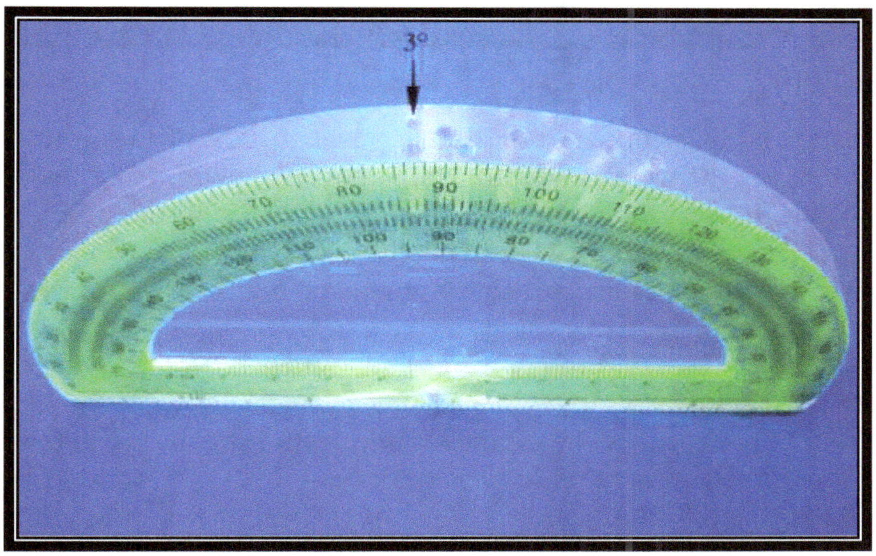

Fig. 5. The jig used to set the degree of taper

Therefore, to achieve a 6° axial taper preparation, the handpiece was secured to the apparatus so that the attached tapered diamond bur was oriented at a 3° angle to the vertical axis of the tooth. This, in additional to the 3° taper of the tapered bur, resulted in a total axial taper angle of 6° corresponding to a convergence angle of 12°.

The axial reduction was performed by rotating the specimen in the fixture against a coarse rotating tapered diamond burs (5856.314.018, chamfer; and 8847KR.314.018, shoulder, Komet GmbH, Lemgo, Germany) as shown in Fig. 6. After that, the preparation surfaces were finished with a fine grit diamond burs (No. 5856.314.018, chamfer; and 8847KR.314.018, shoulder).

Fig. 6. The tooth during axial preparation

After the completion of axial preparation, the occlusal surface of the teeth was cut flat. A pencil was used to mark the prepared tooth 4 mm above the margin and the occlusal surface was flattened with a diamond wheel (Komet No. 909.204.055) to the marked line (Fig. 7a), which resulted in a preparation with 4.0 mm height (Fig. 7b).

The preparation was smoothed and all sharp angles or internal line angles were rounded with a fine abrasive disk (Sof-Lex Discs, 3 M Corp., St. Paul, Minn.) connected to a micromotor handpiece. All preparations were made under copious water irrigation by the same investigator.

These series of reductions resulted in a standardized teeth preparation with 6° axial taper, a 1.2 mm circumferential chamfer/shoulder margin, placed 0.5 mm occlusal to the cemento-enamel junction, and total preparation height of 4.0 mm. The finished preparation is illustrated in Fig. 8a &b.

In summary, 30 premolar teeth were prepared with chamfer margin and divided randomly into 3 groups (n=10) for each of the three ceramic systems used (In-Ceram, Procera and Turkom-Cera). In addition, 10 premolars were prepared with round shoulder margin and used to study the effect of finish line design on the fracture resistance of Turkom-Cera all-ceramic copings.

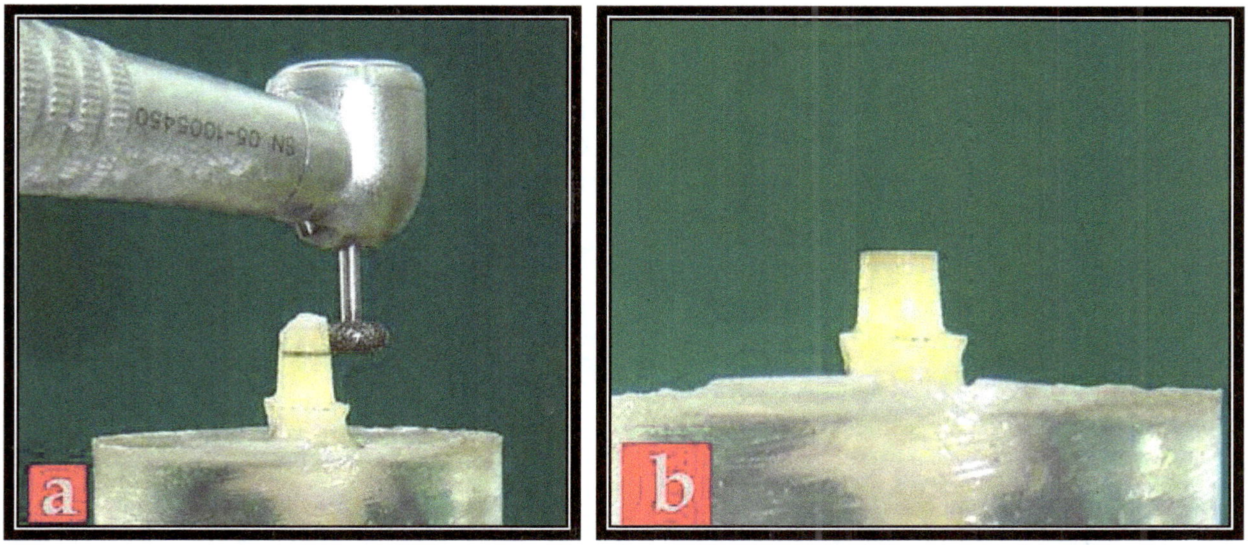

Fig. 7a & b. The tooth before (a) and after (b) oclussal reduction

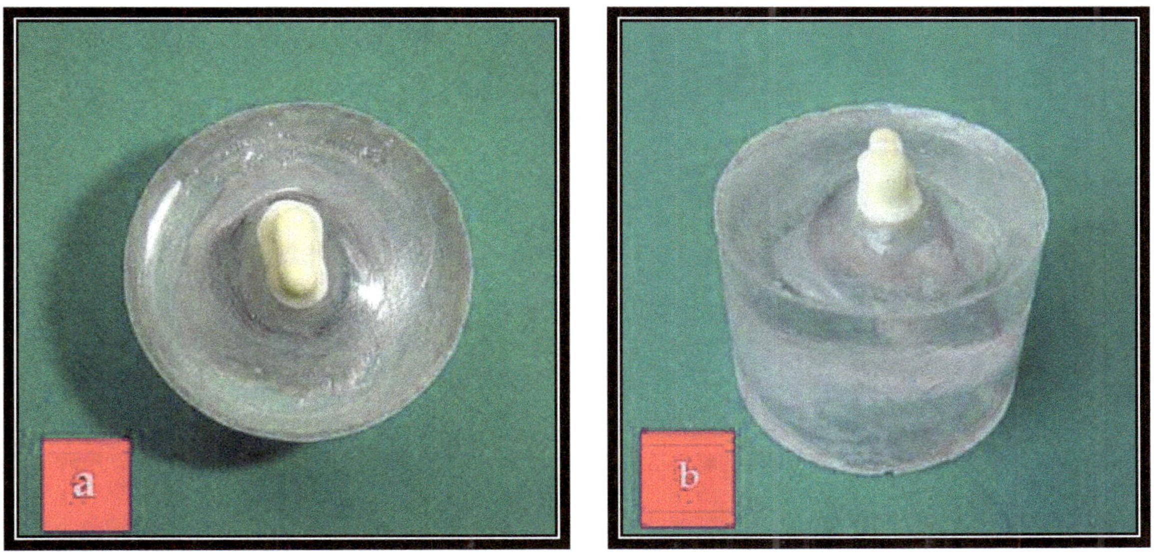

Fig. 8a & b. Occlusal (a) and buccal views (b) of the finished preparation

2.2.3 Impression and die preparation

The teeth were dried with an air/water syringe and impression was taken for each tooth using a bottle cap with pipette which act as an impression tray and silicone impression material (Aquasil Monophase Ultra; Dentsply Caulk, Dentsply International Inc., Milford, Germany) (Fig. 9a & b). The impression material was injected into the cap and placed on the tooth while maintaining finger pressure until setting.

Then, the impressions were boxed (Fig. 10a) using boxing wax (Boxing In Wax, Metrodent Ltd., Huddersfield, Enland). The impressions were then vaporized with a wetting agent and poured in die stone (Densite, Shufo, Kyoto, Japan) (Fig. 10b). The stone was mixed with distilled water in a 20cc liquid to 100 grams of stone ratio as recommended by the manufacturer. After a 4 hours setting time, the dies were trimmed and numbered according to their respective teeth (Fig. 11).

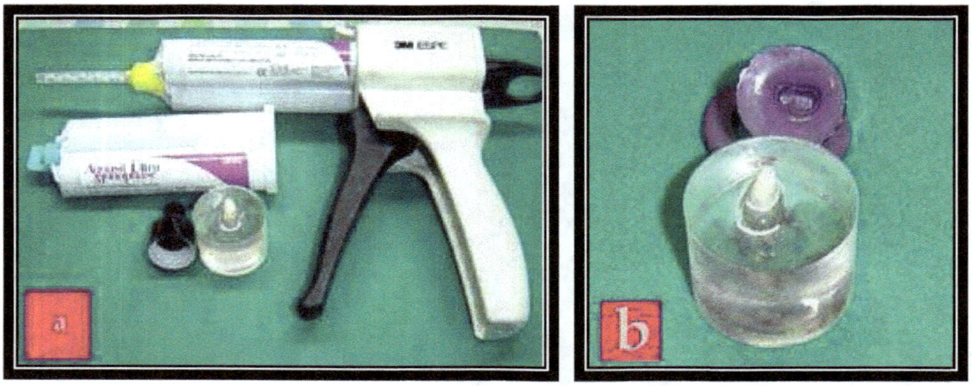

Fig. 9a & b. The impression material, tooth model and plastic cap (a); and impression of the prepared tooth (b)

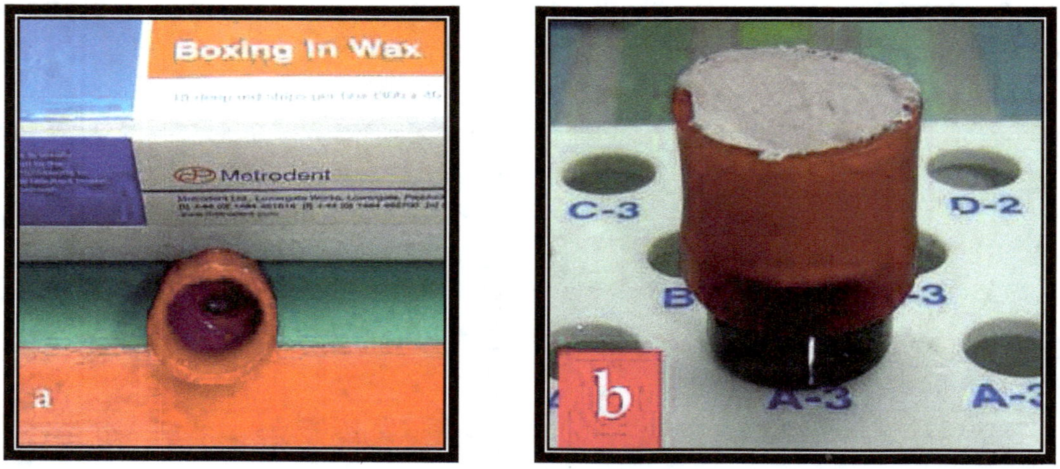

Fig. 10a & b. Boxing of the impression (a) and pouring with die stone (b)

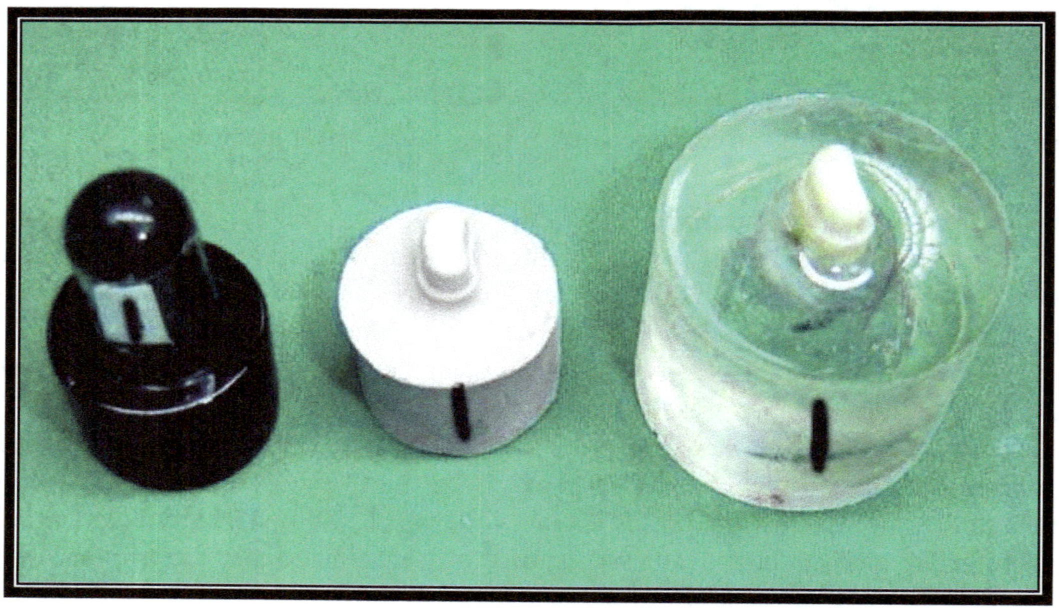

Fig. 11. The die numbered according to its respective tooth

2.2.4 Fabrication of all-ceramic copings

Ten randomly selected stone dies with chamfer margin and the 10 stone dies with shoulder margin were sent to Turkom-Cera dental laboratory and 20 Turkom-Cera copings were fabricated with a thickness of 0.6 mm. The remaining 20 stone dies with chamfer margin were randomly divided into two groups of 10 specimens for In-Ceram (n=10) and Procera (n=10) and sent to other dental laboratories. 10 In-Ceram and 10 Procera copings were fabricated with a thickness of 0.6 mm.

The macroscopic fit of all copings on the corresponding stone dies, and finally on the specimens were visually assessed. Copings that were found to rock or did not seat on the finish line were rejected and refabricated to ensure 10 copings per group.

2.2.5 Cementation

The specimens were identified by their numbers. The teeth were pumiced with a prophylaxis cup mounted on a slow speed handpiece and rinsed with an air/water syringe prior to cementation. The copings were then internally sandblasted with 50 μm aluminium oxide (Al2O3) particles at an air pressure of 2.5 bars for 13s from a distance of 10 mm. After that, the copings were steam cleaned and air dried.

All copings were cemented to their respective teeth using Panavia F resin luting cement. The ED primer was applied to the entire surface of the tooth and allowed to set for 60s before air drying with gentle air flow. The fit surfaces of all copings were silanated with a mixture of Clearfil Porcelain Bond Activator and Clearfil SE Bond Primer. The mixture was applied to the internal surface of the coping and left for 5s before air drying with gentle air flow. Sufficient amount of the Panavia F (one complete turn from each cartridge A &B) were dispensed, mixed for 20s and applied to the internal surface of each coping.

Finger pressure was applied to initially seat each coping on its respective tooth, and each coping was held in place while any excess paste remaining at the margins was removed with a disposable brush. A layer of Oxyguard II (Kuraray) was applied for three minutes around the margins of each specimen. The specimens were then placed in a custom-made vertical loading apparatus (Makramani Load) (AL-Makramani et al., 2008a), for 10 minutes under a 5 kg load. Following cementation, the 30 specimens with chamfer margin and the 10 specimens with shoulder margin were placed in a sealed container of distilled water and left in an incubator at a constant temperature of 37°C for 24 hours.

2.2.6 Testing procedure

The tooth with cemented coping was removed from the storage container, secured in a mounting jig and subjected to testing in a universal testing machine (Shimadzu, Shimadzu Corp., Tokyo, Japan) (Fig. 12). A 3 mm stainless steel bar, mounted on the crosshead of the Shimadzu testing machine was used and applied a compressive load at the centre of the occlusal surface, along the long axis of the cemented copings, at a crosshead speed of 1mm/min until failure occurred (Fig. 13). A piece of tin foil 0.7 mm thick was placed between the loading piston and the specimen to distribute the force over a larger area and to avoid loading stress peaks on the coping surface. The maximum force to produce fracture was recorded in Newtons. The failed copings were examined in order to determine the mode of fracture. The mode of fracture was classified using categories as described by Burke & Watts (1994) (Table 1).

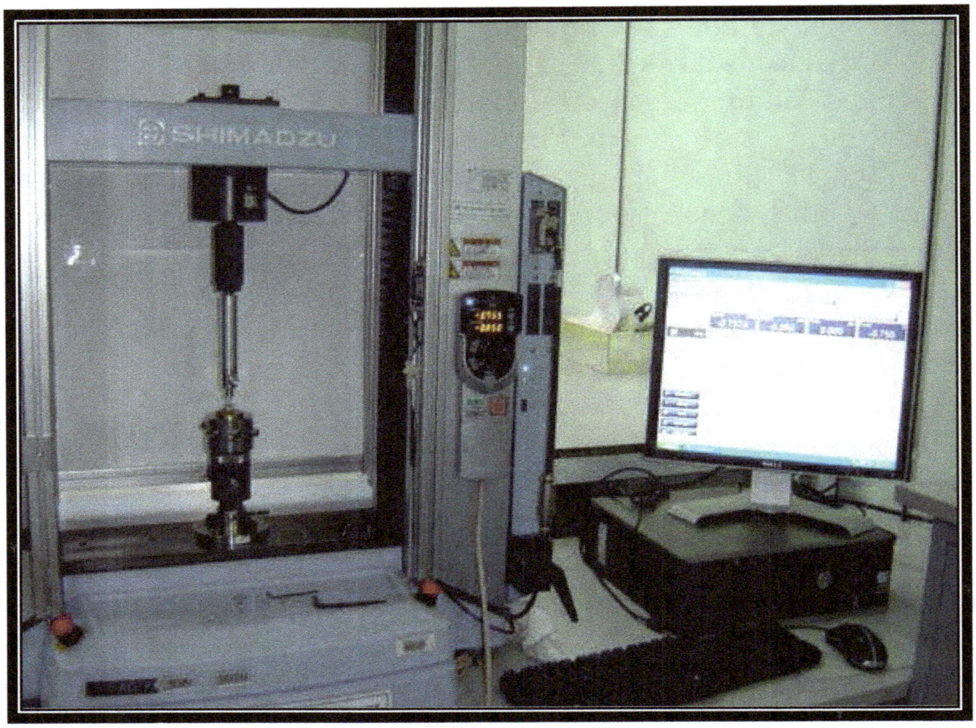

Fig. 12. The universal testing machine used

Fig. 13. The specimen during testing

Mode of fracture	Description
I	Minimal fracture or crack in coping.
II	Less than half of coping lost.
III	Coping fracture through midline (half of coping displaced or lost).
IV	More than half of coping lost.
V	Severe fracture of coping and/or die.

Table 1. Modes of fracture (After Burke and Watts, 1994)

2.2.7 Statistical analysis

The first objective was to determine the occlusal fracture resistance of Turkom-Cera copings compared to In-Ceram and Procera AllCeram copings. Descriptive statistics will be recorded for load at fracture of the three groups. The normally distributed data of load at fracture (histogram and Shapiro-Wilk test) will be analyzed by One Way ANOVA to achieve the objective, provided that equal variances will be assumed (Levene's test). Then, Tukey's HSD test will be carried out for post-hoc comparisons.

Whenever assumptions of normal distribution and equal variances of load at fracture between the three groups will not be met, an equivalent nonparametric Kruskal-Wallis test will be conducted. Subsequently, a post hoc test using Mann-Whitney tests with Bonferroni correction will be performed to test which pair of groups differ from each other significantly.

Regarding the second objective, effect of finish line design on load at fracture of Turkom-Cera copings, descriptive statistics will be recorded for load at fracture of each group. Then, independent samples t-test should be used to determine the significant differences in load at fracture between the two groups of Turkom-Cera copings with chamfer and shoulder finish lines. However, this is dependent on the assumptions of normal distribution and equal variances of load at fracture between each two groups to be met.

Whenever assumptions of normal distribution and equal variances of load at fracture between the two groups will not be met, an equivalent nonparametric Mann-Whitney tests will be performed.

In addition, descriptive statistics for modes of fracture and load at fracture will be recorded and the result will be descriptively analyzed. Statistical analysis will be carried out using a computer program (SPSS, SPSS Inc., Chicago, IL). Statistical significance will be set at $\alpha = 0.05$.

3. Results

3.1 Effect of ceramic material on the fracture resistance

The objective is to test if the mean load at fracture of Procera AllCeram, Turkom-Cera, and In-Ceram all-ceramic copings differ from each other. Descriptive analysis was performed and the mean load at fracture and standard deviation for the three groups are recorded in Table 2.

Ceramic	n	Mean	SD	95% Confidence Interval	
				Lower Bound	Upper Bound
Procera AllCeram	10	975.0	112.7	894.3	1055.6
Turkom-Cera	10	1341.9	216.5	1187.0	1496.8
In-Ceram	10	1151.6	180.1	1022.7	1280.4

Table 2. The mean load at fracture (N) and standard deviation for Procera AllCeram, Turkom-Cera, and In-Ceram copings

Since the assumption of normal distribution was met, the equality of variances (homogeneity) was tested using the Levene's test and showed that there was no significant deviation from homogeneity (p=0.163). Therefore, the parametric One Way ANOVA procedure was used to achieve objective. The results were recorded in Table 3. There was a significant difference in load at fracture between the three groups (p<0.001).

Ceramic	n	Mean (N)	SD	*F* Statistics (*df*)	*P* value[a]
Procera AllCeram	10	975.0	112.7	10.98 (2,27)	<0.001
Turkom-Cera	10	1341.9	216.5		
In-Ceram	10	1151.6	180.1		
[a] One Way ANOVA was used.					
Significant level was set at 0.05.					

Table 3. Comparison of load at fracture between Procera AllCeram, Turkom-Cera and In-Ceram copings by One Way ANOVA

Further analysis using Tukey HSD Post Hoc Test was done to determine the pair of means that differ significantly. Based on Tukey HSD Post Hoc Test (Table 4), the mean load at fracture of Turkom-Cera (1341.9 ± 216.5 N) was significantly more than Procera AllCeram (975.0 ± 112.7 N) (p<0.001). There were no significant differences between the mean load at fracture of In-Ceram (1151.6 ± 180.1 N) and Procera AllCeram (p=0.080) and also between the mean load at fracture of Turkom-Cera and In-Ceram (p=0.056).

Pairewise comparison	Mean (SD)	Mean Difference	*P* value
Procera AllCeram vs Turkom-Cera	975.0 (112.7) 1341.9 (216.5)	-366.93 366.93	<0.001*
Procera AllCeram vs In-Ceram	975.0 (112.7) 1151.6 (180.1)	-176.60 176.60	0.080
Turkom-Cera vs In-Ceram	1341.9 (216.5) 1151.6 (180.1)	190.33 -190.33	0.056
* 2 pairs of means are significantly different by Tukey HSD Test			

Table 4. Multiple pairwise comparisons of fracture load (N) using Tukey HSD Test

3.1.1 Mode of fracture

A cross-tabulation between treatment groups and modes of fracture was obtained (Table 5). The Chi-square test was used to test if there is any association between treatment groups (Procera, Turkom-Cera and In-Ceram) and modes of fracture. Due to unmet assumption of Chi-Square test and non-meaningful combination of different modes the result can be only descriptively analyzed.

Examination of the mode of fracture of specimens revealed that (70 %) of Turkom-Cera and (80 %) Procera AllCeram copings exhibited minimal fracture. Whereas, only 40 % of In-Ceram copings exhibited minimal fracture.

Ceramic	Mode of Fracture				Total
	Minimal Fracture n (%)	Less than half of coping lost n (%)	More than half of coping lost n (%)	Severe fracture of die and/or coping n (%)	
Procera All-ceram	8 (80%)	1 (10%)	1 (10%)	0 (0%)	10
Turkom-ceram	7 (70%)	2 (20%)	1 (10%)	0 (0%)	10
In-ceram	4 (40%)	1 (10%)	2 (20%)	3 (30%)	10

Table 5. Distribution of modes of fracture in each treatment group (Procera, Turkom-Cera and In-Ceram)

Descriptive summary for modes of fracture and mean load at fracture was recorded (Table 6). The identified modes of fracture were: minimal fracture, less than half of coping lost, more than half of coping lost and severe fracture of die and/or coping.

Mode of Fracture	Procera		Turkom-Ceram		In-Ceram	
	n	Mean LAF (SD)	n	Mean LAF (SD)	n	Mean LAF (SD)
Minimal fracture	8	999.72 (102.67)	7	1344.09 (194.26)	4	1158.32 (245.54)
Less than half of coping lost	1	965.36	2	1233.47 (362.48)	1	1077.61
More than half of coping lost	1	786.51	1	1543.40	2	1159.21 (139.41)
Severe fracture of die and/or coping	0	0	0	0	3	1162.12 (206.71)

(LAF: load at fracture)

Table 6. Descriptive summary for modes of fracture and mean load at fracture (N)

As shown in Table 6, the sever fracture mode was not seen within Procera and Turkom-Cera groups. The minimal fracture mode occurred at a higher load with Turkom-Cera group (1344.09 N) followed by In-Ceram (1158.32 N) and Procera AllCeram (999.72 N) groups. Furthermore, the minimal fracture mode occurred at a higher load than other modes of fracture in the Procera group. Whereas, in the Turkom-Cera group the minimal fracture mode occurred at a lower load than the more than half of coping lost mode, and at a lower load than the more than half of coping lost mode. Regarding In-Ceram group, the minimal fracture mode occurred at almost similar load to the other modes of fracture.

3.2 Effect of finish line on the fracture resistance of Turkom-Cera

Specimens were divided into two groups according to finish line used; Group 1 (Turkom-Cera with chamfer finish line) and Group 2 (Turkom-Cera with shoulder finish line). The mean and median load at fracture of the two groups are shown in Table 7.

Finish line	n	Mean (SD) (n=10)	Median (IQR) (n=10)	95% CI	
				Lower Bound	Upper Bound
Turkom-Cera (Chamfer)	10	1341.9 (216.5)	1407.38 (343.99)	1187.0	1496.8
Turkom-Cera (Shoulder)	10	1545.2 (186.6)	1549.35 (318.38)	1411.8	1678.7

Table 7. The mean and median load at fracture (N) of Turkom-Cera (Chamfer) and Turkom-Cera (Shoulder) groups

For testing normality, the histogram and Shapiro-Wilk test were used and showed no normal distribution of the mean load at fracture of the two groups. Since the assumption of normal distribution was not met, comparison of the load at fracture between the two groups was performed using the nonparametric Mann-Whitney U Test (Table 8). There was no significant difference in the load at fracture between the two groups (p=0.059). Therefore, there was no influence of the finish line on the load at fracture of Turkom-Cera all-ceramic copings.

Variable	Turkom-Cera (Chamfer) (n=10)		Turkom-Cera (Shoulder) (n=10)		P value
	Mean (SD)	Median (IQR)	Mean (SD)	Median (IQR)	
Load at fracture (N)	1341.9 (216.5)	1407.38 (343.99)	1545.2 (186.6)	1549.35 (318.38)	0.059*
* 2 pairs of medians are not significantly different by Mann-Whitney Test.					

Table 8. Comparison of load at fracture (N) between Turkom-Cera (Chamfer) and Turkom-Cera (Shoulder) groups by Mann-Whitney Test

4. Discussion

4.1 Methodology

An ideal experimental model of an in-vivo situation to determine the occlusal fracture resistance of all-ceramic crowns is difficult to achieve. The occlusal fracture resistance of a clinical ceramic crown is influenced by several factors, such as method of luting, loading condition and the elastic modulus of the supporting die (Scherrer and de Rijk, 1993, Webber et al., 2003; AL-Makramani et al., 2008a). However, the so called "crunch-the-crown" mechanical test had been widely used to examine the occlusal fracture resistance of sound and restored teeth (Al-Wahadni et al., 2009).

Mechanical tests on ceramic materials are difficult to carry out because of the presence of several limitations related to specimen preparation (Di Iorio et al., 2008). The present study attempted to isolate the ceramic material, artificial ageing and the finish line as the only variables. Natural teeth with comparable size and length were selected for this study to eliminate a possible effect of variations. In addition, a paralleling apparatus was used to prepare the teeth which allowed standardized preparations for all teeth.

In vivo, the teeth are supported by a visco-elastic periodontal ligament, which was not duplicated in the mounting of the specimens in the current study. The ability of the artificial ligament to reproduce the complex visco-elastic properties exhibited by ligament in vivo is limited (King and Setchell, 1990; Gu and Kern, 2006). In the clinical situation, the periodontal ligament may help to dissipate some of the applied load but at high loads simulation of a periodontal ligament in vitro is not useful as previous work has indicated that the root compresses the simulated ligament and impacts against the rigid mounting system and this might not reflect the clinical reality (Fokkinga et al., 2006; Gu and Kern, 2006; Good et al., 2008).

In the current study, two layers of nail varnish were therefore used for coating the root surfaces prior to embedding them in epoxy resin, which helped to avoid external reinforcement of the root by resin (Gu and Kern, 2006).
In the present study, the ceramic copings were cemented to natural teeth to replicate fracture load results more related to clinical situations than using ceramic discs (Wakabayashi and Anusavice, 2000; Ku et al., 2002) or crowns cemented to resin or metal die replicas (Neiva et al., 1998; AL-Makramani et al., 2009). Die replicas made of steel or resin fail to reproduce the actual force distribution at the inner surface of the crown or to reliably produce the characteristics of bonding between crowns and prepared teeth (Kelly, 1999; AL-Makramani et al., 2008a). However, die replicas provide a standardized preparation and identical physical qualities of materials used in comparison with natural teeth (Potiket et al., 2004; AL-Makramani et al., 2008b). Natural teeth show a large variation depending on their age, individual structure, and storage time after extraction, thus, causing difficulties in standardization (Strub and Beschnidt, 1998; Potiket et al., 2004).
Since the effect of the veneering porcelain on the load at fracture of high-strength all-ceramic restorations is still debatable, the copings were not veneered with porcelain (Webber et al., 2003; Beuer et al., 2008). In fact, fracturing of multilayer crowns starts at their weakest part. In the case when a stronger and stiffer core substructure is veneered with weaker porcelain, the failure usually occurs in the weak veneering porcelain or at the bond between the core and veneer (Aboushelib et al., 2006; Zahran et al., 2008). In addition, the veneering procedures could actually introduce factors (such as: flaws, cracks, voids, or internal stresses) that influence the results of mechanical tests (Vult von Steyern et al., 2006; Di Iorio et al., 2008). As stated by Miranda et al., (2001) flaws play a crucial role in the fracture resistance of brittle materials. Therefore, all-ceramic copings without porcelain veneering were loaded until fracture in this study.
The most commonly used artificial ageing technique is long-term water storage. Another widely used ageing technique is thermocycling (De Munck et al., 2005). The ISO/TS 11405 standard (2003) indicates that a thermocycling regimen comprised of 500 cycles in water between 5°C and 55°C is an appropriate artificial ageing test. In order to evaluate the effect of water storage and thermocycling on the fracture resistance of Turkom-Cera copings, 10 Turkom-Cera specimens were stored in distilled water for 30 days and subjected to 500 cycles in water between 5°C and 55°C before testing their fracture resistance.
The present study was conducted to evaluate the occlusal fracture resistance of copings fabricated using three ceramic systems and bonded to prepared teeth using resin luting cement. Such an in vitro study does not require natural teeth as a control group for comparison of results (Al-Wahadni et al., 2009; Komine et al., 2004; Potiket et al., 2004) since the stress distribution in restored teeth is different than in unrestored teeth (Arola et al., 2001). Furthermore, studies have found a large variability in the occlusal fracture resistance of extracted unprepared natural teeth (Attia et al., 2004; Attia et al., 2006).

4.2 Discussion of results

In the present study, the load at fracture of Turkom-Cera, In-Ceram and Procera AllCeram all-ceramic copings cemented to extracted teeth using resin luting cement was evaluated. The data showed that the mean load at fracture for Turkom-Cera, In-Ceram and Procera AllCeram were: 1341.9 N, 1151.6 N and 975.0 N, respectively. Statistical analysis showed that the differences were significant between Turkom-Cera and Procera AllCeram ($p<0.001$).

However, no significant differences were detected between Turkom-Cera and In-Ceram (p=0.056) and also between In-Ceram and Procera AllCeram (p=0.080).

The results of this study for Turkom-Cera and In-Ceram copings were in agreement with those obtained in a previous study (AL-Makramani et al., 2009), which found that Turkom-Cera had a significantly higher load at fracture than Procera AllCeram. Furthermore, the results of this study for In-Ceram and Procera AllCeram copings were in agreement with those obtained in previous studies (Webber et al., 2003; Neiva et al., 1998; Harrington et al., 2003; AL-Makramani et al., 2009), which found no significant differences in load at fracture between Procera AllCeram and In-Ceram copings that were resin cemented.

An in vitro study (AL-Makramani et al., 2009), which evaluated the fracture resistance of Turkom-Cera, In-Ceram and Procera AllCeram copings cemented on a metal master die using resin luting cement, shows that the load necessary to fracture the Turkom-Cera, In-Ceram and Procera AllCeram copings in the current study was less than that reported in that study. In this study, extracted natural teeth were used as abutments. However, metal dies are very rigid and have a higher modulus of elasticity than dentine so that metal dies deform less which results in a lower shear stress at the inner crown surface (Scherrer & de Rijk, 1993). Therefore, the fracture load of all-ceramic restorations may be greater if crowns are supported by dies with a high modulus of elasticity (Scherrer & de Rijk, 1993). This factor should also be considered when interpreting the results of the studies utilizing different die materials.

The results of the present study show that there is no influence of the finish line design on the load at fracture of Turkom-Cera all-ceramic copings. Statistical analysis revealed no significant difference between shoulder (1545.2 N) and chamfer (1341.9 N) margins used in this study (p=0.059). In this study only Turkom-Cera copings were evaluated. Due to this limitation, the load at fracture values obtained in this study should be compared with caution with results obtained in the studies where copings were veneered with feldspathic porcelain.

Results of the present study concurred with other studies on glass-ceramic crowns (Dicor) which did not demonstrate any differences in the loading capacity in relation to the type of finish lines used (Bernal et al., 1993; Malament and Socransky, 1999).

Di Iorio et al., (2008) found that the load at fracture for Procera (alumina-based) crowns with shoulder preparation was significantly higher than the chamfer preparation. In the current study, the load at fracture of Turkom-Cera copings with shoulder margin (1545.2 N) was higher than chamfer margin (1341.9 N), however, statistical analysis revealed no significant difference between them (p=0.059). A possible reason for this may be that the occlusal forces were also borne by the circumferential shoulder margin, and there was less stress concentration on the axial walls compared to chamfer margin (Beuer et al., 2008).

Conversely, a study on ceramic optimized polymer (Ceromer) crowns demonstrated that the fracture resistance of the chamfer finish line specimens was greater than that of the shoulder finish line (Cho et al., 2004).

The classification description of mode of failure by Burke & Watts (1994) was useful in distinguishing between minimal loss of crown material and catastrophic damage (mode II–IV). Examination of the mode of failure of specimens in the current study revealed that Procera AllCeram, Turkom-Cera and In-Ceram copings exhibited (80%), (70%) and (20%) of minimal fracture, respectively. The copings made from Turkom-Cera with shoulder margin and Turkom-Cera with artificial ageing exhibited 50% and 60% of minimal fracture, respectively.

Clinically, dental restorations are subjected to cyclic forces ranging from 60 N to 250 N during normal function and up to 500 N to 800 N for short periods (Zahran et al., 2008; Waltimo and Könönen, 1993; Waltimo and Könönen, 1995). Waltimo & Könönen (1993), reported that the maximum biting force in the molar region was 847 N for men and 597 N for women. The maximum biting force in the premolar region has been reported to be between 181 and 608 N (Widmalm and Ericsson, 1982). However, the normal masticatory forces in human beings have been reported to range from 37% to 40% of the maximum biting force (Lundgren and Laurell, 1986; Gibbs et al., 1981).

Although the results of the present study cannot be directly compared with the in vivo situation, the mean loads at fracture for all groups (ranging from 975.0 to 1545.2 N) exceed the clinically anticipated loads in the molar and premolar regions. However, clinical trials are necessary to validate the results.

Clinically, crown failure usually occurs under a complex type of stresses. However, all specimens in the current study were tested using vertical loads which appear to be appropriate for posterior teeth (Probster, 1992). Therefore, clinical implications of the current study must be limited to that application.

In this study, the specimens were loaded until failure in a single cycle, even though restorations may fail clinically through slow crack growth caused by cyclic fatigue loading (Baran et al., 2001). Subjecting the specimens to cycling fatigue loading could be considered in further investigation to give more information about the longevity and performance of Turkom-Cera crowns in condition relatively resemble the clinical situation.

5. Conclusions

Under the conditions of this study, it was found that:

1. Turkom-Cera copings cemented to extracted teeth using the resin luting cement (Panavia F) provided load at fracture (1341.9 N) that was significantly higher to that obtained by Procera AllCeram (975.0 N) copings tested under the same conditions. However, there was no significant difference between the mean load at fracture of Turkom-Cera and In-Ceram (1151.6 N) and also between In-Ceram and Procera AllCeram copings tested under the same conditions. Thus, the null hypothesis was rejected.
2. There is no influence of the finish line on the load at fracture of Turkom-Cera all-ceramic copings. Thus, the null hypothesis was accepted.

6. Acknowledgment

This study was supported by a grant (P019/2006C); University of Malaya, Kuala Lumpur, Malaysia.

7. References

Aboushelib, M.N.; Kleverlaan, C.J. & Feilzer, A.J. (2006). Microtensile bond strength of different components of core veneered all-ceramic restorations. Part II: zirconia veneering ceramics. *Dental Materials*; Vol. 22, No. 9, (September 2006) 857-863, ISSN 0109-564.

AL-Makramani , B.M.A.; Razak, A.A.A. & Abu-Hassan, M.I. (2008a). Evaluation of Load at Fracture of Procera AllCeram Copings Using Different Luting Cements. *Journal of Prosthodontics*, Vol. 17, No. 2, (February 2008) 120-124, ISSN 1532-849X.

AL-Makramani, B.M.A.; Razak, A.A.A. & Abu-Hassan, M.I. (2008b). Effect of luting cements on the compressive strength of Turkom-Cera all-ceramic copings. *The Journal of Contemporary Dental Practice*, Vol. 9, No. 2, (February 2008), 33-40, ISSN 1526-3711.

AL-Makramani, B.M.A.; Razak, A.A.A. & Abu-Hassan, M.I. (2009). Comparison of the load at fracture of Turkom-Cera to Procera AllCeram and In-Ceram all-Ceramic restorations. *Journal of Prosthodontics*, Vol. 18, No. 6, (August 2009) 484-488, ISSN 1532-849X.

Al-Wahadni, A.M.; Hussey, D.L.; Grey, N. & Hatamleh, M.M. (2009). Fracture Resistance of Aluminium Oxide and Lithium Disilicate-based Crowns using Different Luting Cements: An in vitro Study. *J Contemp Dent Pract*, Vol. 10, No. 2, (March 2009) 051-058, ISSN 1526-3711.

Arola, D.; Galles, L.A. & Sarubin, M.F. (2001). A comparison of the mechanical behavior of posterior teeth with amalgam and composite MOD restorations. *Journal of Dentistry*, Vol. 29, No. 1, (January 2001) 63-73, ISSN 0300-5712.

Attia, A. & Kern, M. (2004). Fracture strength of all-ceramic crowns luted using two bonding methods. *The Journal of Prosthetic Dentistry*, Vol. 91, No. 3, (March 2004) 247-252, ISSN 0022-3913.

Attia, A.; Abdelaziz, K.M.; Freitag, S. & Kern, M. (2006). Fracture load of composite resin and feldspathic all-ceramic CAD/CAM crowns. *The Journal of Prosthetic Dentistry*, Vol. 95, No. 2, (February 2006) 117-123, ISSN 0022-3913.

Baran, G.; Boberick, K. & McCool, J. (2001). Fatigue of restorative materials. *Critical Reviews in Oral Biology & Medicine*, Vol. 12, No. 4, (July 2001) 350-360, ISSN 1544-1113.

Bernal, G.; Jones, R.M.; Brown, D.T.; Munoz, C.A. & Goodacre, C.J. (1993). The effect of finish line form and luting agent on the breaking strength of Dicor crowns. *The International Journal of Prosthodontics*, Vol. 6, No. 3, (May-June 1993) 286-290, ISSN 0893-2174.

Beuer, F.; Aggstaller, H.; Edelhoff, D. & Gernet, W. (2008). Effect of Preparation Design on the Fracture Resistance of Zirconia Crown copings. *Dental Materials Journal*, Vol. 27, No. 3, (May 2008) 362-367, ISSN 0287-4547.

Blatz, M.B.; Richter, C.; Sadan, A. & Chiche, G.J. (2004). Critical appraisal. Resin bond to dental ceramics, Part II: high-strength ceramics. *Journal of Esthetic and Restorative Dentistry*, Vol. 16, No. 5, (September 2004), 324-328, ISSN 1708-8240.

Burke, F.J.T. & Watts, D.C. (1994). Fracture resistance of teeth restored with dentin-bonded crowns. *Quintessence International*, Vol. 25, No. 5, (May 1994) 335-340, ISSN 0033-6572.

Cho, L.; Choi, J.; Yi, Y.J.; Park, C.J. (2004). Effect of finish line variants on marginal accuracy and fracture strength of ceramic optimized polymer/fiber-reinforced composite crowns. *The Journal of Prosthetic Dentistry*, Vol. 91, No. 6, (June 2004) 554–560, ISSN 0022-3913.

De Munck, J.; Van Landuyt, K.; Peumans, M.; Poitevin, A.; Lambrechts, P.; Braem, M. & Van Meerbeek, B. (2005). A critical review of the durability of adhesion to tooth tissue: methods and results. *Journal of Dental Research*, Vol. 84, No. 2, (February 2005) 118-132, ISSN 1544-0591.

Della Bona, A.; Borba, M.; Benetti, P. & Cecchetti, D. (2007). Effect of surface treatments on the bond strength of a zirconia-reinforced ceramic to composite resin. *Brazilian Oral Research*, Vol. 21, No. 1, (January-March 2007) 10-15, ISSN 1806-8324.

Di Iorio, D.; Murmura, G.; Orsini, G.; Scarano, A. & Caputi, S. (2008). Effect of Margin Design on the Fracture Resistance of Procera® All Ceram Cores: An in vitro Study. *The Journal of Contemporary Dental Practice*, Vol. 9, No. 2, (February 2008), 001-008, ISSN 1526-3711.

Doyle, M.G.; Goodacre, C.J.; Munoz, C.A. & Andres, C.J. (1990a). The effect of tooth preparation design on the breaking strength of Dicor crowns: 3. *The International Journal of Prosthodontics*, Vol. 3, No. 4, (July-August 1990) 327-340, ISSN 0893-2174.

Doyle, M.G.; Munoz, C.A.; Goodacre, C.J.; Friedlander, L.D. & Moore, B.K. (1990b). The effect of tooth preparation design on the breaking strength of Dicor crowns: 2. *The International Journal of Prosthodontics*, Vol. 3, No. 3, (May-June 1990) 241-248, ISSN 0893-2174.

Fokkinga, W.A.; Kreulen, C.M.; Le Bell-Rönnlöf, A.M.; Lassila, L.V.; Vallittu, P.K. & Creugers, N.H. (2006). In vitro fracture behavior of maxillary premolars with metal crowns and several post-and-core systems. *European Journal of Oral Sciences*, Vol. 114, No. 3, (June 2006) 250-256, ISSN 0909-8836.

Gibbs, C.H.; Mahan, P.E.; Lundeen, H.C.; Brehnan, K.; Walsh, E.K, and Holbrook, W.B. (1981). Occlusal forces during chewing and swallowing as measured by sound transmission. *The Journal of Prosthetic Dentistry*, Vol. 46, No. 4, (October 1981) 443-449, ISSN 0022-3913.

Good, M.L.; Orr, J.F. & Mitchell, C.A. (2008). In vitro study of mean loads and modes of failure of all-ceramic crowns cemented with light-cured or dual-cured luting cement, after 1 and 30 d of storage. *European Journal of Oral Sciences*, Vol. 116, No. 1, (February 2008) 83-88, ISSN 0909-8836.

Goodis, H.E.; Marshall, G.W. Jr.; White, J.M.; Gee, L.; Hornberger, B. & Marshall, S.J. (1993). Storage effects on deiitin permeability and shear bond strengths. *Dental Materials*; Vol. 9, No. 2, (March 1993) 79-84, ISSN 0109-564.

Gu, X.H. & Kern, M. (2006). Fracture resistance of crowned incisors with different post systems and luting agents. *Journal of Oral Rehabilitation*, Vol. 33, No. 12, (December 2006) 918-923, ISSN 13652842.

Harrington, Z.; McDonald, A. & Knowles, J. (2003). An in vitro study to investigate the load at fracture of Procera Allceram crowns with various thickness of occlusal veneer porcelain. *The International Journal of Prosthodontics*, Vol. 16, No. 1, (January-February 2003) 54-58, ISSN 0893-2174.

International Organization for Standardization. ISO/TS 11405:2003. *Dental materials–testing of adhesion to tooth structure*, 2nd ed. Geneva, Switzerland: ISO; 2003.

Kelly, J.R. (1999). Clinically relevant approach to failure testing of all-ceramic restorations. *The Journal of Prosthetic Dentistry*, Vol. 81, No. 6, (June 1999) 652-661, ISSN 0022-3913.

King, P.A. & Setchell, D.J. (1990). An in vitro evaluation of a prototype CFRC prefabricated post developed for the restoration of pulpless teeth. *Journal of Oral Rehabilitation*, Vol. 17, No. 6, (November 1990) 599-609, ISSN 13652842.

Komine, F.; Tomic, M.; Gerds, T. & Strub, J.R. (2004). Influence of different adhesive resin cements on the fracture strength of aluminum oxide ceramic posterior crowns. *The Journal of Prosthetic Dentistry*, Vol. 92, No. 4, (October 2004) 359-364, ISSN 0022-3913.

Ku, C.W.; Park, S.W. & Yang, H.S. (2002). Comparison of the fracture strengths of metal-ceramic crowns and three ceromer crowns. *The Journal of Prosthetic Dentistry*, Vol. 88, No. 2, (August 2002) 170-175, ISSN 0022-3913.

Lundgren, D. & Laurell, L. (1986). Occlusal force pattern during chewing and biting in dentitions restored with fixed bridges of cross-arch extension. I. Bilateral end abutments. *Journal of Oral Rehabilitation*, Vol. 13, No. 1, (January 1986) 57-71, ISSN 13652842.

Malament, K. & Socransky, S. (1999). Survival of Dicor glassceramic dental restorations over 14 years. Part II: Effect of thickness of Dicor material and design of tooth preparation. *The Journal of Prosthetic Dentistry*, Vol. 81, No. 6, (June 1999) 662-667, ISSN 0022-3913.

Miranda, P.; Pajares, A.; Guiberteau, F.; Cumbrera, F.L. & Lawn, B.R. (2001). Role of flaw statistics in contact fracture of brittle coatings. *Acta Materialia*, Vol. 49, No. 18, (October 2001) 3719-3726, ISSN 1359-6454.

Neiva, G.; Yaman, P.; Dennison, J.B.; Razzoog, M.E. & Lang, B.R. (1998). Resistance to fracture of three all-ceramic systems. *Journal of Esthetic and Restorative Dentistry*, Vol. 10, No. 2, (March 1998) 60-66, ISSN 1708-8240.

Ozcan, M.; Alkumru, H.N. & Gemalmaz, D. (2001). The effect of surface treatment on the shear bond strength of luting cement to a glass-infiltrated alumina ceramic. *The International Journal of Prosthodontics*, Vol. 14, No. 4, (July-August 2001) 335-339, ISSN 0893-2174.

Potiket, N.; Chiche, G. & Finger, I.M. (2004). In vitro fracture strength of teeth restored with different all-ceramic crown systems. *The Journal of Prosthetic Dentistry*, Vol. 92, No. 5, (November 2004) 491-495, ISSN 0022-3913.

Probster L. (1992). Compressive strength of two modern all-ceramic crowns. *The International Journal of Prosthodontics*, Vol. 5, No. 5, (September-October 1992) 409-414, ISSN 0893-2174.

Scherrer, S.S. & de Rijk, W.G. (1993). The fracture resistance of all-ceramic crowns on supporting structures with different elastic moduli. *The International Journal of Prosthodontics*, Vol. 6, No. 5, (September-October 1993) 462-467, ISSN 0893-2174.

Shillingburg, H.T.; Hobo, S.; Whitsett, L.D.; Jacobi, R. & Brackett, S.E. (1997). *Fundamentals of fixed prosthodontics*. 3rd ed., Quintessence Publishing Co, Inc, 0-86715-201-X, Chicago.

Shimada, Y.; Yamaguchi, S. & Tagami, J. (2002). Micro-shear bond strength of dual-cured resin cement to glass ceramics. *Dental Materials*; Vol. 18, No. 5, (July 2002) 380-388, ISSN 0109-564.

Strub, J.R. & Beschnidt, S.M. (1998). Fracture strength of five different all-ceramic crown systems. *The International Journal of Prosthodontics*, Vol. 11, No. 6, (November-December 1998) 602-609, ISSN 0893-2174.

Valandro, L.F.; Ozcan, M.; Bottino, M.C.; Bottino, M.A.; Scotti, R. & Della Bona, A. (2006). Bond strength of a resin cement to high-alumina and zirconia-reinforced ceramics: the effect of surface conditioning. *The Journal of Adhesive Dentistry*, Vol. 8, No. 3, (June 2006) 175-181, ISSN 1461-5185.

Vult von Steyern, P.; Ebbesson, S.; Holmgren, J.; Haag, P. & Nilner, K. (2006). Fracture strength of two oxide ceramic crown systems after cyclic pre-loading and thermocycling. *Journal of Oral Rehabilitation*, Vol. 33, No. 9, (September 2006) 682-689, ISSN 13652842.

Wakabayashi, N. & Anusavice, K. (2000). Crack initiation modes in bilayered alumina/porcelain disks as a function of core/veneer thickness ratio and supporting substrate stiffness. *Journal of Dental Research*, Vol. 79, No. 6), (June 2006) 1398-1404, ISSN 1544-0591.

Waltimo, A. & Könönen, M. (1993). A novel bite force recorder and maximal isometric bite force values for healthy young adults. *Scandinavian Journal of Dental Research*, Vol. 101, No. 3, (June 1993) 171-175, ISSN 0029-845X.

Waltimo, A. & Könönen, M. (1995). Maximal bite force and its association with signs and symptoms of craniomandibular disorders in young Finnish non-patients. *Acta Odontologica Scandinavica*, Vol. 53, No. 4, (August 1995) 254-258, ISSN 1502-3850.

Webber, B.; McDonald, A. & Knowles, J. (2003). An in vitro study of the compressive load at fracture of procera allceram crowns with varying thickness of veneer porcelain. *The Journal of Prosthetic Dentistry*, Vol. 89, No. 2, (February 2003) 154-160, ISSN 0022-3913.

Widmalm, S.E. & Ericsson, S.G. (1982). Maximal bite force with centric and eccentric load. *Journal of Oral Rehabilitation*, Vol. 9, No. 5, (September 1982) 445-450, ISSN 13652842.

Zahran, M.; El-Mowafy, O.; Tam, L.;Watson, P.A. & Finer, Y. (2008). Fracture strength and fatigue resistance of all-ceramic molar crowns manufactured with CAD/CAM technology. *Journal of Prosthodontics*, Vol. 17, No. 5, (July 2008) 370-377, ISSN 1532-849X.

3

Mechanical Properties of New Ceramic Materials Obtained from Granular Solid Residuals Coming from Mines and Diatomaceous Earth

Jaime Vite-Torres, María del Carmen Carreño de León,
Manuel Vite-Torres and Juan Rodrigo Laguna-Camacho
Instituto Nacional de Investigaciones Nucleares (ININ),
SEPI-ESIME-Campus Zacatenco-IPN,
UniversidadVeracruzana,Campus Poza Rica, Veracruz
México

1. Introduction

Granular solid residuals are produced by different industries in Mexico. Some of these are considered dangerous because they are stored in open areas contaminating and affecting the flora, fauna, and aquifers, and occasionally to the human being. These solid residuals coming from mines and diatomaceous earth are known as "jales" (náhualt derivated from xalli, it means sand). The "jales" have a high content of cyanide and heavy metals. These high contents are commonly out of the limits permitted by the Mexican standards. On the other hand, the "diatomita" (diatomaceous earth) is not as dangerous as the "jales", however, this needs a treatment to eliminate the occluded organic material. This is normally used by producers of gelatin, antibiotics, industrial grease and oil, sugar refining, etc.

The leaching process of metals using the sodium cyanide was introduced in 1887 (Li, 1993). This revolutionizes the techniques to obtain gold and silver. This is a method usually employed in the world to recover metals. Two international patents were obtained by developing equipment and also a process to extract toxic metals and wastes from foundry sands (Vite, 1994a, 1994b). In addition, a national patent was achieved by the leaching of heavy metals from foundry sands (Vite, 1997).

Approximately, 330,000 tons of these solid residuals are generated daily in Mexico. In some states, such as Guanajuato, Zacatecas and cities as Pachuca in the state of Hidalgo, the waste deposits are found in open areas. The chemical composition of these solid residuals is varied. Figure 1 shows a photograph of the pipe used to transport the solid residuals until a "jales" dam.

An important case is related to "jales" when they have high pyrite content (FeS_2). This affects considerably the chemical composition due to their weathering. Due to this fact, it is necessary to process the "jales" before being used at applications such as the construction of landfills in the leveling of roads, additives for asphalt mixtures or dam construction (Botz & Stevenson, 1995, Orozco & Orozco, 1992, Sengupta, 1993, Vite et al., 2003, Wills, 1994).

In this study, a multidisciplinary work combining different mechanical, chemical and physical aspects was conducted. Industrial solid residuals coming from mines were leached to eliminate the toxic metals present in their chemical matrix. This process was carried out by using special equipment as mentioned and described above. It was developed and patented in Mexico and United States of America. The resultant solid material was mixed with solid residuals coming from diatomaceous earth and mines, and binders were employed to obtain new compact ceramic materials.

Fig. 1. Pipe used to transport the solid residuals until the jales dam

The mechanical properties of these ceramic materials, such as the compressive strength were determined in accordance to ASTM C170-50 and ASTM C-36 standard. The final results showed that these ceramic materials are lighter and exhibit higher compression strength than conventional materials such as bricks.

1.1 Background of the invention

The invention refers to an apparatus and process for extracting metals from foundry sands and, more specifically, to an apparatus and process for extracting these metals by making use of a thermostated column.

In the field of metal extraction based on foundry sands, the extraction techniques most utilized to date in a liquid-solid system involve mechanically-stirred leaching tanks, the Pachuca (State of Mexico) tank and Bonotto, Kennedy or Rotocel extractors. None of this equipment however has been specifically designed to extract small quantities of metals from foundry sands, which represents a problem not, so far, resolved.

Of the conventional apparatus mentioned above, the Pachuca tank is that which most resembles the apparatus object of the present application. However, the thermostated column of the new apparatus differs from the Pachuca tank insofar as regards its air dispersion system, the use of decarbonated and humidified air, because it consists of a heating jacket and because temperature control of the heating fluid is by means of a thermostat. These features of an apparatus which is desired to protect, allow a suitable pulp suspension, prevent the formation of a silicate and carbonate layer at the bottom of the column, permit better temperature control and a finer control of the process variables, such as the pH, the air flow and the solution´s oxidation-reduction potencial.

1.2 Summary of the apparatus

The apparatus and the process of the present invention allow low concentrations of the metals contained in the mining tailings to be simultaneously extracted, thus achieving two main goals which are 1) the recovery of metal values which can be recycled and 2) An important reduction in the toxicity of the industrial solid wastes (Mining Tailings) which once treated, can be handled as normal industrial waste, which can be used to make ceramic materials or construction materials, with important savings. From among the metals which are possible to recover, one can mention, among others platinum, gold, cobalt, germanium, zinc, manganese, silver and cooper, metals with many industrial uses, and some others such as chromium, vanadium and arsenic, which are highly toxic. Figure 2 shows a schematic diagram of the patented equipment ((Vite, 1994a, 1994b).

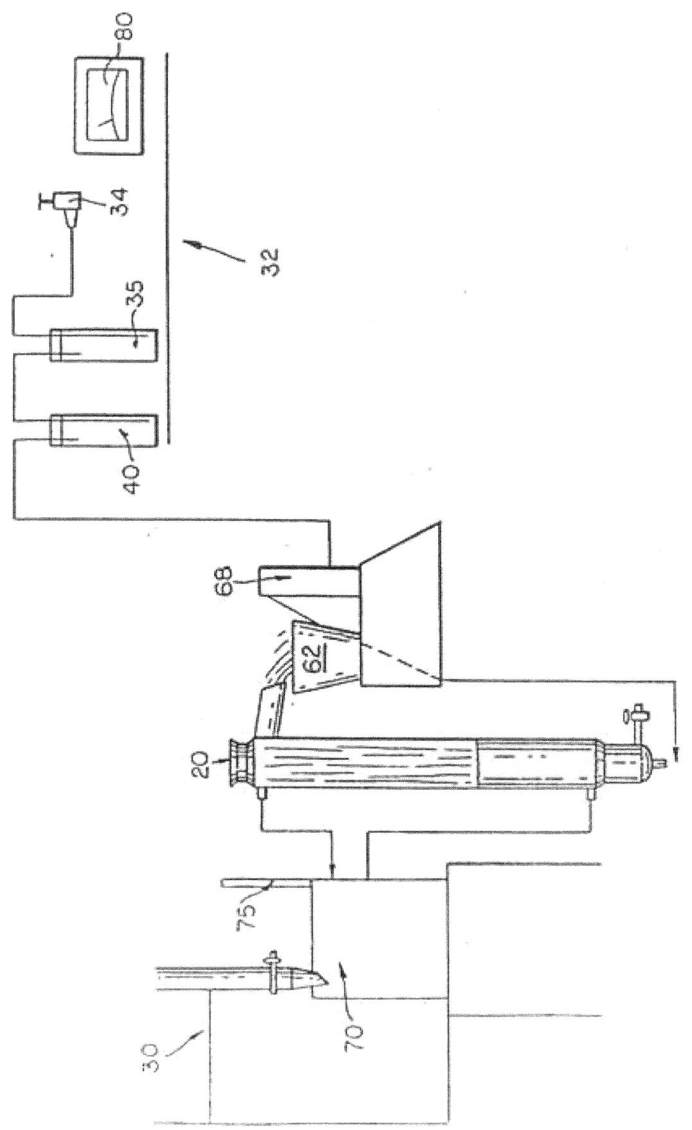

Fig. 2. Schematic diagram of the patented equipment (Vite, 1994a, 1994b)

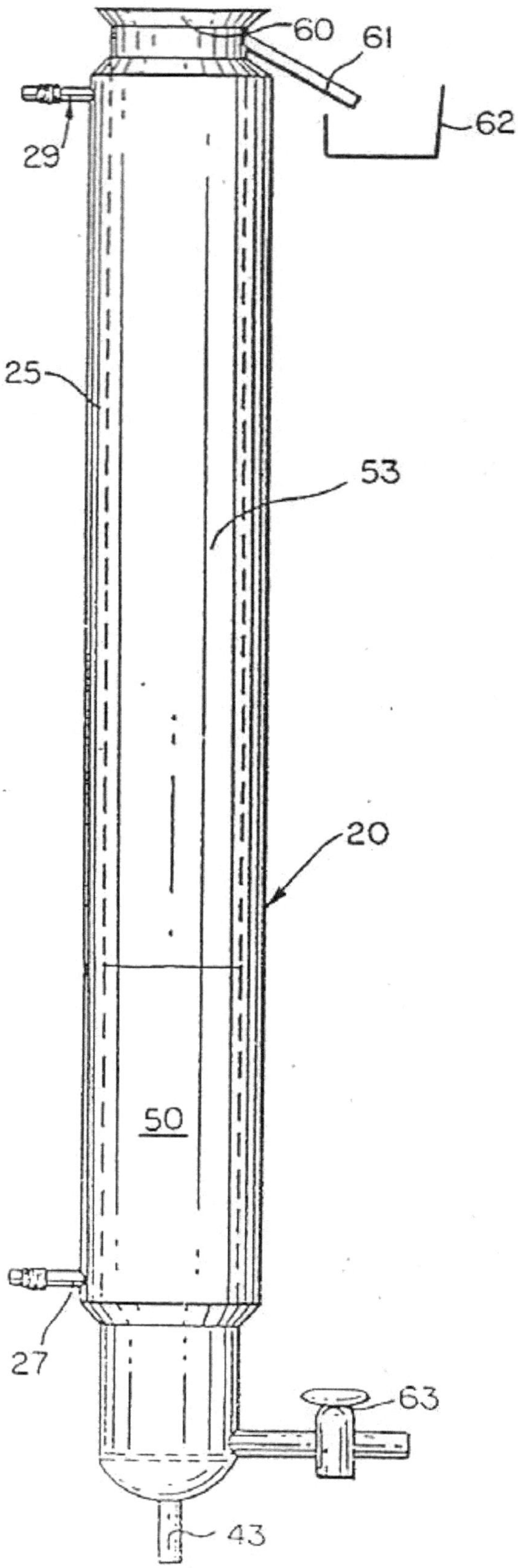

Fig. 3. Side view of the thermostated column (Vite, 1994a, 1994b)

In order to achieve its purpose, the apparatus consists of a thermostatic includes column which is connected to a decarbonated and humidified air supply source, and with a controlled heating system. This column includes an inlet for introduction of the components to form an aqueous pulp, and device for sparging, it in such away that a suitable suspension and homogenization of same is obtained. For it's part , the process is characterized by the introduction into the column of some reagents, distilled water and industrial solid wastes (mining tailings) in order to form an aqueous pulp which is processed in same for 2 hours under controlled temperature , air flow and pH conditions. Upon termination of the above mentioned period, the pulp in the column is extracted and once again heated for 20 minutes, after which it is filtered in order to obtain liquor with the metals in solution.
Another distinctive characteristic of this process is that, as distinguished from the existing system, it permits the extraction of metals from metallic alloys; the thermodynamic stability is higher than covalent bonds.

1.3 Detailed description of the invention

In the structural aspect of the invention (Figure 2, 3, 4 and 5) this consists of an apparatus comprising a thermostated column 20 having a double wall defining a heating jacket 25 through which a heating fluid circulates coming from a heating system 30 which is in flow communications with said jacked by means of the heating fluid inlet and outlet means 27, 29 disposed in element 25 for connection with said system 30. The purpose of said heating system is to maintain the interior of the column 20 at an approximate temperature of 60° C., in accordance with one of the variables of the process for extracting metals from foundry sands. To this end, the heating system 30 includes thermal control devices such as a thermostat 70 and thermometer 75 to control the temperature of the heating fluid.
The apparatus also includes a source 32 of supply of air to the column, where said source comprises a valve 34 for the admission of air and a carbonating device 35 and humidifier 40, the latter two placed in the path which communicates said source with de column. The air is passed through the device 35 and humidifier 40 before it reach the column, to which it is introduced by an inlet 43 and crosses to a sparging device 45 placed next to said inlet which server to finely divide the air in order to foster suitable suspension and homogenization of the pulp. Device 45 consist of a membrane with multiple orifices 46 of a diameter such as not to allow the pulp to pass through same towards the airflow inlet 43. The upper end of the column 20 includes an opening 48 through which the different components forming the aqueous pulp 50 are introduced, said pulp containing the metals which it is desired to extract in accordance with the aim of the apparatus and process invented. Pulp 50 is placed in the inner chamber 53 of the column, which has been previously heated by the heating system 30 and supplied with air from supply source 32, with the characteristics mentioned hereinabove, wherein said pulp is sparged by the air flow crossing device 45 in order to provide it with the suspension and homogenization properties required for the process. A frosted glass cap 60 is placed in opening 48 in order to avoid the escape of air and foam. However, the column is provided with means 61 through which the foam may leave into a reservoir 62. For extracting samples from the pulp during the process, the column has a valve 63, through which the necessary samples are taken to verify that the pulp being processed retains the ideal conditions for obtaining the desired results. In addition, a device 68 for controlling and measuring the airflow supplied to the column is located at a point

before the air inlet 43. With regard to the `process the tailings are analyzed in order to verify their compatibility with said process and in order to subsequently carry out the corresponding metallurgic balances. Once analyzed the tailings are finely ground and suspended in an aqueous pulp with a solid-liquid ratio of 1 to 4 and the necessary reagents added; that from 10 to 40 g , of a mineral acid in order to reach a pH of 2 , from 4 to 15 of a compound of the family of the bisulphites and from 0.3 to 1.5 g of mono and dialkylamines amineoxides and ethoxylated alkyl amines, all these reagents being of analytical grade. The addition of surfactant is preferred because it is a cationic surfactant which reduces the surface tension and promotes foam formation , thus obtaining a greater efficiency in the leaching process.

Figure 3 and 4 displayed a side view of the thermostated column and the membrane device of the column, respectively. Additionally, Figure 5 shows a design of diagram of a preliminary pilot plant for leaching metal values and toxic metals from industrial solid wastes. Finally, Figure 6 presents a block diagram of the process for the extraction of metal values from foundry sands.

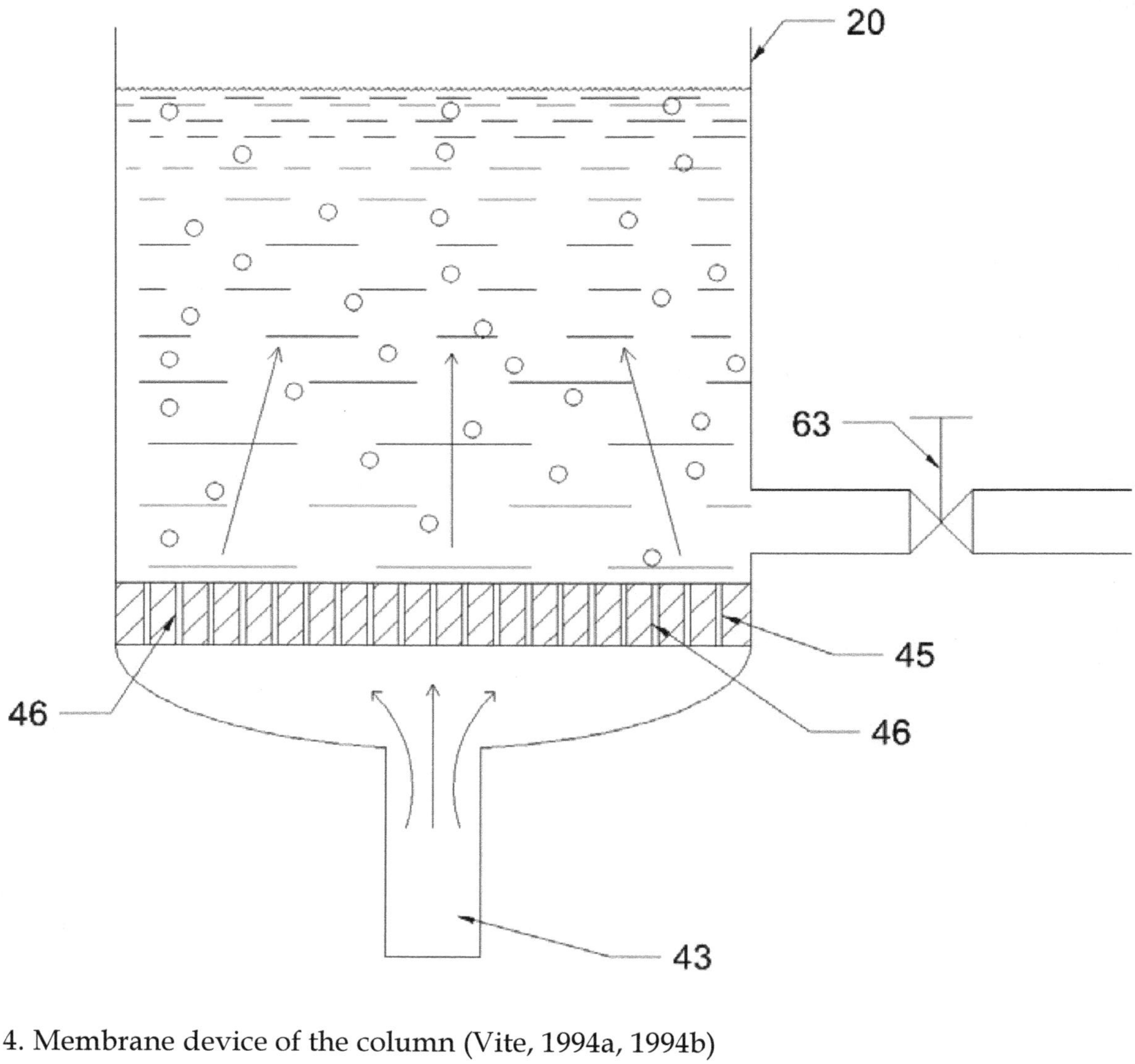

Fig. 4. Membrane device of the column (Vite, 1994a, 1994b)

2. Experimental details

2.1 Preparation of granular solid industrial residuals

2.1.1 Treatment of granular solid industrial residuals coming from diatomaceous earth

The diatomite (diatomaceous earth) is a marine sedimentary rock. This can be found in lake deposits and consist mainly of the accumulation of shells or hydrated silica secreted by diatoms which are microscopic organisms. Chemically consists of silicon dioxide (SiO_2), which is essentially inert; is attacked only by strong alkalis and hydrofluoric acid. Other components of the diatomite are: Sand, clay, volcanic ashes, magnesium carbonate, soluble salts and organic matter. The types and amounts of impurities are highly variable and depend on the conditions of sedimentation and deposition time of the diatomite. Diatomite is used primarily as filters, insulation or filler material. In the first case, the substance is used to filter sludge. Another application is in sugar refining and industrial plants in various solvents antibiotics, grass and oils. In many other cases diatomaceous earth are used for separation process. The diatomite is contaminated with organic material, after being used in the processes described above.

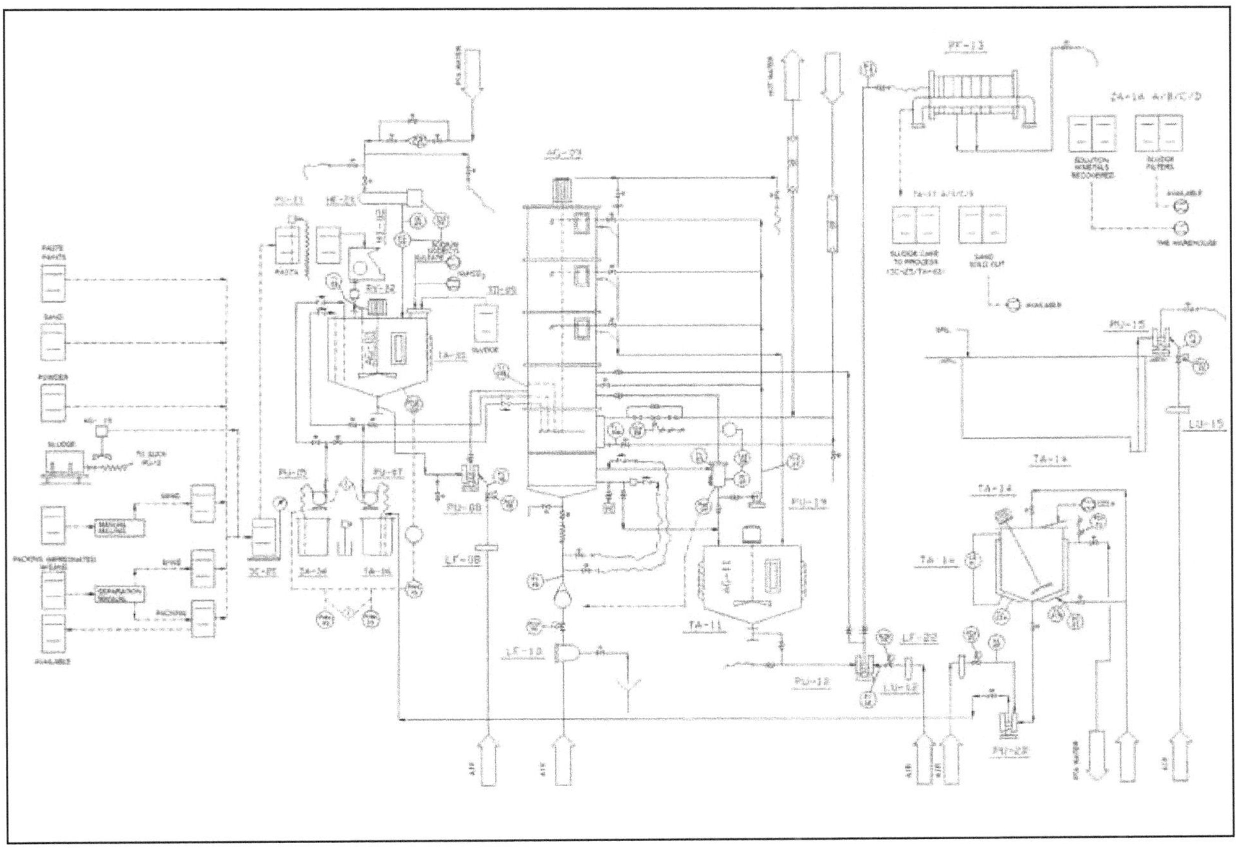

Fig. 5. Design plant for leaching of metal for values and toxic metals from industrial solidwastes (Vite, 1994a, 1994b)

In the laboratory, the organic material adhered to the diatomaceous earth was removed by a physical-chemical process developed here, thus obtaining innocuous material. Subsequently, the sample was sieved to achieve a grain size between 100-200 μm, likewise, a representative sample was taken, using the counting method and quartering, to analyze qualitatively and quantitatively. Figure 7 shows the morphology of one of the samples obtained from granular solid industrial wastes coming from diatomaceous earth.

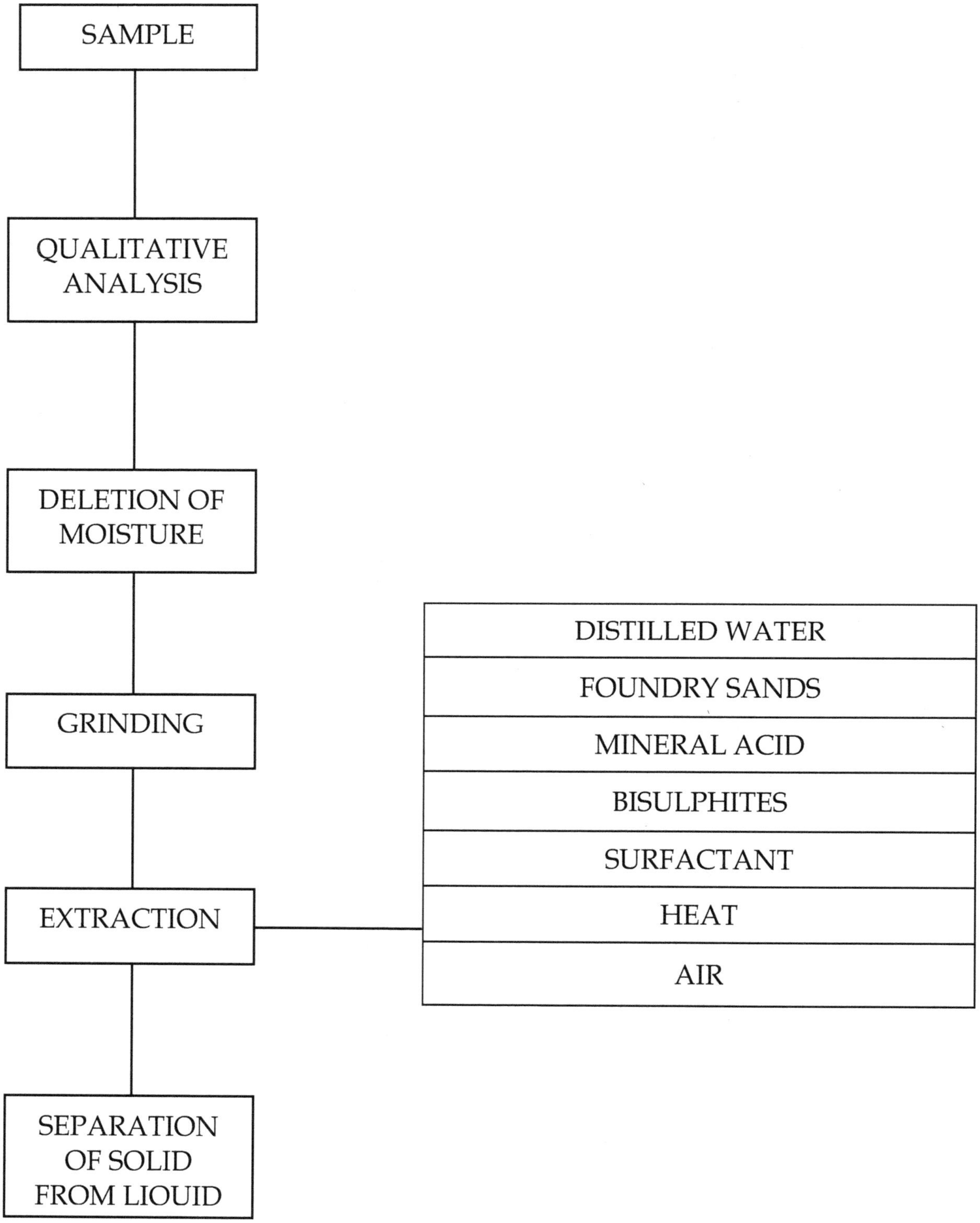

Fig. 6. Block diagram of the process for the extraction of metal values from foundry sands (Vite, 1994a, 1994b)

Fig. 7. Morphology characteristics of industrial residual coming from diatomaceous earth

2.1.2 Treatment of the mining industrial residuals samples (MIR)

To determine the presence and concentration of cyanide and heavy metals, representative samples were analyzed qualitative and quantitatively using conventional and nuclear techniques. Then, the content of cyanide was removed using thermostated coupled columns. This process and the equipment was developed and patented in the United States and Mexico (Vite, 1994a, 1994b). Samples are placed into the equipment, where sodium hypochlorite is added, the temperature in the columns is set at 28 ° C, during 20 min. Once the process is completed, the separation solid-liquid is performed. Then, the solid sample is dried and placed back in the thermostated coupled columns in order to leach heavy metals. The sample of "jales" is mixed with reducers, surfactants and complexing agents to perform the leaching process at different pH. Then, an acid or a base is added, la pulp is stirred by injecting air and each sample is processed for 120 min. at 60 ° C. After all this process, the solid-liquid separation is made. The solid was sieved to obtain a grain size between 100-200 µm and was analyzed qualitative and quantitatively, in order to verify the decrease in the concentration of cyanide and heavy metals content in the samples, if there is any evidence of a toxic component, the operation is repeated until obtain innocuous samples. The treatment of solid residuals to leach cyanide and heavy metals was successful, as it eliminates 100% of the cyanide content, as well as 80-100% of the heavy metals content occluded in the chemical matrix of the sample.

2.2 Specimens

The materials used to conduct the tests were ceramic materials identified as "JADIAR" due to a mixture of Jale-Diatomaceous earth-Clay. The samples were named as 1A, 1B and 1C. In the sample 1A, 50% was material coming from diatomaceous earth, 40% of mining industrial residual (Jale) and 10% clay. In sample 1B, was used 50% of diatomaceous earth, 30% solid residual (jale) and 20% clay. Finally, in sample 1C, the contents were 50% diatomaceous earth, 20% mining industrial residuals and 30% clay. Table 1 presents the material contents. Each sample was mixed separately and was heat- treated first to remove moisture in an oven at a temperature of 100°C for 120 min., then the sample was treated in an oven at 950 °C for 180 min. The diagram shown in Figure 8, presents a flow chart of processing of materials coming from mining industrial residuals and diatomaceous earth.

Sample	Diatomaceous earth (%)	Mining industrial residuals (jales) (%)	Clay (%)
1A	50	40	10
1B	50	30	20
1C	50	20	30

Table 1. Material contents

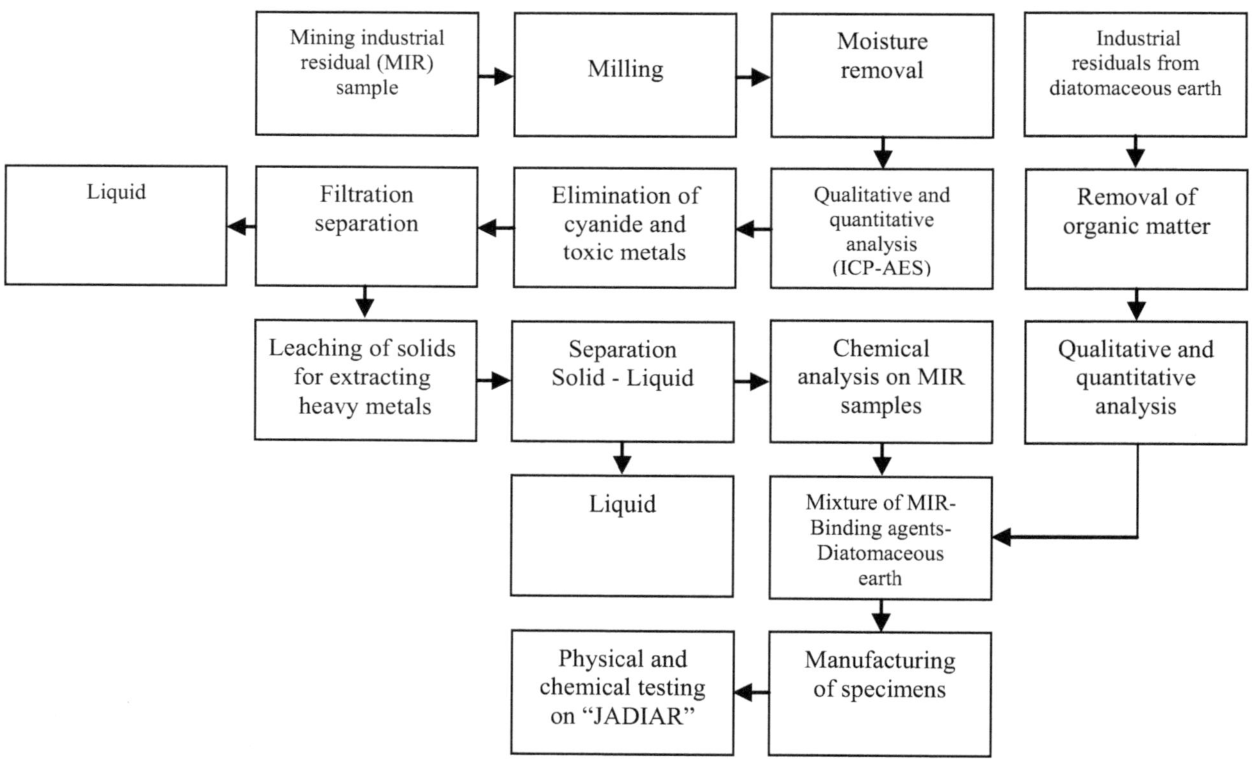

Fig. 8. Flowchart of processing of materials

3. Results and discussion

3.1 Methodology for the production of construction materials (lattice and bricks)

The main objectives pursued with these building blocks are: Ease of construction, increased insulation and ease of operation. Light inert materials are generally chosen. The manufacture may be carried out by mortars and concretes conveniently dispensed with fluid consistencies, paste or dried, depending on the system followed and the shape of the final piece.

3.1.1 Pasta preparation

Pastas are prepared from the mixture of granular solid industrial residuals coming from mines and diatomaceous earth with binding agents such as cement or clay mixed with water conveniently without exceeding their plasticity.

3.1.2 Molding

The molding is made by direct compression or by vibration, with mortar or concrete, the materials has to be practically dry, if the process is done by casting, the material must be more or less fluid. The procedures may be manual or mechanical, depending on the production, cost and product quality.

In this particular case, the manual procedure was used because the pieces were made solely for laboratory analysis. Metallic and Wood molds were used for the lattices. The results of the materials obtained using the methodology mentioned above, can be seen in the bricks and lattices in Figure 9 and Figure 10 where it is possible to observe the manufactured samples for tribological tests.

Fig. 9. Bricks and lattices obtained from granular solid residuals coming from diatomaceous, mining industrial residuals and binders

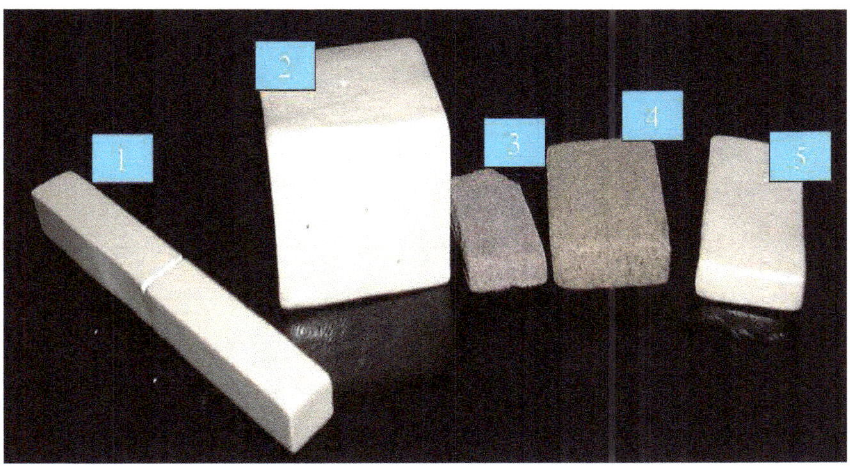

Fig. 10. Manufactured samples with clay and materials coming from mines and diatomaceous earth, 1) Impact and flexion testing, 2) Compression testing, 3) Erosive wear tests (Vite et al., 2010) 4) wet abrasion tests, 5) dry abrasion tests (Vite et al., 2011)

These bricks and lattices were obtained from solid industrial residuals coming from diatomaceous earth, which were mixed with mining industrial residuals and clay. These materials duplicate in tensile strength compared to conventional bricks, according to NOM-C-101, also, the percentage of absorption in the material obtained is smaller (about 17.1%) than required by the NOM-C-101, whose maximum value is 25%.
Note: It should be mentioned in the case of the lattice there are no regulations that specify the quality requirements to be met. The procedure for manufacturing was based on the proportions of the materials that give to a brick, the optimal resistance conditions.

3.2 Mechanical testing

3.2.1 Compressive strength according to ASTM C-170-50 and ASTM C-36

Compression tests are required to subject the sample to a compressive load, until it breaks or fractures. These can be made on most materials according to ASTM C-170-50 standard. The compression tests are conducted on materials such as cement, concrete, rubber, wood, plastic and clay (ceramics). This test consists of applying a variable vertical load to the material, in order to determine whether it complies or not the preset compressive load according to the type of the material studied. The manufacture of these materials can be carried out by mortars and concretes conveniently dispensed. The samples tested were named as 1A, 1B and 1C as mentioned previously. The dimensions of the test samples were 50.8 X 50.8 mm (a 2 inch cube). Figure 11 shows that the compressive strength of the samples analyzed had a maximum when the content of the binder agent was 30% in volume and such resistance was 160.3 kg/cm^2 and a minimum of 119.7 kg/cm^2 was obtained when the binder content was 10%. In both cases, the resistance was higher than that of a conventional brick whose resistance varies between 80-100 Kg/cm^2, also, the weight of the material obtained was 20-25% lower than the weight of conventional bricks, had less porosity and for this reason these materials were considered as a sustainable because the costs of processing were 25% lower, that the materials available on the market. Table 2 presents the compressive strength results for each testing sample.
The parameters obtained were:

σ max - Maximum stress
σ yield - Yield stress
E - Young Modulus
Є max - Maximum strain

Sample	Binder agent (%)	σ max (Kg/cm^2)	σ yield	E	Є max (%)
1A	10	119.7	--------------	---------------	2.7
1B	20	151.9	--------------	---------------	2.5
1C	30	160.3	--------------	---------------	3.0

Table 2. Mechanical test results

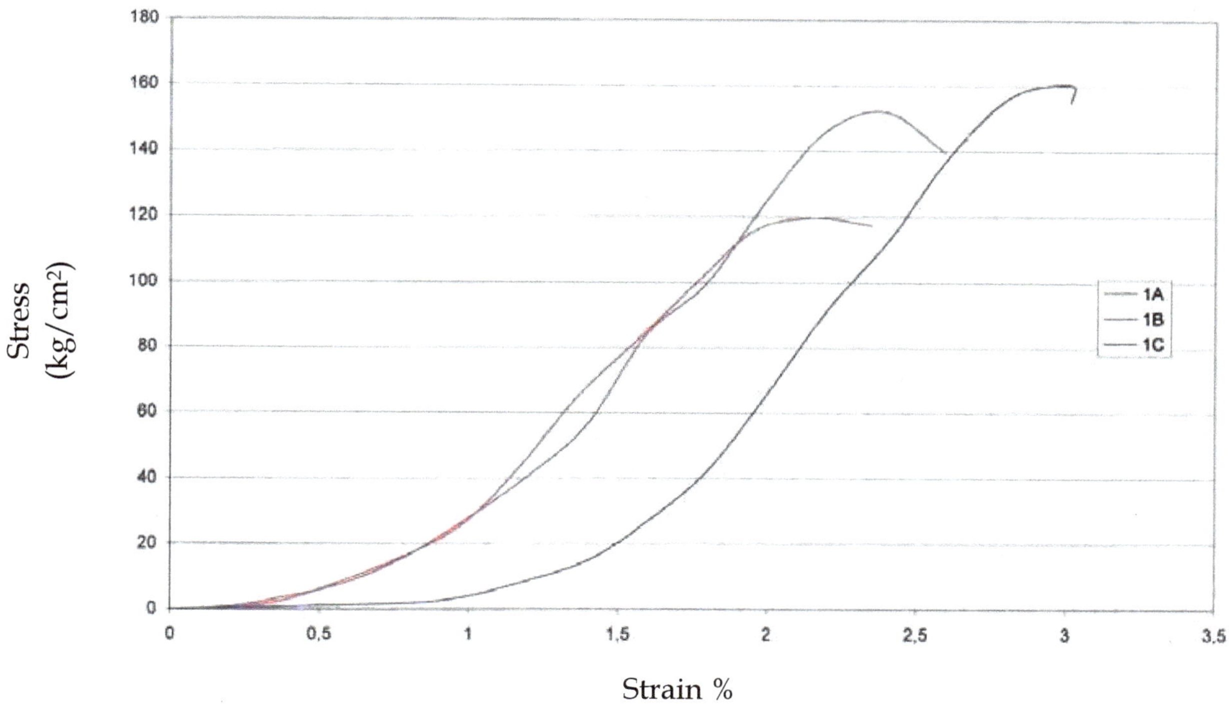

Fig. 11. Graph of the compressive resistance of test samples 1A, 1B and 1C

There are a variety of granular solid industrial residuals in all industrial countries, of anthropogenic origin and produced by nature, with which it is possible to generate proposals based on technical feasibility, economic and social development for the implementation of processes that help to the production of ceramic materials [10] or inputs for the construction industry. Due to this fact, the use of the patented equipment and the process employed in this work can provide sustainable economic and environmental benefits.

The management and implementation of solid and semisolid industrial residuals coming from different industrial sectors is important because it is possible to recover its usability by incorporating the production cycle as materials applicable to the construction industry. In addition, the impact to the environment and health of the population caused by proper handling of industrial solid residuals, which are often stored in open areas and dispersed by wind and rainwater.

4. Conclusions

Using thermostated columns can completely remove the cyanide content of "Jales" present in a recent process. After this treatment, the solid material is dried, characterized and processed with a binder agent, the final result provides a material with excellent mechanical properties for application in construction industries.

The bricks obtained had excellent physical and mechanical properties, such as greater resistance to compression than conventional bricks.

The materials obtained are 20-25% lighter than conventional materials, are also less porous and are sustainable, because development costs are 25% lower compared to other materials used in the construction industry.

There is a wide variety of possibilities of using solid residuals coming from mining and granular residuals from diatomaceous earth to be applied in the construction industry or to the development of ceramic materials.

5. References

Botz, M., Stevenson, J. (1995). Cyanide Recovery and Destruction. *Engineering and Mining Journal.*, Vol.196, pp. 44-47, ISSN 0095-8948.

Li, J., Wadsworth, E.M. (1993). Indirect Manipulation of a Sphere on a Flat Disk Using Force Information. *J. Electrochemical Soc.*, Vol.140, No.7, pp. 1921-1927, ISSN 0013-4651.

Orozco, R.V., Orozco, Y.Y. (1992). Las Presas de Jales en México, Criterios Básicos para su Proyecto, Construcción y Operación, XVI Reunión Nacional de Mecánica de Suelos, México, 1992.

Sengupta, M. (1993). *Environmental Impacts of Mining: Monitoring, Restoration and Control,* Lewis Publishers, ISBN 0873714415, United States of America.

Vite, J. (1994a). Apparatus for Extracting Metal Values from Foundry Sands. U.S. Patent 5376000.

Vite, J. (1994b). Process for Extracting Metal Values from Foundry Sands. U.S. Patent 5376001.

Vite, J. (1997). Aparato para extraer Metales Valiosos de Arenas de Fundición. Mexican Patent 186702.

Vite, J., Vite, M., Castillo, M., Laguna-Camacho, J.R., Soto, J., Susarrey, O. (2010). Erosive wear on ceramic materials obtained from solid residuals and volcanic ashes. *Tribology International,* Vol.43, pp. 1943-1950, ISSN 0301-679X.

Vite, M., Vite, J., Laguna-Camacho, J.R., Castillo, M., Marquina-Chávez, A. (2011). Abrasive wear on ceramic materials obtained from solid residuals coming from mines. *Wear,* Article in press, ISSN 0043-1648.

Vite-Torres, J., Díaz-Calva, A., Vite-Torres, M., Carreño de León, C. (2003). Application of Coupled Thermostated Columns in Civil Engineering and for Leaching Heavy Metals of Wastes from Foundry Sands and Mining. *International Journal of Environment and Pollution,* Vol.19, No. 1, pp. 46-65, ISSN 0957-4352.

Wills, B. (1994). *Tecnología del Procesamiento de Minerales,* Editorial Limusa, México.

4

Spectroscopic and Dielectric Characterization of Plasma Sprayed Titanates

Pavel Ctibor[1] and Josef Sedlacek[2]
[1]*Institute of Plasma Physics, ASCR, v.v.i., Prague,*
[2]*Department of Electrotechnology, Faculty of Electrical Engineering, Czech Technical University, Prague*
Czech Republic

1. Introduction

Synthetic perovskite $CaTiO_3$ (CT), geikielite $MgTiO_3$ (MT) and their mixture $MgTiO_3$-$CaTiO_3$ (MCT) are materials well known and widely used as dielectrics in a sintered state. CT is high-permittivity linear dielectric material whose structure is not influenced by plasma spraying - neither chemical nor phase composition, as demonstrated earlier [Ctibor, 2003]. MT belongs to the family of low-loss microwave dielectrics and MT-CT solution is known by its temperature stability of permittivity. Namely the composition $(Mg_{0.95}Ca_{0.05})TiO_3$ is used in connection with this feature.

In recent decades plasma spraying has become a well accepted technology as the coating method for metallic and ceramic materials and has been used in a variety of fields including electrical engineering. The coatings have lamellar character of a body formed with porosity aligned with respect to the lamellas. Spraying does not require heating of the substrate to the melting point of the deposited material.

Often the research is focused on the microstructure of plasma sprayed coatings and features like porosity, unmelted particles, cracks and residual stress. Above listed characteristics are responsible for the behavior of coatings. However in the case of titanates phenomena taking place on atomic level and single crystal cell level are also important. Raman spectroscopy, infrared spectroscopy and near-field microwave microscopy are suitable techniques for this characterization.

Ca and Mg have the same charge but different ionic radii (rCa^{2+} = 0.134 nm and rMg^{2+} = 0.103 nm) [Hirata, 1996]. A mixture of CT and MT melts and forms an eutectic liquid at 1462°C, which, under proper solidification conditions, can be used to achieve a highly dense product. During processing presence of intermediate phases of $MgTi_2O_5$ and Mg_2TiO_4 was noted, and they were difficult to eliminate completely from the reaction products [Zhang, 2006; Zheng, 2003; Huang, 2002].

MCT exhibits differences from MT in the metal-oxygen bond lengths which are relevant to the stability of the compounds. While the infrared and Raman spectra of CT and MT have been measured [Jiang, 1998; Cavalcante, 2008], other cations at the *A* and/or *B* sites alters the vibrational properties of ABO_3 studied by these spectroscopic techniques. This encourages the author to measure the Raman and infrared spectra. In the plasma sprayed MCT we have earlier recognized also $MgTi_2O_5$ and Mg_2TiO_4 phases [Ctibor, 2003].

Besides the mentioned dielectric titanate materials we have sprayed also one representative of ferroelectrics - barium titanate. $BaTiO_3$ (BT) is an interesting multifunctional oxide that exhibits complex phase appearance. Between 120°C (393K) and 1457°C (1730K) $BaTiO_3$ has a cubic perovskite structure that consists of corner linked oxygen octahedra containing Ti^{4+}, with Ba^{2+}. Cooling below 120°C results in small displacements in the positions of the cations in the unit cell resulting in polar ferroelectric phase existing in the temperature interval between 5°C (278K) and 120°C [Boutinaud, 2006].

$BaTiO_3$, due to its high dielectric constant, is used frequently as multilayer capacitor components and sensors. However, it has been found that with respect to the electrical properties $BaTiO_3$ in the form of thin-films does not reach the qualities of bulk material. This difference was explained by a combination of the intrinsic dead layer effect, a stress effect, an effect of the microstructure within the thin film, and an effect of the stoichiometry [Zhao, 2008]. In particular, the relative permittivity of films decreases when the film thickness is reduced [Setter, 2000]. The optimal dielectric characteristics are obtained for sintered $BaTiO_3$-based sample with bulk density of about 5300 $kg.m^{-3}$ [Jin, 2003]. Any deviation from the stoichiometric Ba/Ti ratio leads to suppression of the high relative permittivity of the ferroelectric barium titanate [Mitic, 2001]. To detect the tetragonal $BaTiO_3$ phase by X-ray diffraction, the split of peaks of (002) and (200) reflection is a well-established indication [Waser, 1999; Yu, 2009; Simon-Seveyrat, 2007].

In general there are differences of the behavior of barium titanate in the form of a single-crystal, sintered bulk material and thin film [Boutinaud, 2006; Mitic, 2001]. Plasma spraying enables to create layers with 'bulk-like' thickness but adhering on a metallic substrate of various shapes. Free-standing parts of titanate ceramics can be fabricated as well by plasma spraying [Wu, 2009]. $BaTiO_3$ itself was up to now very seldom plasma sprayed and the understanding of its behavior in the form of sprayed coating is not satisfactory. For coatings with the thickness of about 100 μm the values of relative permittivity 50 and loss factor 0.08 were reported [Dent, 2001]. The dielectric properties of the plasma sprayed $BaTiO_3$ were related to the degree of crystallinity [Dent, 2001]. The coatings containing more crystalline material have higher relative permittivity. The relative permittivity was affected also by cracks and splat interfaces within the coating [Dent, 2001]. The reported value of relative permittivity is however surprisingly low, because one and more orders higher values are typical for bulk $BaTiO_3$ [Buchanan, 2004].

In frame of the presented chapter we are focused on selected aspects of the dielectric characteristics of the as-sprayed barium titanate coatings and we provide comparison of them with other plasma sprayed titanates.

2. Experimental

2.1 Feedstock materials

All materials were obtained in the form of tablets of industrial purity, produced by the sintering of micropowders. The sintering was carried out by companies Epsilon (Librice, Czech Republic), Ceramic Capacitors (Hradec Kralove, Czech Rep.) and Teceram (Hradec Kralove, Czech Rep.).

The synthetic form of perovskite $CaTiO_3$ was produced by reactive sintering of CaO and TiO_2. $CaTiO_3$ powder used for experiments was sintered without any additives (like ZnO), normally used for decreasing the sintering temperature. Tablets were crushed and sieved into a powder of the correct size for spraying. $MgTiO_3$ and MCT were sintered using MgO, $CaCO_3$ and TiO_2. After sieving the size distribution of the feedstocks was 63–125 microns for all three materials.

$BaTiO_3$ feedstock powder was obtained by crushing and sieving of sintered coarse agglomerates. Those agglomerates were prepared by a reactive sintering of micrometer-sized powders of $BaCO_3$ and TiO_2 used as starting materials. After sieving the size distribution of the BT feedstock for spraying was between 20 and 63 µm with an average at 40 µm, whereas the bulk density measured by helium pycnometry was 5721 kg.m^{-3}.

2.2 Plasma spraying

The CT, MT and MCT samples were manufactured using a high throughput Water-Stabilized Plasma (WSP) spray system WSP500® at Institute of Plasma Physics (Prague, Czech Republic) at ambient atmosphere. The WSP system operates at about 160 kW arc power and can process high amounts of material. This system can be used to fabricate deposits similar but not identical to those prepared by means of conventional atmospheric plasma-spray systems based on gas-stabilized torches. As substrates flat carbon steel coupons (Euronorm S355) were used whereas the powder was fed in by compressed air through two injectors. Just before spraying, the steel was grit blasted with Al_2O_3 with a mean diameter of 650 µm. The deposited thickness was about 1.5 mm for self-supporting deposits. Thick deposits were stripped from the substrate by a releasing agent or by thermal cycling between +200 and -70°C.

For manufacturing of BT samples a Gas-Stabilized Plasma gun (GPS) was used to perform Atmospheric Plasma Spraying (APS) process. The conventional d.c. GPS gun F4 consisted of a thoriated tungsten cathode of 10 mm in diameter with a conical tip and a copper anode/nozzle. The plasma gas mixture used was argon/hydrogen with the total flow rate 60 slm. The powder was injected perpendicularly to the plasma jet axis with argon as a carrier gas (at constant flow rate 5 slm at pressure 0.3 MPa for all spray experiments) through an injector located 3 mm downstream (called external injection) of the torch nozzle exit. The system can process 1 to 5 kg/hour of a ceramic powder. Barium titanate was sprayed at arc power around 30 kW. Spray distance was 100 mm and plasma spraying deposition time about five minutes to reach the thickness 0.9 to 1 mm. Substrates, rectangular shaped (120x20 mm^2) 3 mm thick, were made of carbon steel (Euronorm S355). Just before spraying, they were grit blasted with Al_2O_3 with a mean diameter of 400 µm. The substrates were disposed on a rotating sample holder which diameter was 90 mm. This substrate holder was rotated (tangential speed of 1 m/s) with a horizontal axis and simultaneously translated back and forth orthogonally to the plasma jet axis at a velocity of 24 mm/sec, with an excursion of 160 mm, the plasma torch being stationary.

2.3 Characterization techniques

X-ray diffraction (XRD) was performed as a phase identification with SIEMENS D5000™ equipment *allowed* identifying *phases* present within powders and coatings. For estimation of the crystallinity of the plasma sprayed BT coating relative peak areas have been used. These have been calculated from relative ratios of the areas of the three main peaks (101, 111 and 200) from the tetragonal titanate phase.

Raman spectra were collected by the FORAM 685 apparatus which is equipped by a 685 nm laser with output power up to 40 mW. Spectral resolution 8 cm^{-1} and various integration times from 10 to 60 seconds were used. Raman spectroscopy of BT was performed using a Lambda Solutions P1 apparatus – laser wavelength 785 nm, objective 50 x, integration time 25 s.The surface of the coating was polished before the test.

Infrared reflective spectra of CT samples were measured by Bruker IFS 113v Fourier transform spectrometer.

Microwave microscope Agilent AFM 5400 was used for simultaneous monitoring of the surface profile by AFM and dielectric response on external field applied at resonant frequency (approx. 2.6 MHz). This relatively new technique - near-field scanning microwave microscopy (SMM) permits characterization of the effects of inhomogeneities and defects in crystals, films, and compacts on the local dielectric behavior. By moving the tip/cavity assembly over a surface, one can map the microwave cavity resonant frequency *fr* and quality factor *Q* as a function of position and generate images of the sample. In addition to qualitative images, the microscope can provide quantitative characterization of local dielectric properties [Cheng, 2003].

The complex dielectric permittivity of CT was studied in the 440 Hz – 1 MHz frequency range and 10–270K temperature interval using HP 4192A impedance analyzer with a Leybold He-flow cryostat (operating range 5–300 K). The impedance of the cylindrical sample with Au electrodes sputtered on the cylinder ends was recorded on cooling rate of 2 K/min.

All other electric measurements were performed at room temperature. The deposits were stripped of from the substrates. The surfaces were ground after spraying to eliminate surface roughness. A thin layer of aluminum as the electrode was sputtered in a reduced pressure on the ground surface [Ctibor, 2003]. A three-electrode system was used with a guarded electrode, whereas an unguarded electrode was sputtered on the entire surface of the sample opposite side. The electric field was applied parallel with the spraying direction (i.e., perpendicular to the substrate surface). Capacity was measured in the frequency range from 120 Hz to 1 MHz using the impedance analyzer 4284A (Agilent, CA, USA). Applied voltage was 1V AC whereas the stabilized electric source was equipped with a micrometric capacitor type 16451A (Agilent, CA, USA). Relative permittivity ε_r was calculated from measured capacities C_P and specimen dimensions (Eq. 1)

$$C_P = \varepsilon_0 \times \varepsilon_r \times 1/k \qquad (1)$$

where $\varepsilon_0 = 8.854\times10^{-14}$ F cm^{-1}; 1/k [cm] is defined as the ratio between the guarded surface and the thickness of the sample [Morey, 2003].

This same arrangement and equipment was used for the loss factor measurement at the same frequencies as capacity.

Electric resistance was measured with a special resistivity adapter – Keithley model 6105. The electric field was applied from a regulated high-voltage source and the values read by a multi-purpose electrometer (617C, Keithley Instruments, USA). The magnitude of the applied voltage was 100±2V DC. Volume resistivity was calculated from the measured resistance and specimen dimensions. Typically 4 - 5 specimens were measured and the average calculated.

3. Results

3.1 Spectroscopic measurements

Figure 1 shows the Raman spectra of plasma sprayed CT in comparison with the sintered sample. SD means stand-off distance, in millimeters, of the as-sprayed samples and "an" denotes annealed samples. Annealing details are given elsewhere [Ctibor, 2003].

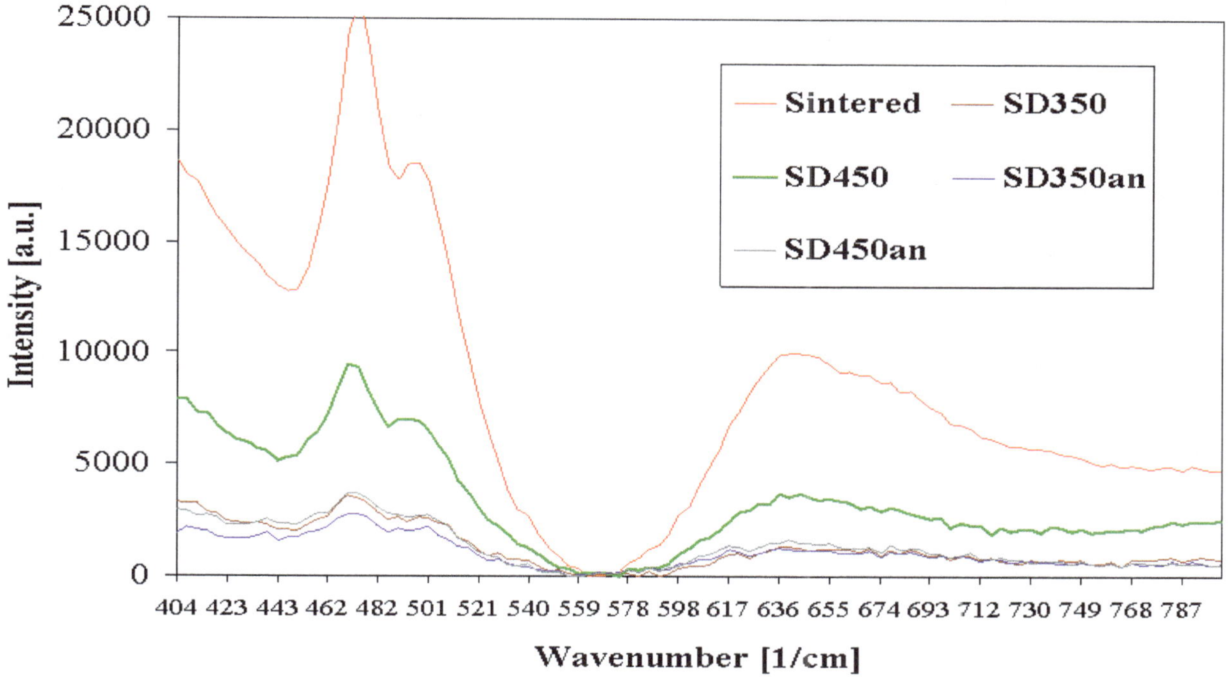

Fig. 1. Raman spectrum of $CaTiO_3$

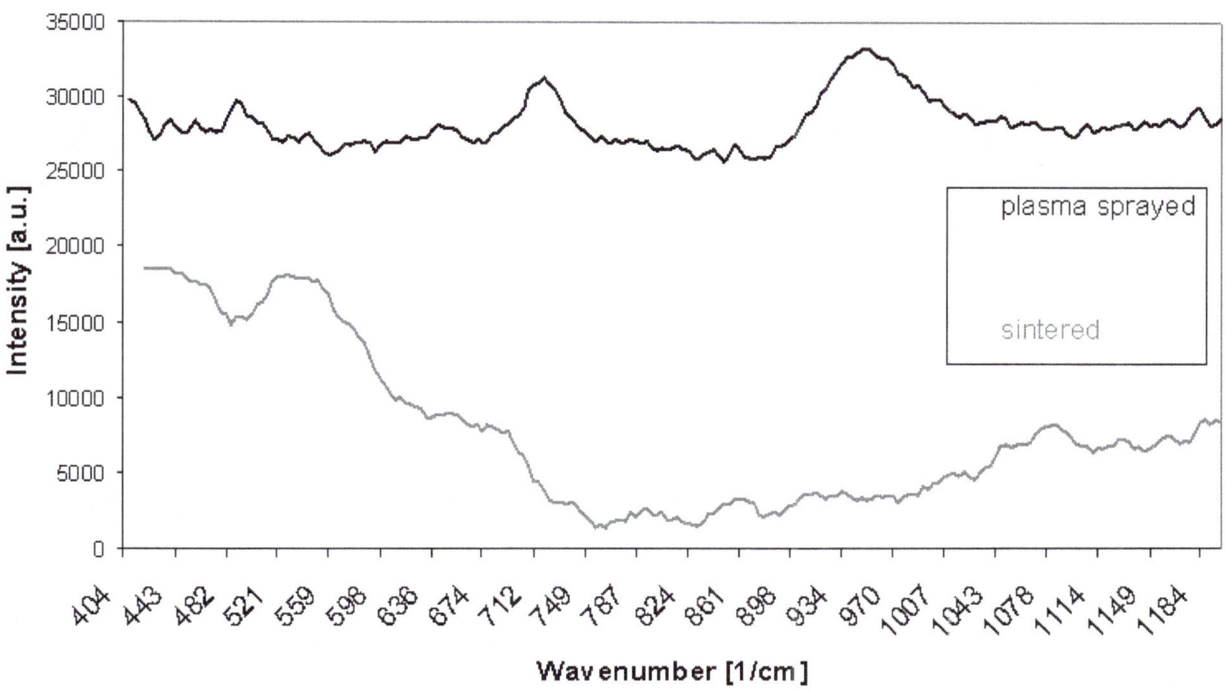

Fig. 2. Raman spectrum of $MgTiO_3$

The absolute values of the intensity of the reflection are associated with surface roughness and could be omitted. From the graph we see that the positions of all peaks are identical for all samples. Wavenumbers of all three main peaks observed at 471, 495, and 640 cm^{-1} are in

agreement with [Cavalcante, 2008]. The bands at 471 and 495 cm^{-1} are assigned to Ti–O torsional (bending or internal vibration of oxygen cage) modes [Hirata, 1996; Zheng, 2003]. The Ti–O stretching mode is centered at 640 cm^{-1} [Boutinaud, 2006]. Two small peaks at about 650 cm^{-1}, suggesting the simultaneous presence of [TiO_6] and [TiO_5] clusters, however in our case they are even less pronounced compare to [Cavalcante, 2008]. Also in our plasma sprayed samples incomplete organization of the $CaTiO_3$ lattice can be attributed to the defects in the covalent bond due to the oxygen vacancies ($VO^{\bullet\bullet}$) between the clusters [TiO_6–$TiO_5 \cdot VO^{\bullet\bullet}$] as in [Cavalcante, 2008]. This is associated with slightly reducing atmosphere in the plasma stream based on H and O atoms (water stabilization).
The Raman spectrum of MT is displayed on the **Fig. 2** in which the active modes at 478 cm^{-1} and 501 cm^{-1} [Hirata, 1996] are observed only on the coating. Next active modes are at 641 and 712 cm^{-1} [Hirata, 1996] and also were detected only on the coating, whereas in the sintered sample a shoulder centered at about 530 cm^{-1} is present.

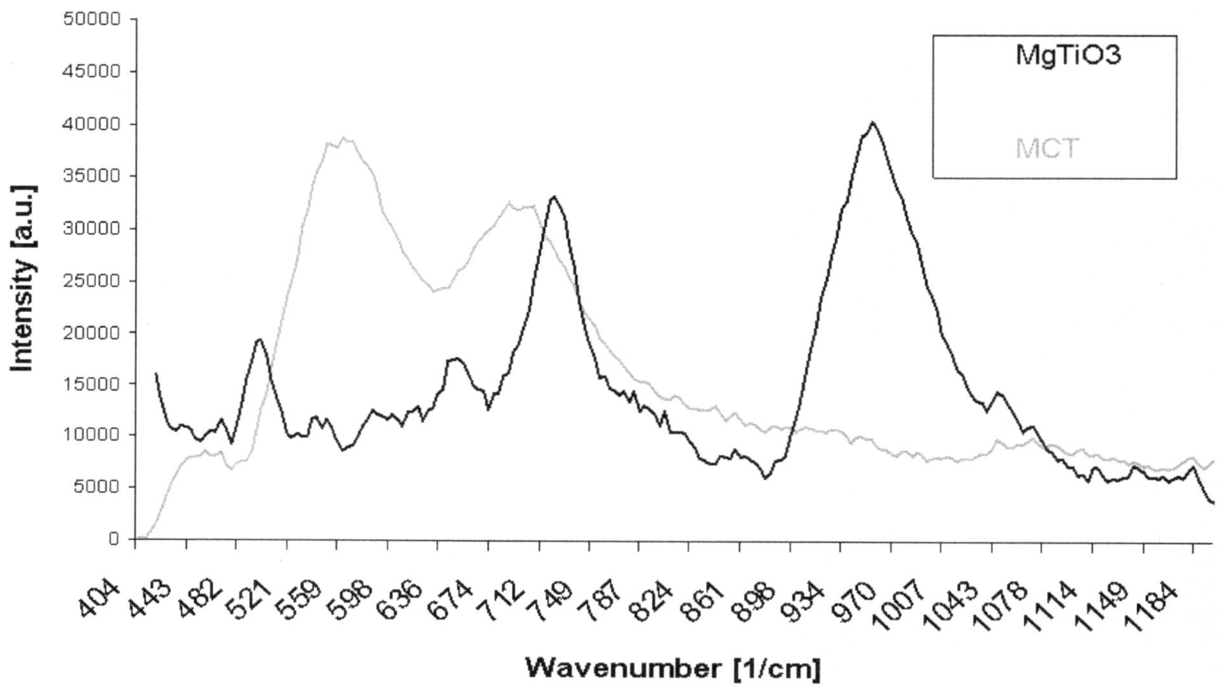

Fig. 3. Raman spectrum of MCT

The Raman spectrum of plasma sprayed MCT is displayed on the **Fig. 3** in comparison with MT measured at exactly the same conditions. We can see for MCT a red shift of the peak centered at 712 cm^{-1} in the case of MT. Other pronounced maximum is at 565 cm^{-1}. The Raman pattern of MCT is different compare to MT and CT and moreover it is not a simple combination of both of them. This is because of $MgTi_2O_5$ and Mg_2TiO_4 origin during the spraying, as confirmed by XRD [Ctibor, 2003]. However only two phases with different relative permittivity were detected by scanning microwave microscopy, see **Fig. 4**, similarly as in [Zhang, 2006]. Permittivities of $MgTi_2O_5$, Mg_2TiO_4 and MT are very similar together and very different from CT. **Figure 4** is a superposition of the AFM contact mode image (roughness) and scanning microwave microscopy capacitance mode image (colors). The image is artificially colored – blue and red zones represent different relative permittivity.

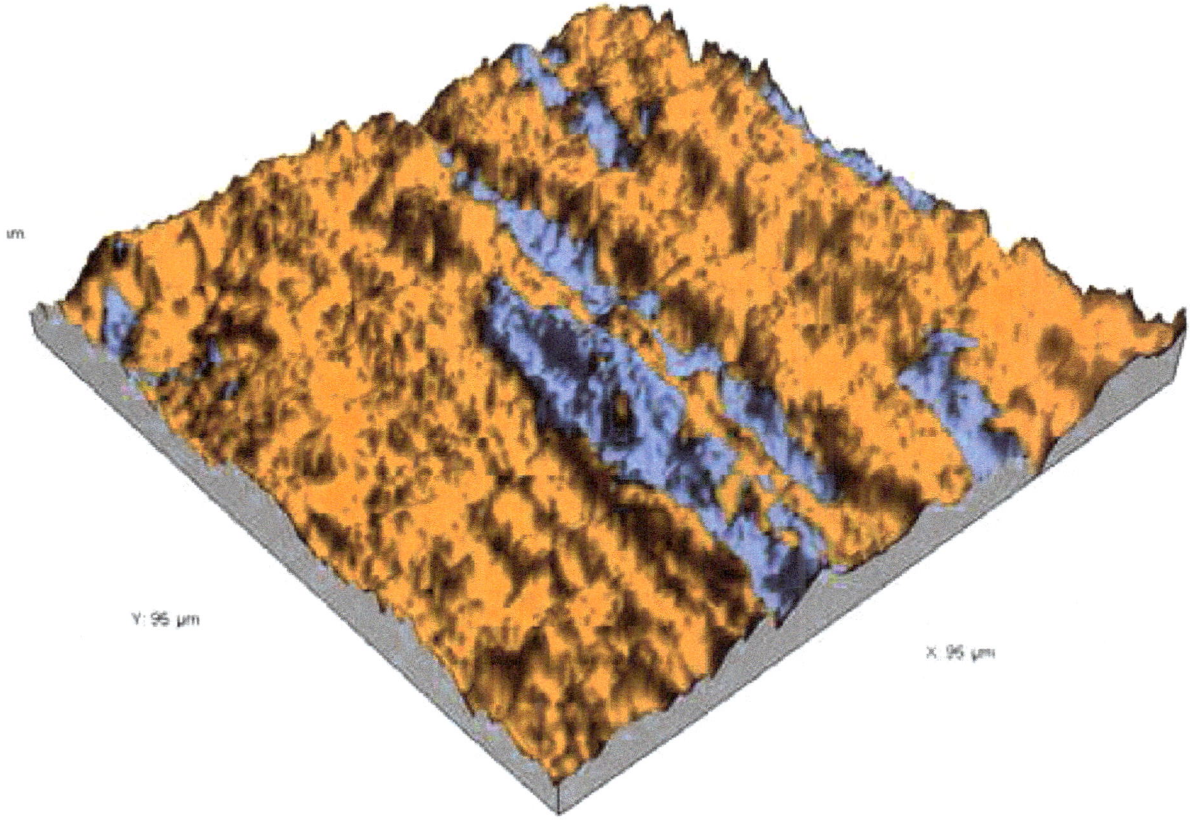

Fig. 4. Scanning microwave microscopy image of MCT as-sprayed surface (artificially colored – blue and red zones represent different relative permittivity)

CT (perovskite) is a high permittivity material $\varepsilon_r' = 170$ [Ferreira, 1997], $MgTiO_3$ (giekielite) has a low $\varepsilon_r' = 17$ [Zeng, 1997] and Mg_2TiO_4 (qandilite) also has even lower $\varepsilon_r' = 12$ [Haefie, 1992]. However SMM setup is not able to distinguish well the two last phases [Wing, 2006], also in our **Fig. 4** the blue lamellas correspond to CT, c.f. **Fig. 5**, and the red background to all other phases.

Raman spectrum of BT coating is displayed on **Fig. 6**. The spectrum with peaks at 311 and 507 cm^{-1} corresponds to tetragonal phase of $BaTiO_3$ [Souza, 2006]. At low oxygen pressure which is the condition relevant for plasma spraying, the density of oxygen vacancies is higher, and then the expansion of the lattice volume is greater [Souza, 2006]. This is why the Raman modes shift to lower frequencies: 507 cm^{-1} in our case instead of 518 cm^{-1} [26] or 532 cm^{-1} in [Souza, 2006] for the A_1 torsion mode. Raman spectrum of the long-SD coating is practically identical.

The modes further split into longitudinal (LO) and transverse (TO) components. The spectrum in **Fig. 6** shows the stretching mode of $A1(TO_1)$, $A1(TO_2)$ and $A1(TO_3)$ at around 163, 259 and 507 cm^{-1}, respectively [Mattsson, 2010; Guo, 2005]. The stretching mode of E (TO_2) appeared at 311 cm^{-1}, while $A1(LO_1)$ stretching modes at 188 cm^{-1} and $A1(LO_2)$ at about 470 cm^{-1} [Mattsson, 2010], however the last one was not very pronounced in our case.

By the Raman spectroscopy presence of TiO_2 in anatase form was mentioned - the peak at about 645 cm^{-1} was observed in TiO_2 film [Giolli, 2007] or coating [Buralcov, 2007]. Elsewhere [Ostapchuk, 2005] such a peak was shown without comments. In our case also week peak at 631 cm^{-1} was detected, which can correspond to anatase-TiO_2 individual phase in the $BaTiO_3$ coating.

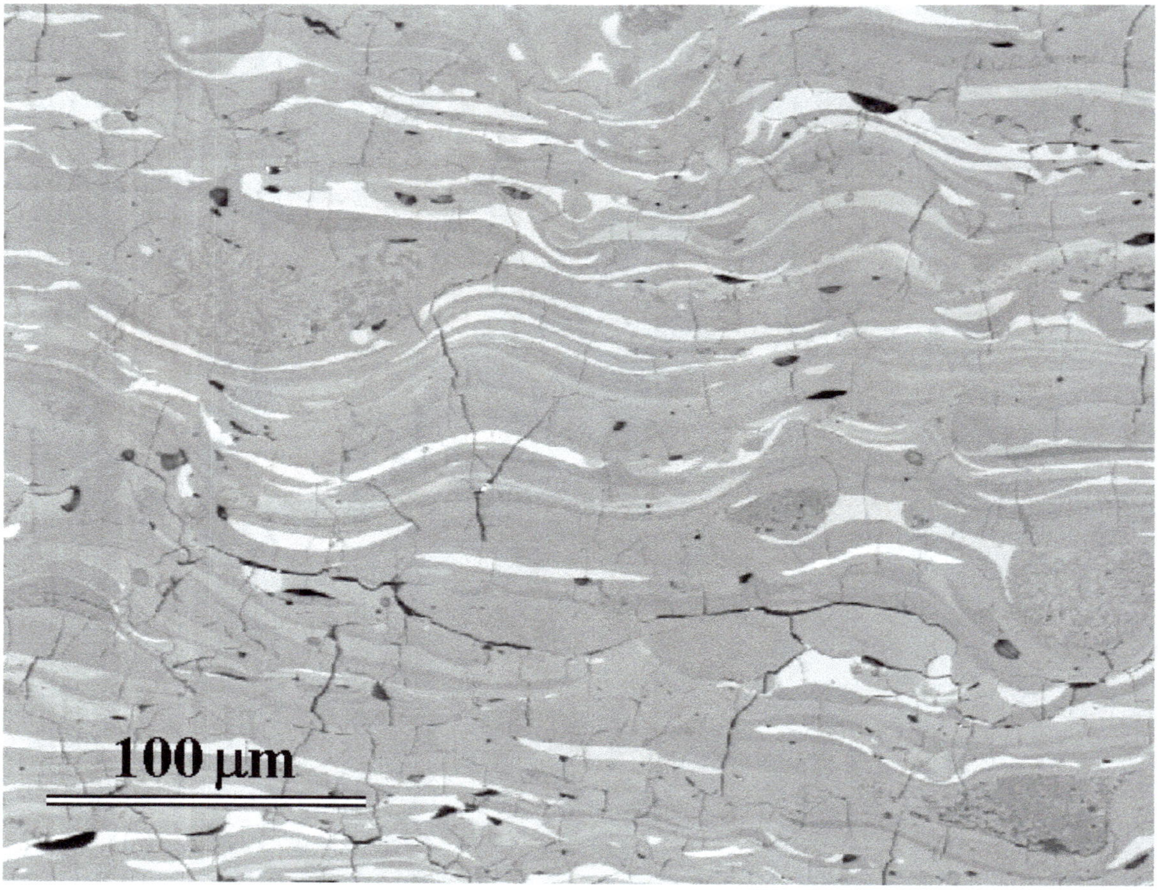

Fig. 5. SEM-BE image showing the lamellar microstructure

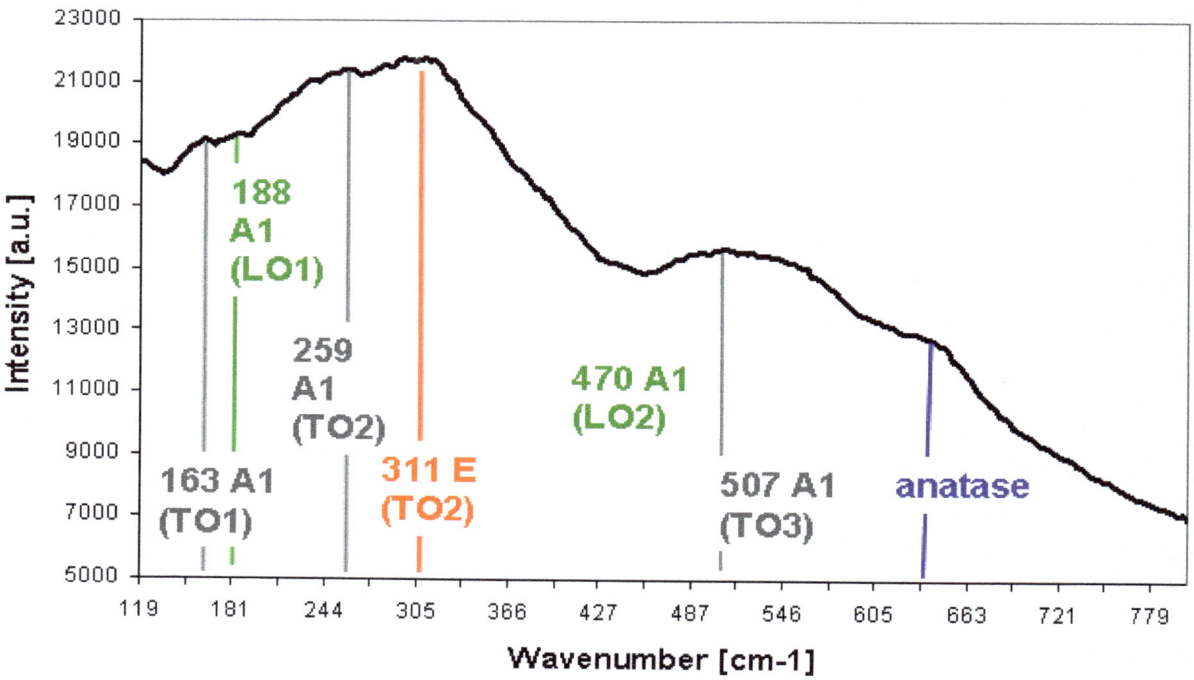

Fig. 6. Raman spectrum of plasma sprayed $BaTiO_3$

Figure 7 shows the infrared spectra of plasma sprayed CT in comparison with the sintered sample. The infrared-active mode at 575 cm^{-1}, assigned to the Ti–O stretch, and also the mode at 455 cm^{-1}, assigned to the Ti–O_3 torsion, are present in both samples with the same intensity. The slightly more pronounced local valley in the case of plasma sprayed sample – localized at about 680 cm^{-1} is the only subtle distinction.

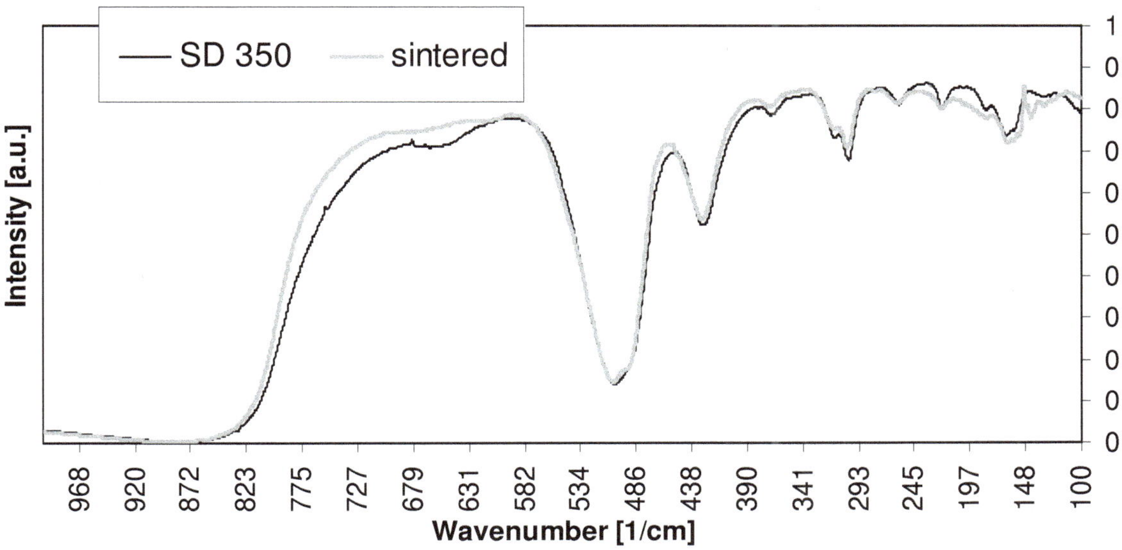

Fig. 7. Infrared spectrum of $CaTiO_3$

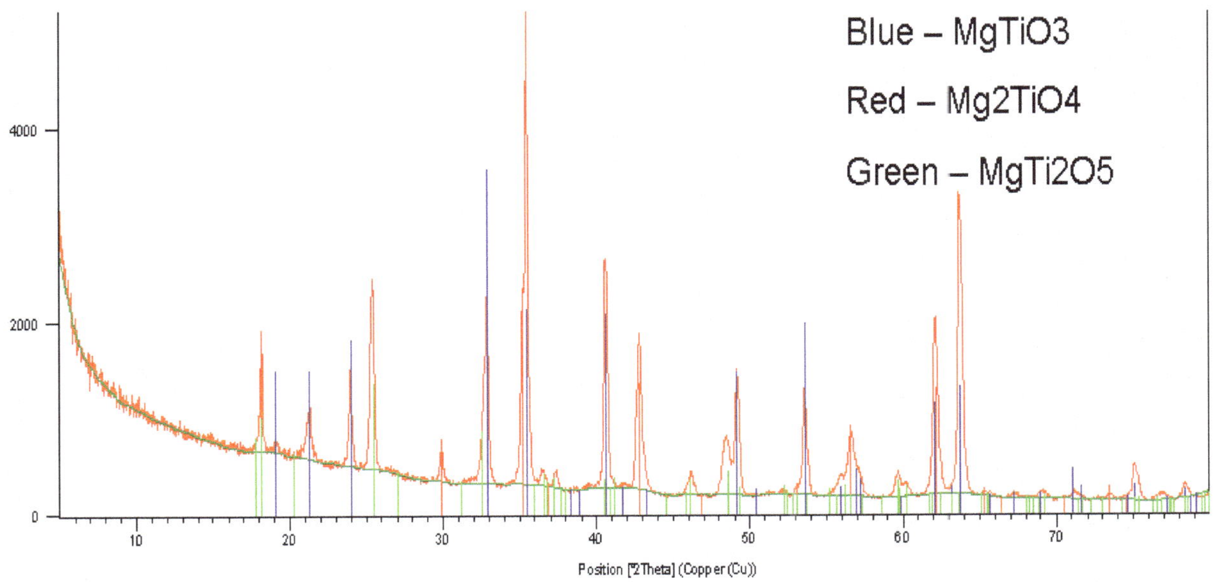

Fig. 8. XRD pattern of plasma sprayed $MgTiO_3$

According X-ray diffraction measurement of CT coating, the phase composition is the same as in the feedstock powder - pure $CaTiO_3$ (PDF2 card No. 00-022-0153). The XRD pattern of MT, **Fig. 8**, shows that the original metatitanate partly decomposed during the spray process on Mg_2TiO_4 and $MgTi_2O_5$. Al these components are present also in MCT coating, **Fig. 9**, whereas also $CaTiO_3$ is present as an individual phase.

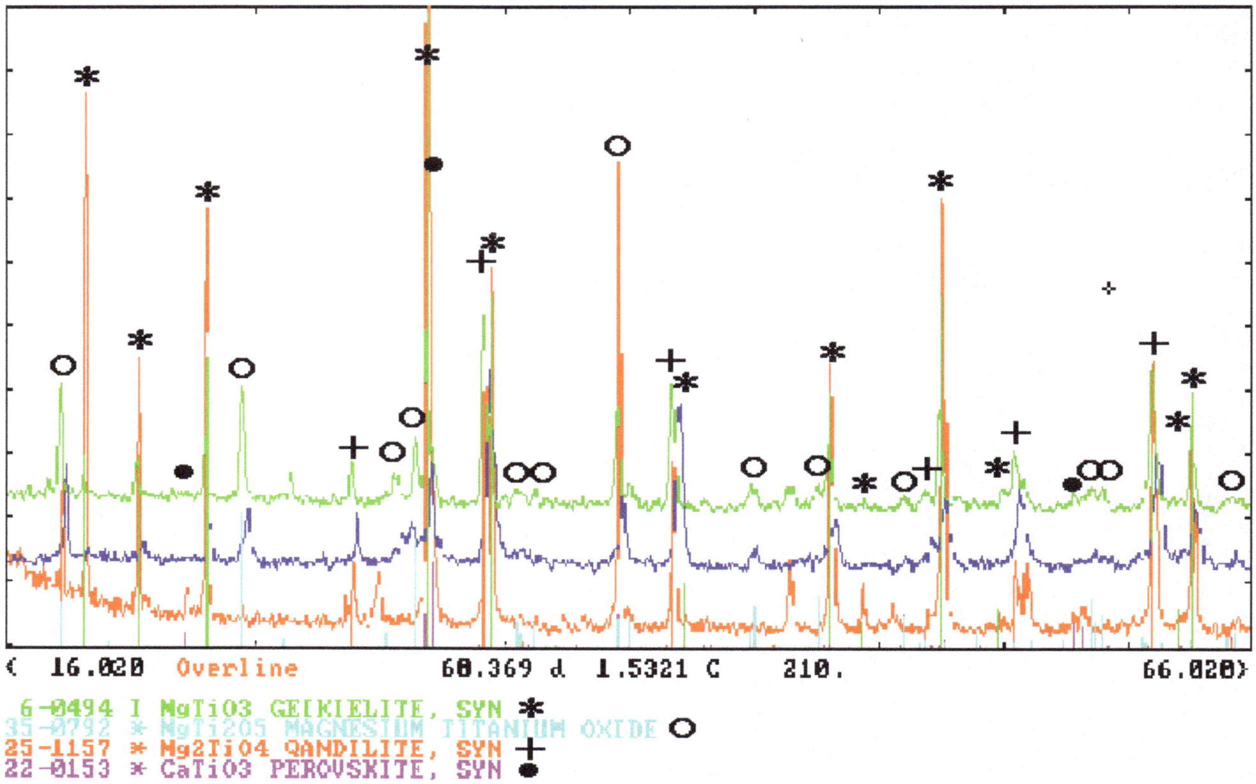

Fig. 9. XRD pattern of plasma sprayed MCT

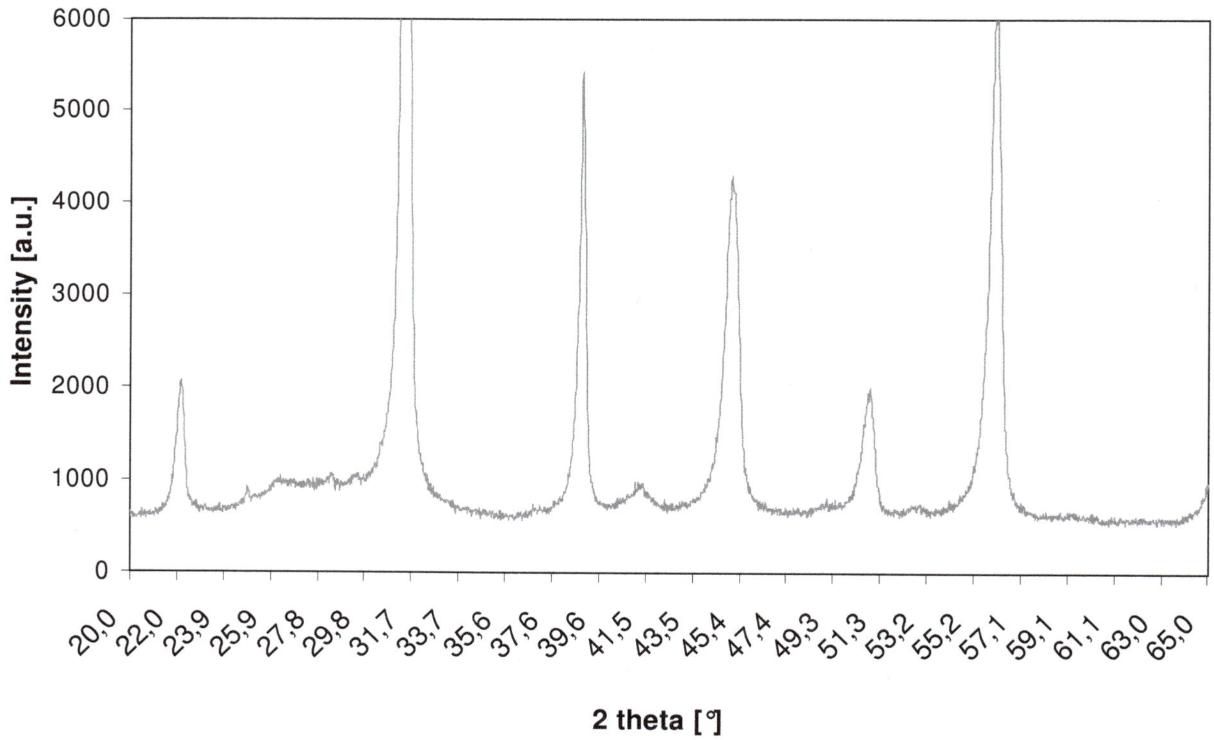

Fig. 10. XRD pattern of plasma sprayed BT coating

The XRD pattern of BT, **Fig. 10**, corresponds strictly to the tetragonal phase (PDF2 card No. 01-081-2204), which was confirmed to be the constituent of the feedstock powder as well as of the coatings. Intensity ratio of individual peaks of the feedstock as well as of the coatings is very similar to patterns reported in the literature [Wang, 2008]. In the coating, certain quantity of amorphous phase is present, manifested itself by a halo centered on 28° 2θ in the pattern, and quantified as 10 %.

3.2 Dielectric measurements

Dielectric properties of plasma sprayed $CaTiO_3$ between 10 K and room temperature are displayed on the **Figures 11 to 13**. The incipient ferroelectricity of CT [Lemanov, 1999; Sudheendran, 2008] is suppressed in the case of plasma sprayed coatings, see **Fig. 11**. Rather relaxor type of behavior takes place. Relaxor features are believed to be connected with a situation when more than one type of ions occupies equivalent crystallographic position. One of the most important features of relaxors is their low frequency dispersion at maximum dielectric permittivity. The other features of this include dependence of phase transition temperature on the method of its determination. The origin of such properties is probably connected with a presence of microscopic polar regions (nano-domains) in these materials [Bak, 2009]. With temperature falling down also the ε'' values decrease to minimum (for a given frequency), cf. **Fig. 12**.

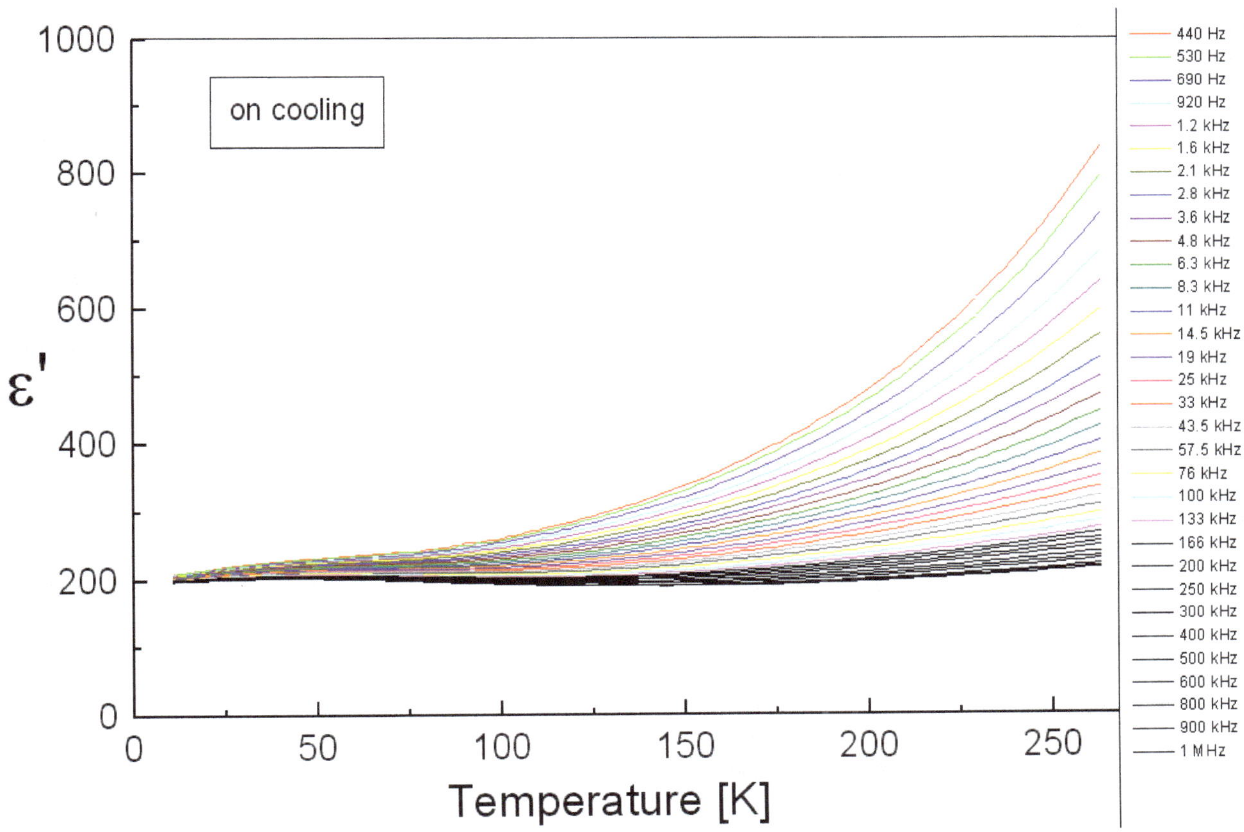

Fig. 11. Real part of permittivity of plasma sprayed $CaTiO_3$ between 10 K and room temperature

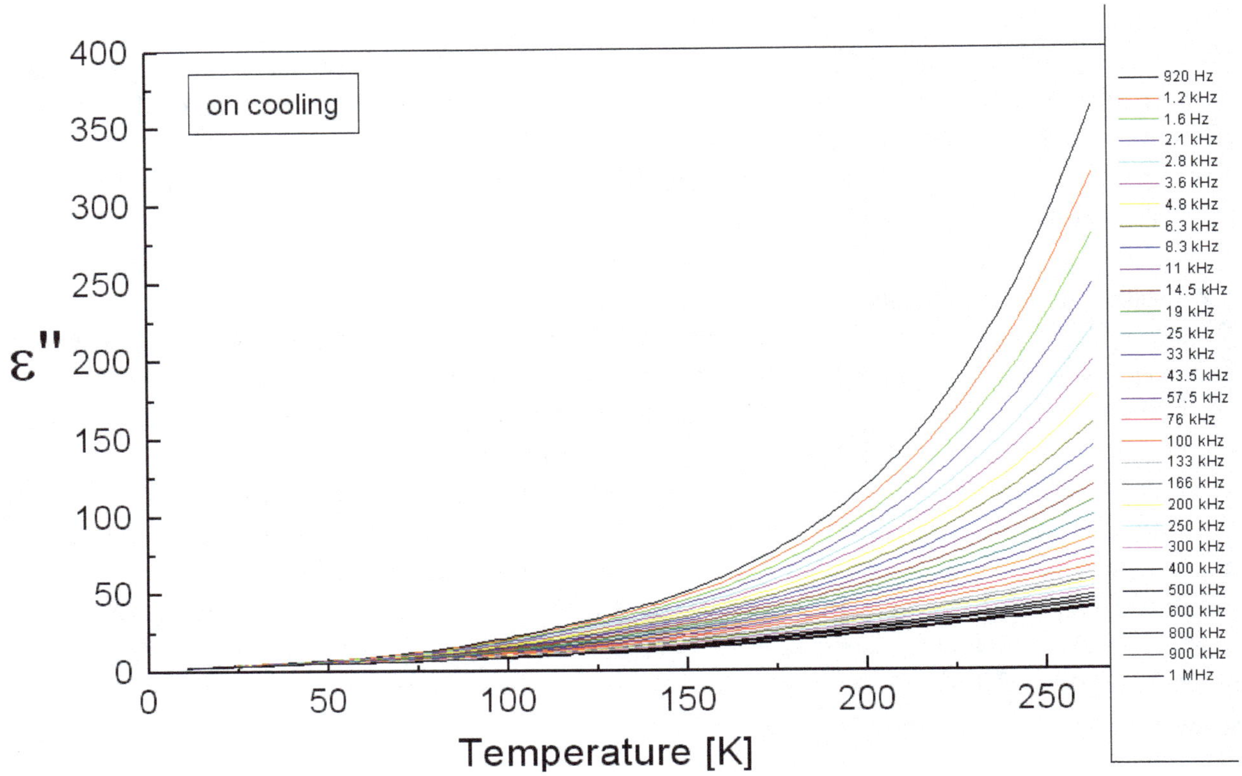

Fig. 12. Imaginary part of permittivity of plasma sprayed $CaTiO_3$ between 10 K and room temperature

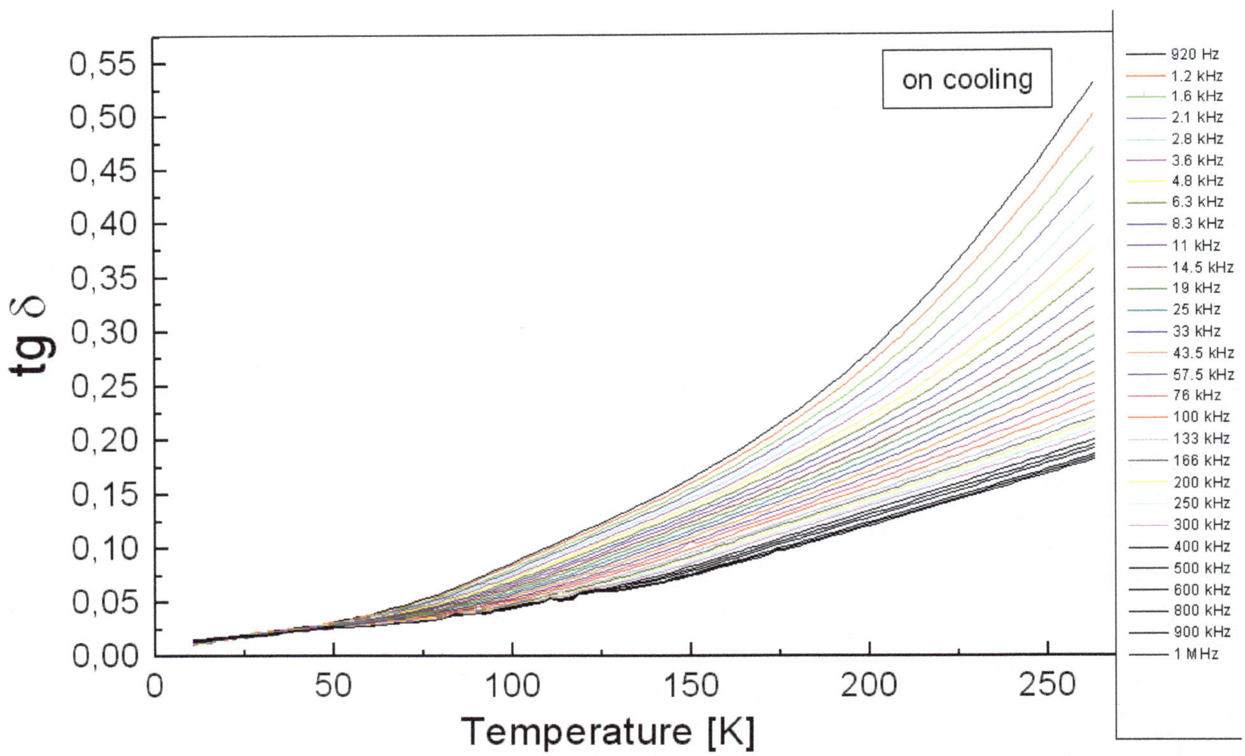

Fig. 13. Loss tangent of plasma sprayed $CaTiO_3$ between 10 K and room temperature

In the studied frequency range, the relative permittivity of all specimens exhibits significant ionic relaxation process [Dervos, 2004]. Strong relaxation effects is shown by the ε_r' reduction with the operating frequency, see **Fig. 14**. In Ti-based materials, the concentrations of intrinsic defects are determined by oxygen partial pressure and, additionally, variations in defect concentrations can be detected by conductivity measurements. At low oxygen partial pressures the materials show n-type conduction with oxygen vacancy as the dominant defect while at high oxygen partial pressures the materials show p-type conduction with the cation vacancy as the dominant defect. For the different materials the transition from n-type conduction to p-type conduction occurs at different oxygen partial pressures [Hu, 2011]. The partial pressure of oxygen required to reduce for example TiO_2 to Ti_2O_3, Ti_3O_5 or Ti_4O_7 is of the order of 10^{-5} Pa at around 2000°C, while during plasma spraying in the air, oxygen partial pressure does not go below 1 Pa [Ctibor, 2010]. Dielectric losses of our titanate samples, namely BT and CT are however high, see **Fig. 15**. Electric conductivity is responsible for it with dominating contribution of dc-conductivity component. The ε'' dependence can be treated as typical for materials with dc-conductivity which dominates over relaxation processes [Bak, 2008].

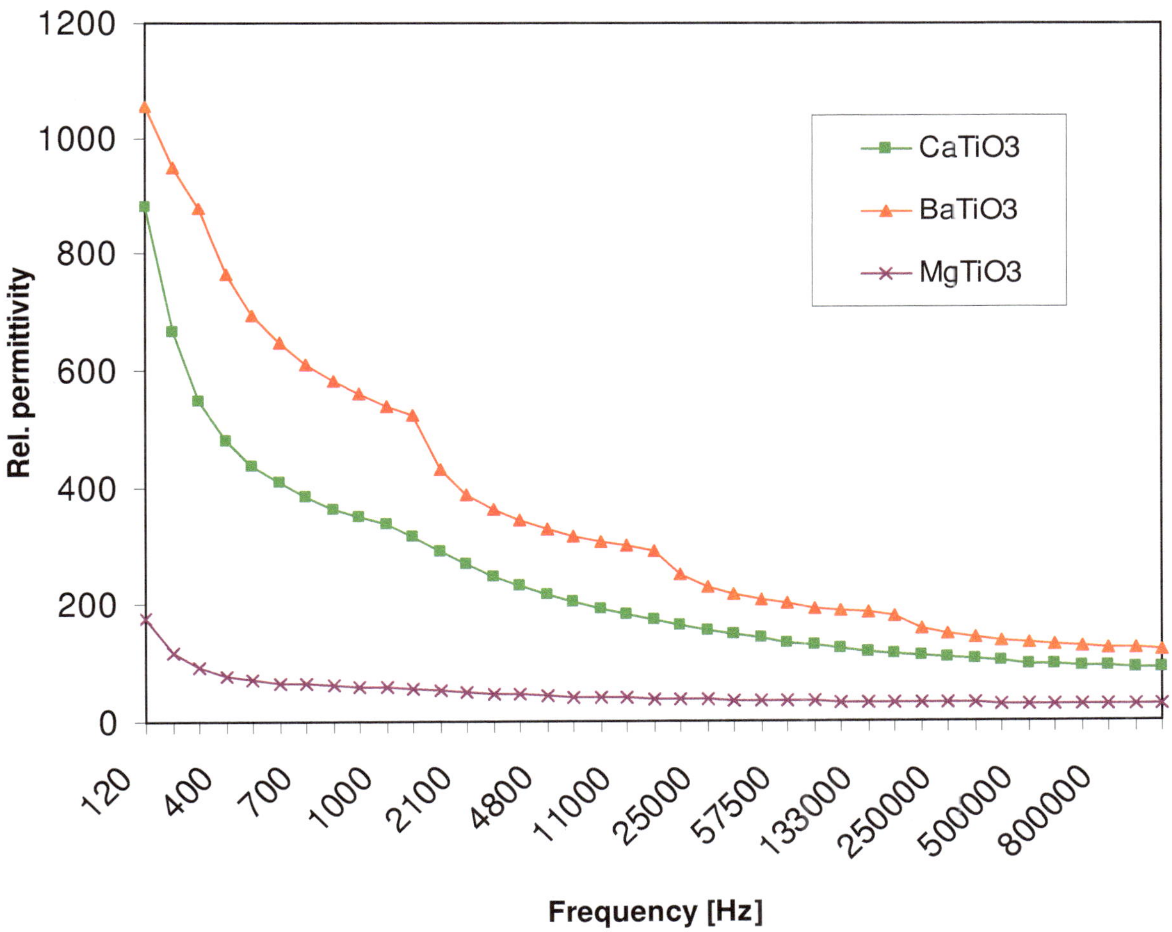

Fig. 14. Frequency dependence of the relative permittivity of plasma sprayed $BaTiO_3$, $CaTiO_3$, and $MgTiO_3$ at room temperature

Table 1 shows the volume resistivity and **Fig. 16** corresponding conductivity of plasma sprayed and sintered BT, CT, MT and MCT materials. We see the different values with typically several orders of magnitude lower resistivity values for plasma sprayed specimen compare to sintered materials.

Sample	BT (P)	BT (S)	CT (P)	CT (S)	MT (P)	MT (S)	MCT (P)	MCT (S)
ρ [Ωm]	6.08×10^4	2.42×10^9	7.51×10^9	1.41×10^{12}	1.15×10^8	6.90×10^{11}	1.17×10^7	7.54×10^{11}

Table 1. Volume resistivity of studied materials (P – plasma sprayed, S – sintered)

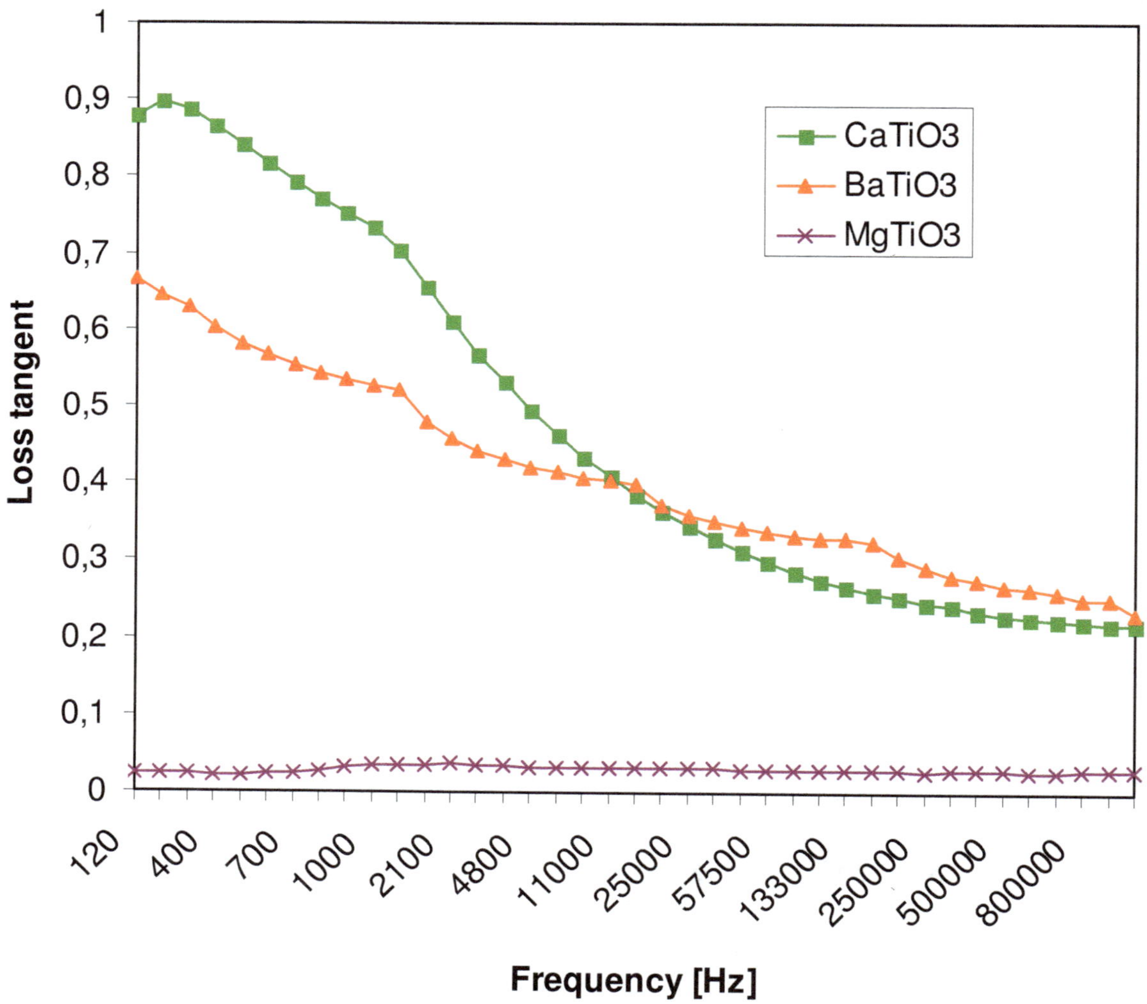

Fig. 15. Frequency dependence of the loss tangent of plasma sprayed $BaTiO_3$, $CaTiO_3$, and $MgTiO_3$ at room temperature

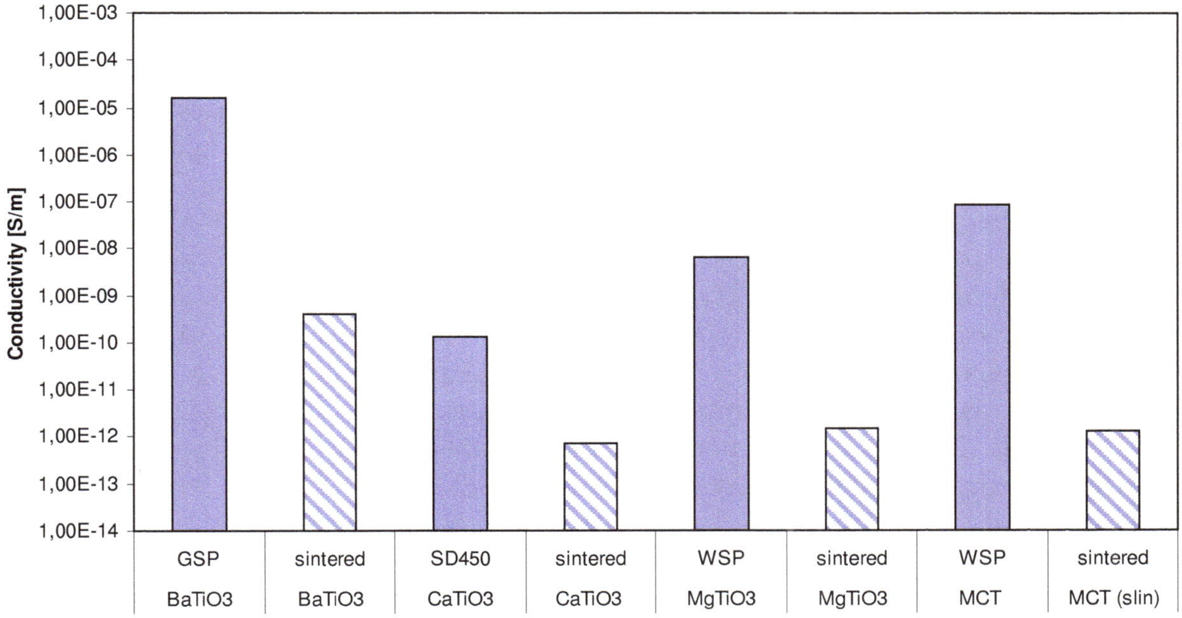

Fig. 16. Conductivity of plasma sprayed and sintered BT, CT, MT and MCT

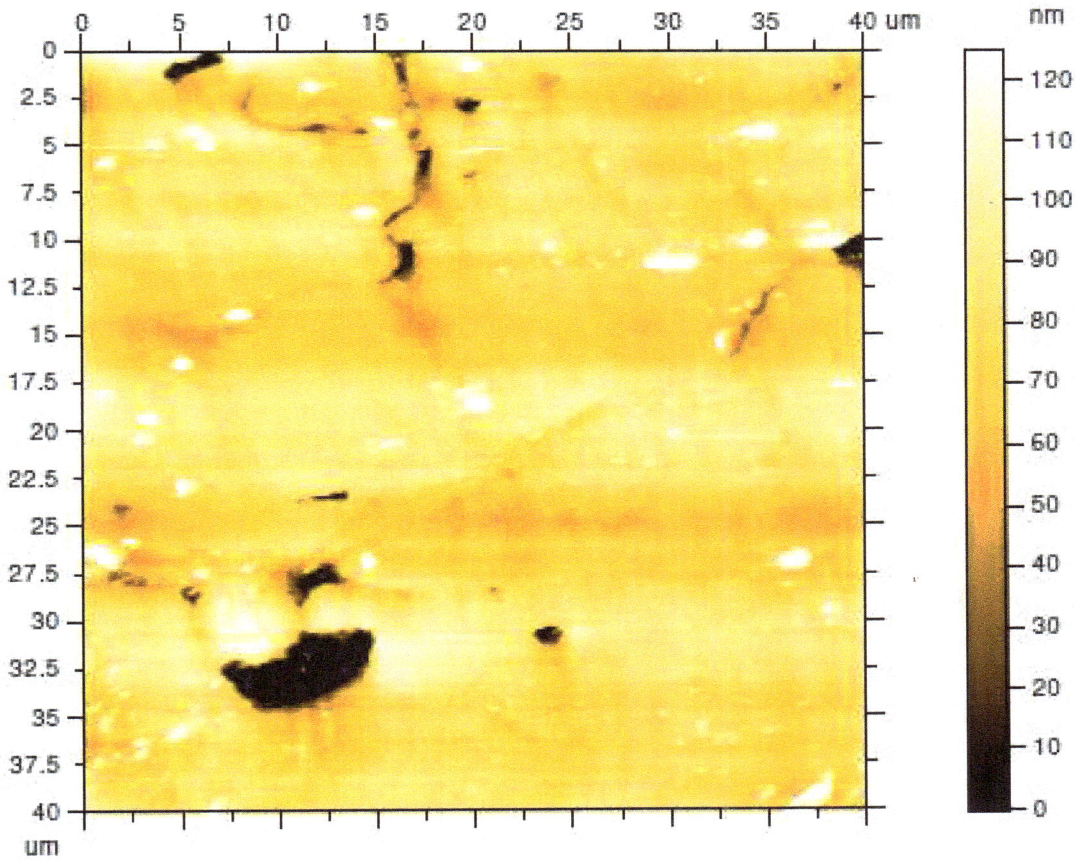

Fig. 17. AFM contact mode image of $BaTiO_3$ (in plane orientation). Roughness representation: peaks are lighter and valleys darker (black color represents pores)

3.3 Scanning microwave microscopy

Scanning microwave microscopy images of BT are given on **Figs. 17** and **18**. AFM images represent surface roughness. Since the surface was materialographically polished, the subtle differences in the roughness are associated with hardness of various phases. These phases are concentrated predominantly in individual lamellas, cf. MCT SEM micrograph in the **Fig. 5**. In the case of BT, the difference between lamellas is mainly in terms of Ba/Ti ratio [Ctibor 2, 2010]. Such a feature is, as expectable, more easily visible on cross sections, **Fig. 19,** then on in-plane (i.e. spray direction) sections **Fig. 18.** The lamellas, red and yellow on capacitance mode images, are more Ti-rich (higher permittivity) whereas blue areas are Ba-rich (lower permittivity). The Ba-rich phase has slightly lower hardness (hardness measured by us: $BaTiO_3$ is about 6.5 GPa versus TiO_2 about 11.0 GPa; BaO not reported as a layer). The Ba-rich phase is darker on AFM contact mode image, since it is more easily worn-out by polishing process. These "steps" on border of lamellas are about 80 nm high, see also **Fig. 20**, which height is consistent with observations made by SEM.

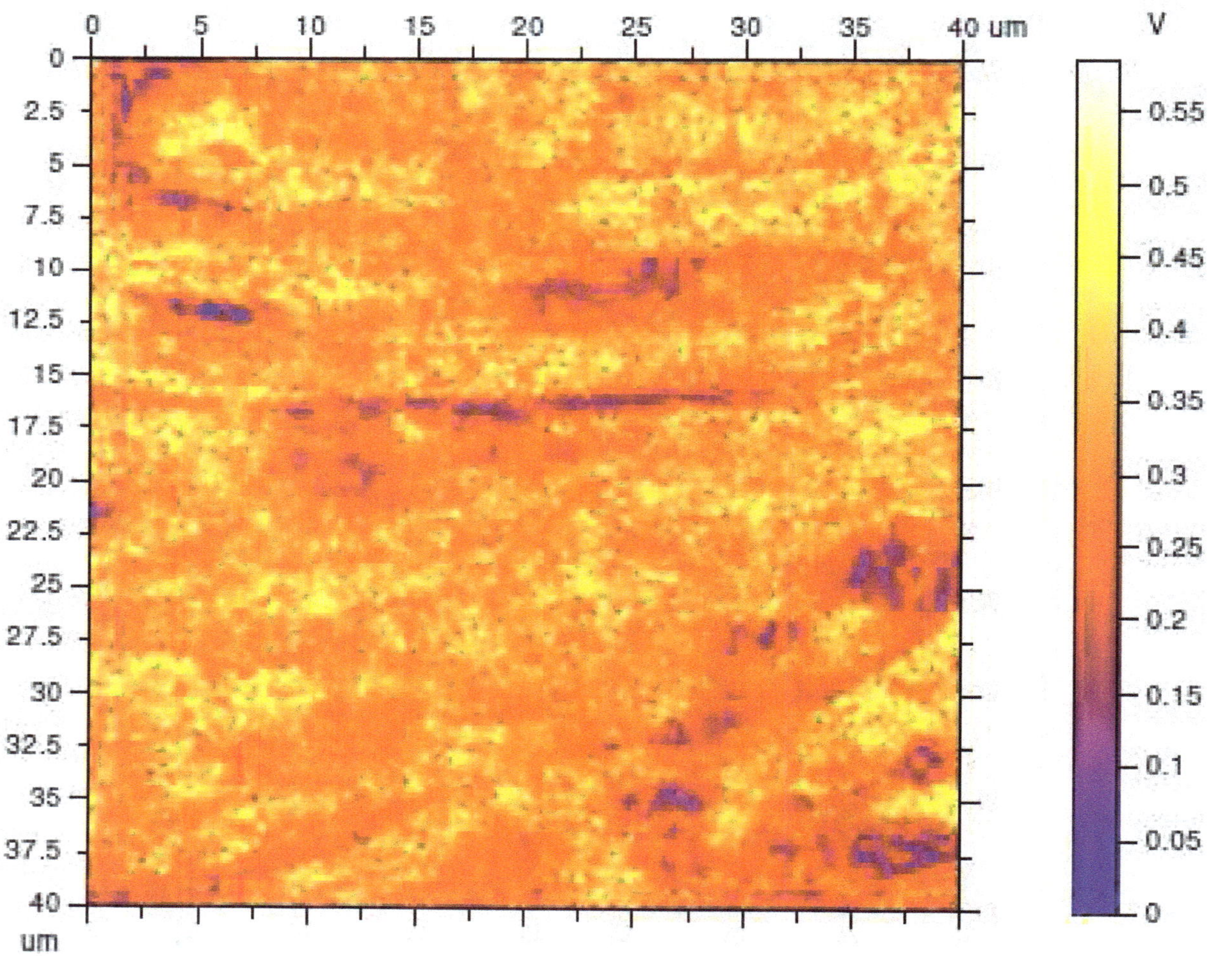

Fig. 18. Scanning microwave microscopy image of $BaTiO_3$ (in plane orientation) - artificially colored: blue and red zones represent different relative permittivity

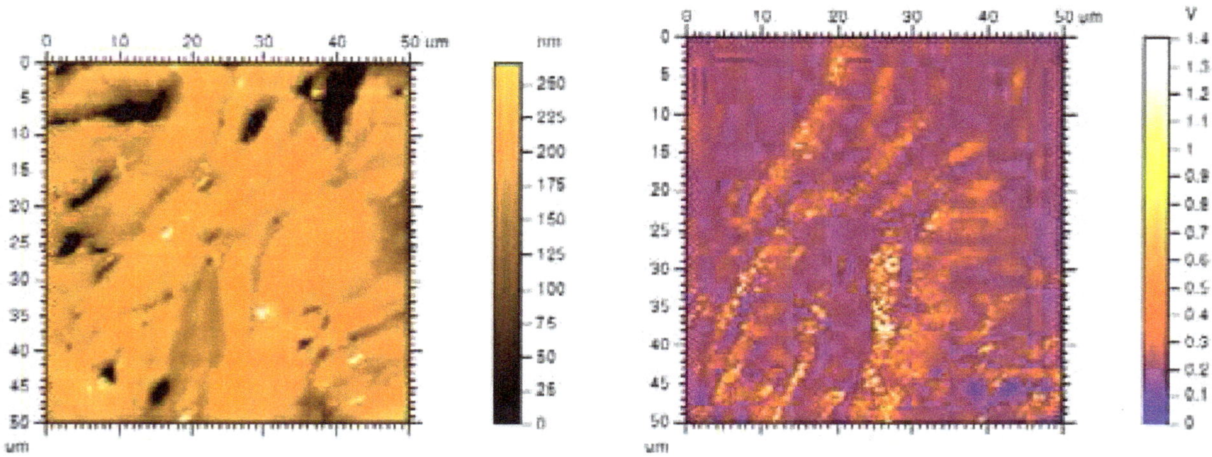

Fig. 19. AFM contact mode image of $BaTiO_3$ (cross sectional orientation). Left: Roughness representation: peaks are lighter and valleys darker (black color represents pores). Right: Scanning microwave microscopy image of $BaTiO_3$ (cross sectional orientation) - artificially colored: blue and red zones represent different relative permittivity

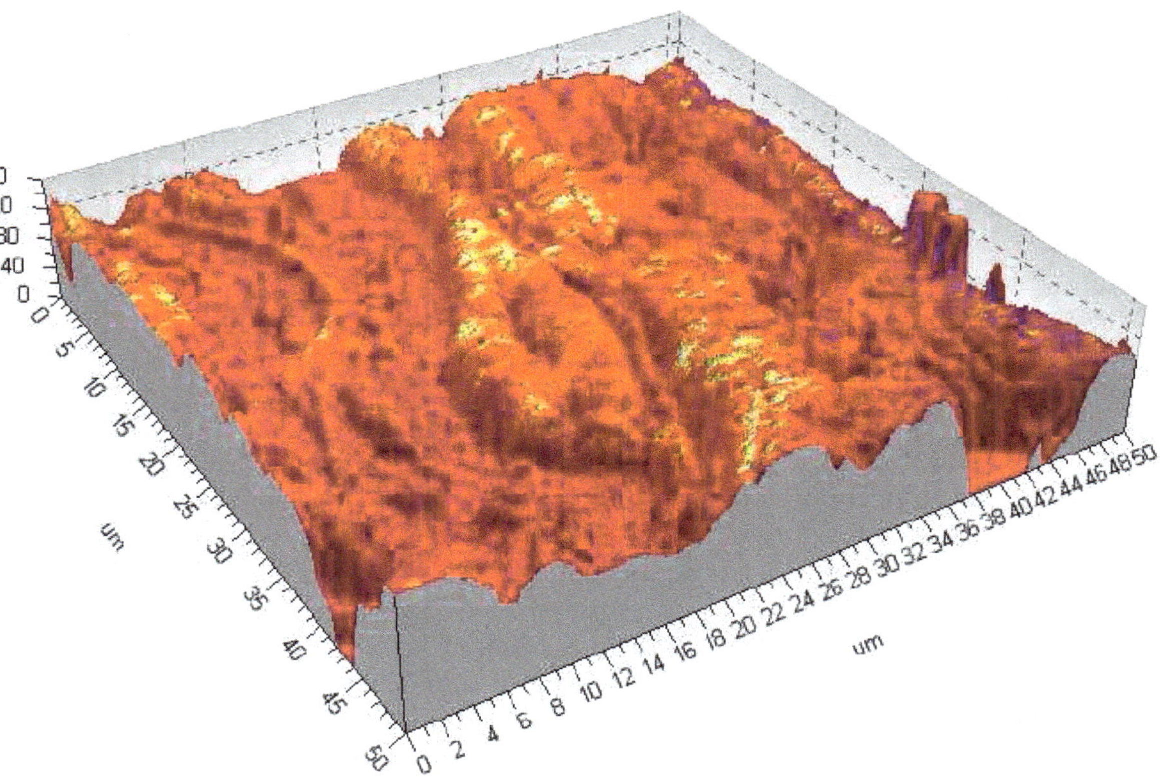

Fig. 20. Superposition of the AFM contact mode image and SMM image - BaTiO3 cross section. Hard and high-permittivity Ti-rich lamellas are yellow

4. Conclusions

$BaTiO_3$, $CaTiO_3$, $MgTiO_3$ and a mixture of the last two materials with 5 wt.% of $CaTiO_3$ in $MgTiO_3$ were plasma sprayed by two complementary spray techniques. Raman and infrared spectroscopic measurements were carried out on plasma sprayed titanates. Coating samples were tested as mono-block capacitors. Volume dc-resistivities of the materials were summarized. Microwave microscopy was used for mapping of the dielectric response of selected samples.

$CaTiO_3$ exhibits anomalous dielectric losses because of intrinsic conductivity. Relaxor type of behavior takes place at dielectric measurements, whereas dc-conductivity is responsible for the high dielectric losses. All coatings are slightly oxygen deficient - however under detection limits of XRD materials - the color of all of them is darker compare to sintered samples.

Polarization of grain boundaries combined with n-type conduction with oxygen vacancy as the dominant defect contributes to the observed pseudo-relaxor response of $BaTiO_3$ and $CaTiO_3$ to AC electric field. $MgTiO_3$ as a low-loss material has the above mentioned effects slightly suppressed and is more similar to sintered bulk. The existence of irregularities in the crystal lattice was confirmed by spectroscopic measurements. Plasma spraying of applicable ceramic dielectrics able to substitute bulk parts will need further research. In the other hand, plasma spraying, with its possibility to cover wide range of substrate materials with thick coatings having very versatile thicknesses and areas, offers promising way to future development.

5. Acknowledgment

The authors thank to J. Dubsky (IPP ASCR) for XRD measurements, to M. Savinov (Inst. of Physics, ASCR) for the temperature measurement of dielectric properties of $CaTiO_3$ and to V. Zelezny (Inst. of Physics, ASCR) for doing the infrared spectroscopy. Scanning microwave microscopy was done by M. Fenner (Agilent, Germany). The company ELMES, Lt.D. is acknowledged for lending of the Foram device. $BaTiO_3$ was sprayed at University of Limoges, France.

This work was supported by IPP ASCR under project AV0Z20430508.

6. References

Bak W.; Starzyk, F.; Kajtoch, C. & Nogas-Cwikiel, E (2008). Elevated temperature induced dispersion phenomena in $Ba_{1-x}Na_xTi_{1-x}Nb_xO_3$. *Archives of Materials Science and Engineering*, Vol. 29, No. 1, pp. 5-9

Bak W. (2009). Study of the relaxor behaviour in $Ba_{0.68}Na_{0.32}Ti_{0.68}Nb_{0.32}O_3$ ceramic. *Journal of Achievements in Materials and Manufacturing Engineering*, Vol. 37, pp. 24-27

Boutinaud, P.; Tomasella, E.; Ennajdaoui A. & Mahiou R. (2006), Structural characterization and luminescent properties of $CaTiO_3:Pr^{3+}$ thin films deposited by radio frequency sputtering.*Thin Solid Films*, Vol. 515, pp. 2316–2321

Buchanan, R.C. (2004). Ceramic materials for electronics (3rd edition), M. Dekker, New York

Burlacov, I.; Jirkovsky, J.; Kavan, L.; Ballhorn, R. & Heimann, R.B. (2007). Cold gas dynamic spraying (CGDS) of TiO_2 (anatase) powders onto poly(sulfone) substrates: Microstructural characterisation and photocatalytic efficiency, *Journal of Photochemistry and Photobiology A: Chemistry*, Vol. 187, pp. 285–292

Cavalcante, L.S. et al. (2008). Synthesis, structural refinement and optical behavior of $CaTiO_3$ powders: A comparative study of processing in different furnaces. Chemical *Engineering Journal*, Vol. 143, pp. 299–307

Ctibor, P.; Sedlacek, J.; Neufuss, K. & Chraska, P. (2003). Dielectric relaxation in calcium titanate-containing ceramics prepared by plasma spraying. *Ceramics International*, Vol. 29, pp. 955-960

Ctibor, P. & Hrabovsky, M. (2010). Plasma sprayed TiO_2: The influence of power of an electric supply on particle parameters in the flight and character of sprayed coating. Journal of the European *Ceramic Society*, Vol. 30, pp. 3131-3136

Ctibor, P.; Ageorges, H.; Sedlacek, J. & Ctvrtlik, R. (2010). Structure and properties of plasma sprayed $BaTiO_3$ coatings, *Ceramics International*, Vol. 36, pp. 2155-2162

Cheng H.-F. et al. (2003). Study of second-phases in $Ba(Mg_{1/3}Ta_{2/3})O_3$ materials by microwave near-field microscopy. *Journal of the European Ceramic Society*, Vol. 23, pp. 2667–2670.

Dent, A.H. et al. (2001). High velocity oxy-fuel and plasma deposition of $BaTiO_3$ and $(Ba,Sr)TiO_3$. Materials Science and Engineering B, Vol. 87, pp. 23-30.

Dervos, C.T.; Thirios, E.; Novacovich, J.; Vassiliou, P. & Skafidas, P. (2004). Permittivity properties of thermally treated TiO_2. *Materials Letters*, Vol. 58, pp. 1502– 1507

Ferreira, V. M.; Azough, F.; Freer R. & Baptista J. L. (1997). The Effect of Cr and La on MgTiO3 and MgTiO3–CaTiO3 Microwave Dielectric Ceramics, *Journal of Materials Research*, Vol. 12, pp. 3293–3299

Giolli C. et al. (2007). Characterization of TiO_2 coatings prepared by a modified electric arc-physical vapour deposition system. *Surface & Coatings Technology*, Vol. 202, pp. 13–22

Guo H.Z. et al. (2005). Structure dynamics of strongly reduced epitaxial $BaTiO_{3-x}$ studied by Raman scattering. *Journal of the European Ceramic Society*, Vol. 25, pp. 2347-2352.

Haefie H. et al. (1992). Mg2TiO4 as a novel substrate for high-temperature superconducting thin films, *Applied Physics Letters*, Vol. 61, pp. 9-19

Hu P. et al. (2011). Influence of thermal treatments on the low frequency conductivity and microwave dielectric loss of $CaTiO_3$ ceramics, *Materials Science and Engineering* B, Vol. 176, No. 5, pp. 401-405.

Huang, C.-L.; Pan, C.-L. & Shium S.-J. (2002). Liquid phase sintering of $MgTiO_3$-$CaTiO_3$ microwave dielectric ceramics.

Materials Chemistry and Physics, Vol. 78, pp. 111–115

Hirata, T.; Ishioka K. & Kitajima M. (1996). *Journal of solid state chemistry*, Vol. 124, pp. 353–359

Jiang, Y.; Guo, R. & Bhalla, A. S. (1998). LHPG grown crystal fibers of $MgTiO_3$-$CaTiO_3$ eutectic system. *Journal of Physics and Chemistry of Solids*, Vol. 59, pp. 611-615

Jin, H.Z. et al. (2003). An interfacial defect layer observed at $(Ba,Sr)TiO_3$-Pt interface. *Thin Solid Films*, Vol. 429, pp. 282-285.

Lemanov, V.V.; Sotnikov, A.V.; Smirnova, E.P.; Weihnacht, M. & Kunze, R. (1999). Perovskite $CaTiO_3$ as an incipient ferroelectric, *Solid State Communications*, Vol. 110, pp. 611–614

Mattsson A. & Oesterlund L. (2010). Adsorption and photoinduced decomposition of acetone and acetic acid on anatase, brookite, and rutile TiO_2 nanoparticles, *Journal of Physical Chemistry*, Vol. 114 C, pp.14121–14132

Mitic, V.V. & Mitrovic I. (2001). The influence of Nb_2O_5 on $BaTiO_3$ ceramics dielectric properties, *Journal of the European Ceramic Society*, Vol. 21, pp. 2693-2696

Morey, O.; Goeuriot P.; Juve D. & Treheux D. (2003). Dielectric investigations on 'MgAlON' compounds: Role of nitrogen content. *Journal of the European Ceramic Society*, Vol. 23 pp. 345-350.

Ostapchuk T. et al. (2005). Soft-mode spectroscopy of $BaTiO_3$ thin films. *Journal of the European Ceramic Society*, Vol. 25, pp. 3063–3067

Setter, N. & Waser, R. (2000). Electroceramic materials, *Acta Materialia*, Vol. 48, pp.151-178

Simon-Seveyrat, L.; Hajjaji, A.; Emziane, Y.; Guiffard, B. & Guyomar, D. (2007). Re-investigation of synthesis of $BaTiO_3$ by conventional solid-state reaction and oxalate coprecipitation route for piezoelectric applications, *Ceramics International*, Vol. 33, pp. 35-40

Souza I.A. et al. (2006). Theoretical and experimental study of disordered $Ba_{0.4}$ $Sr_{0.55}TiO_3$ photoluminescence at room temperature, *Chemical Physics*, Vol. 322, pp. 343–348

Sudheendran, K. & James Raju K.C. (2008). Temperature dependent impedance and dielectric properties of $0.7CaTiO_3$–$0.3NdAlO_3$ ceramics. *Indian journal of engineering & material sciences*,Vol. 15, p. 133-136

Waser, R. (1999). Modeling of Electroceramics - Applications and Prospects, *Journal of the European Ceramic Society*, Vol. 19, pp. 655-664

Wang, X.; Zhang, L.; Liu, H.; Zhai J. & Yao X. (2008). Dielectric nonlinear properties of $BaTiO_3$–$CaTiO_3$–$SrTiO_3$ ceramics near the solubility limit, *Materials Chemistry and Physics*, Vol. 112, pp. 675-678

Wing, Z. N.; Halloran J. W.; Zhang, Q.; McGinn, P. J. (2006). Variable Dielectrics in the Calcium Magnesium Titanate System Characterized with Scanning Microwave Microscopy, *Journal of the American Ceramic Society*, Vol. 89, No. 5, pp. 1610–1614

Wu, L. et al. (2009). Dielectric properties of barium titanate ceramics with different materials powder size, *Ceramics International*, Vol. 35, pp. 957-960

Yu, P.; Cui, B. & Chany, Z. (2009). Preparation and characterization of Ag-doped $BaTiO_3$ based X7R ceramics, *Materials Research Bulletin*, Vol. 44 , pp. 893-897

Zeng, J.; Wang, H.; Shang, S.; Wang, Z. & C. Lin (1997). Preparation of Textured Mg_2TiO_4 Thin Films on Si Substrate by Atmospheric Pressure Metallorganic Chemical Vapour Deposition, *J. Mater. Sci. Mater. Electron.*, Vol. 8, pp. 159–162

Zhao, M.H.; Bonnell, D.A. & Vohs, J.M. (2008). Effect of ferroelectric polarization on the adsorption and reaction of ethanol on $BaTiO_3$, *Surface Science*,Vol. 602, pp. 2849-2855

Zhang Q. & McGinn P. J. (2006). Characterization of Calcium Titanate–Magnesium Titanate Eutectic by Scanning Microwave Microscopy, *Journal of the American Ceramic Society*, Vol. 89, pp. 3817 - 3823

Zheng,H. et al. (2003). Raman spectroscopy of B-site order-disorder in $CaTiO_3$-based microwave ceramics. *Journal of the European Ceramic Society*, Vol. 23, pp. 2653–2659

5

Oxygen Isotope Exchange in Nanocrystal Oxides

Anatoly Fishman[1], Tatyana Kurennykh[2],
Vladimir Vykhodets[2] and Evgeniya Vykhodets[3]
[1]*Institute of Metallurgy, Ural Division, Russian Academy of Sciences,*
[2]*Institute of Metal Physics, Ural Division, Russian Academy of Sciences,*
[3]*Ural Federal University, Yekaterinburg*
Russia

1. Introduction

The subject of investigation in this work is oxygen isotope exchange (OIE) between oxides and oxygen-containing gases $^{18}O_2$ and $C^{18}O_2$. OIE studies yield information about the rate of processes on the gas – solid interface and oxygen self-diffusion in oxides. In turn, the surface processes can involve some elementary stages, in particular, physical and chemical adsorption. Several types of diffusion processes can be observed also inside oxides, for example, volume and grain boundary diffusion. OIE investigations are of much practical interest, for example, in connection with the problems of catalytic oxidation of metals and development of materials for chemical and electrochemical devices (fuel cells, electrolyzers, sensors, hydrogen storage devices, devices for separation of gas and isotope mixtures, etc.).
Two experimental approaches to OIE examination are known. The pioneer approach was based on measurements of the isotope composition of a gas mixture interacting with oxide. In the alternative approach developed in the last decades, the isotope composition of the oxygen subsystem of oxides was measured, which was studied usually by secondary ion mass spectrometry (SIMS) and nuclear microanalysis (NRA). The present work is devoted to the examination of the isotope composition of solid-state samples. The overwhelming majority of studies by means of this approach were performed on bulk samples. A distinguishing feature of this work is its orientation toward isotope exchange examination in nanoscale oxides. As will be shown below, this results in novel or supplementary data on the surface reaction rates during isotope exchange and the rates of diffusion processes in oxides. Moreover, the investigations into oxygen isotope exchange in oxide nanomaterials are of great practical importance. It is not improbable that the use of nanomaterials may lead to favorable changes in the functional properties of oxides when it is necessary to increase the quantity of absorbed gas or to increase rates of gas absorption or extraction etc.

2. Investigations on bulk samples

Most often, OIE studies on bulk samples involve isothermal annealing of samples in oxygen (or other oxygen-containing gas) enriched with ^{18}O isotope and measurement of concentration profiles (depth distribution) of ^{18}O tracers in the sample. In case of bulk diffusion described by one value of oxygen self-diffusion coefficient D, the concentration profile $C(x,t)$ is described by the expression (Ruiz & Kilner, 1997)

$$C(x,t) - C_G = C_S^{\max}\left[erfc\,(Z) - \exp\left(Hx + H^2Dt\right) erfc\left(Z + \frac{Hx}{2Z}\right)\right], \; Z = \frac{x}{2\sqrt{Dt}}, \qquad (1)$$

where H is a parameter characterizing the surface energy barrier; C_S^{max} is the maximal permissible concentration of ^{18}O atoms on the surface of a sample; t is the diffusion annealing time; C_G is the concentration of ^{18}O in a sample prior to annealing. According to (Ruiz & Kilner, 1997), the parameter H in expression (1) is connected with the parameter k characterizing the surface reaction rate during isotope exchange by the relation

$$k = HD\,. \qquad (2)$$

It is seen from expressions (1)-(2) that OIE studies on bulk samples provide measurements both of diffusion coefficients D and surface reaction rates k.

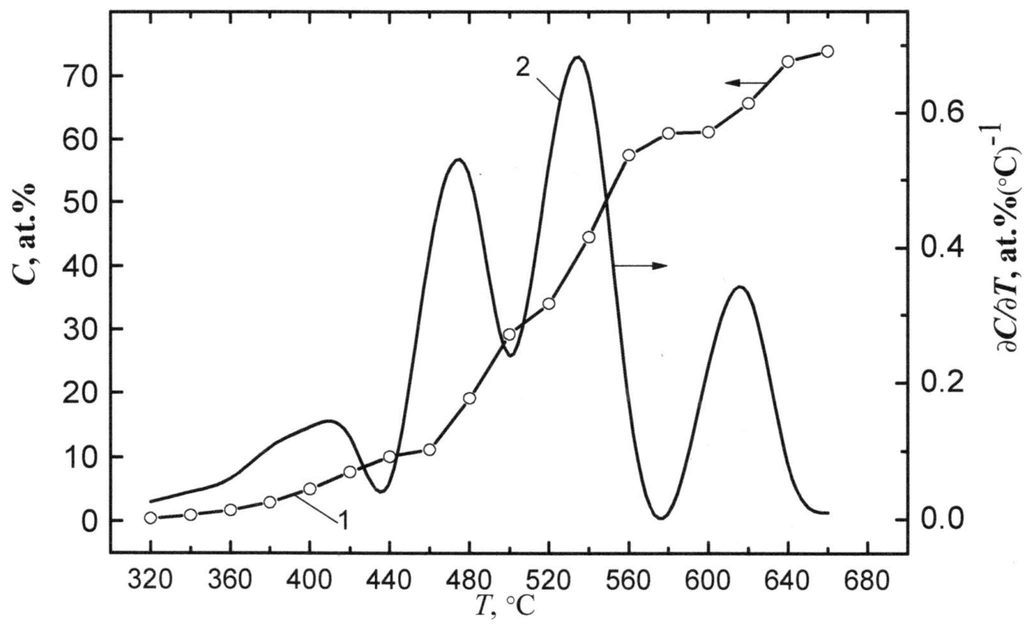

Fig. 1. Temperature dependence of concentration C and derivative $\partial C/\partial T$ during step-wise isochronous annealing of $YBa_2Cu_3O_{7-\delta}$ (Vykhodets et al., 1994). The concentration is given in percents of the number of positions in the oxygen sublattice of the oxide

For strongly anisotropic diffusion systems (HTSC oxides and other oxides with complicated crystal structure and chemical composition), a different route of OIE experiments can be chosen. In some works, step-wise isochronous annealing of samples was carried out in oxygen enriched with ^{18}O isotopes. The isochronous annealing conditions can be selected such that the systems with strong anisotropy of oxygen diffusion coefficients can exhibit a site-plane effect. It consists in successive substitution of ^{18}O atoms for crystallographic oxygen planes or positions occupied by ^{16}O atoms before annealing.

By way of example, Fig. 1 demonstrates the isochronous-annealing curve $C(T)$ and its derivative $\partial C/\partial T$ characterizing the variation in ^{18}O isotope concentration C in $YBa_2Cu_3O_{7-\delta}$ samples as a function of experiment temperature T (Vykhodets et al., 1994). The results displayed in Fig. 1 show that there are several types of energetically non-equivalent sites for oxygen atoms in the oxide lattice.

The data on the site -plane effect make it possible to obtain information about the fraction of each type of sites in the oxygen subsystem of the oxide, to estimate the activation energy and the frequency factor for oxygen atom transition from one site to another, as well as to examine the effect of phase transformations in oxides on the distribution of oxygen atoms in the lattice sites, etc.

3. Investigations on nanocrystalline powders

The concentration profiles of ^{18}O atoms in a nanoparticle can not be determined experimentally since neither SIMS nor NRA methods ensure the locality of measurements commensurable with the size of nanopowder particles. Therefore, in our OIE studies on nanopowders we measured the average concentration of ^{18}O atoms in a large assembly of powder particles (Gizhevskii et al, 2008); the same assembly was annealed in oxygen enriched with ^{18}O isotopes. In view of the destructive character of SIMS measurements, this technique can scarcely be used to determine the average oxygen isotope concentration in powders. Here, the NRA method, which permits measurement of ^{18}O and ^{16}O atom concentration in the depth to several microns without sample failure, is certainly advantageous.

At first, we shall consider the isotope exchange model, in which only oxygen ions from the external atomic layer of the oxide particle participate. Let only $^{18}O_2$ molecules be present in the gaseous phase, while the oxide particle contains no ^{18}O atoms at all in the start time. In that case, the differential equation for isotope exchange has the form

$$dN_{18} = N_{16}\Gamma dt \quad , \tag{3}$$

where dN_{18} is the number of ^{18}O atoms that entered into the oxide during time dt; N_{16} is the number of ^{16}O atoms located in the instant time t on the surface of the oxide atomic layer; Γ is the frequency, with which ^{16}O atom in the oxide surface atomic layer is replaced by ^{18}O atom. Generally, this parameter is a function of the annealing temperature and the pressure in the gaseous phase. During isotope exchange, when neither oxidation nor reduction of the oxide take place, for the numbers of atoms ^{16}O and ^{18}O the relation $N_{16} + N_{18}$ = const is fulfilled. Then equation (3) for spherical particles is solved as

$$C(t) = C_0 \frac{3\Delta}{r}(1 - \exp(-\Gamma t)) \quad , \tag{4}$$

where $C(t)$ is the average concentration of ^{18}O atoms in the oxide; Δ is the monolayer thickness in the oxide; r is the particle radius; C_0 is the atomic concentration of oxygen in the oxide. For arbitrarily shaped particles it is easy to obtain an analogous "relaxation"-type dependence

$$C(t)=C_0 S\rho\Delta(1-\exp(-\Gamma t)) \ , \tag{5}$$

where S is the powder specific surface determined usually with the BET technique and ρ is the oxide density. Frequency Γ measured in OIE experiments on oxide powders is an analogue of the parameter k in formula (2). Based on the results (Vykhodets et al., 2000), it is possible to establish a correlation between Γ and k

$$k = l\Gamma \ , \tag{6}$$

where l is an atomic jump distance.
A fundamentally different "diffusion" model for the description of OIE (Gizhevskii et al, 2008) is used for processes characterized by the very high surface reaction rate and by comparatively slow volume diffusion of oxygen in the oxide particle. Here, for spherically shaped particles it is easy to obtain

$$C(t)=C_0\left(1-\frac{3\Delta}{r}\right)\left(1-\frac{6}{\pi^2}\sum_{s=1}^{\infty}\frac{1}{s^2}\exp\left(-\left(\frac{s\pi}{r}\right)^2 Dt\right)\right)+C_0\frac{3\Delta}{r} \ . \tag{7}$$

Finally, the OIE model version providing for simultaneous implementation of the relaxation and diffusion mechanisms (Gizhevskii et al, 2008) was considered. In this case, the $C(t)$ dependence is found by numerical calculations.
Expressions for relaxation-type isotope exchange analogous with (4)-(5) will be valid not only at very slow oxygen diffusion in oxide, when a typical diffusion length is $(Dt)^{1/2} < d$ (d is the oxide particle size). It can be shown (Fishman et al, 2009) that they are correct also at very fast diffusion when $(Dt)^{1/2} \gg d$. In that case, all oxygen ions of the oxide particle take part in isotope exchange, and the concentration of ^{18}O atoms becomes similar in the bulk of the particle more quickly than it grows on the boundary with the gaseous phase. Here we have

$$C(t)=C_0\left[1-\exp\{-(3\Delta/r)\Gamma t\}\right] \tag{8}$$

or

$$C(t)=C_0\left[1-\exp\{-(S\rho\Delta)\Gamma t\}\right] \ . \tag{9}$$

From the aforesaid it is seen that the examination of isotope exchange on nanopowders may give data on their dimensional characteristics. Otherwise, the information received from isotope exchange studies on powders and bulk samples is in many respects analogous. This concerns both the diffusion characteristics of oxide materials and the processes on the surface of oxides.
OIE studies on nanopowders attract particular interest since they permit measuring low D values. Analysis of expression (7) shows that diffusion coefficients of oxygen tracers as small as 10^{-23} - 10^{-24} m^2/s or ever lower can be measured in such studies. This level of measurements is due to the small size of the powder nanoparticles, and is not typical of OIE investigations on bulk samples.
Expressions (4),(7),(8) are written for spherically shaped particles with a single value of the particle radius r. Certainly, for real powders, dimension distribution of powder particles

should be taken in account (Fishman et al, 2009). Corresponding relaxation-type dependences for an assembly of different-size particles can be easily obtained. For example, for $(Dt)^{1/2} \ll d$ we have

$$C(t) = C_0 3\Delta\left[1 - \exp\{-\Gamma t\}\right]\sum_i r_i^2 \Big/ \sum_i r_i^3 \quad , \tag{10}$$

and for $(Dt)^{1/2} \gg d$

$$C(t) = C_0 \sum_i r_i^3 \left[1\text{-}\exp\{\text{-}(3\Delta / r_i)\Gamma t\}\right] \Big/ \sum_i r_i^3 \quad . \tag{11}$$

The expediency and informativity of experimental studies of isotope exchange on nanopowders is not a trivial problem. First, there may be an unacceptable diversity of annealing conditions for individual particles contained in the powder assembly. Agglomeration of powders aggravates the situation. Second, analysis of expressions (10)-(11) shows that the presence in a nanopowder of large, for example, micron-size particles will result in a very strong decrease in the ^{18}O concentration as compared with that for a powder containing no large particles. Even with a small part of large particles, isotope exchange experiments may turn out senseless since the experimental data on $C(t)$ will characterize to a greater extent the isotope exchange on large particles rather than the assembly of particles. Virtually, this means that the considered technique will be effective only for nanopowders with a narrow size distribution function of powder particles.

Taking into consideration the above observations, in this work we devote much attention to the optimization of the technique for experimental studies on nanopowders, in particular, we examine powders obtained by different technologies and analyze the quantitative correspondence between the theoretical and experimental $C(t)$ values.

4. Oxygen isotope exchange with oxide powders

In this work we present OIE data for oxide nanopowders produced by mechanoactivation (grinding), laser sputtering of a ceramic target, and electrical explosion of wire. In some cases, comparative studies on micropowders were also performed. These technologies are characterized by the following features. The powders obtained by laser sputtering of a ceramic target or by electrical explosion of wire are traditional nanopowders with particles ranging within the nanoscale interval. For the nanopowders produced by grinding, the term *nanopowder* is not conventional. The size of their particles is usually within the micron range. At the same time, such powders have also nanoscale characteristics: the average size of X-ray coherent scattering domains is often of the order of 10 nm.

4.1 Oxygen isotope exchange with oxide powders $LaMnO_{3+\delta}$

The oxide $LaMnO_{3+\delta}$ is a convenient model object for optimization of the OIE technique (Gizhevskii et al, 2008) on nanopowders. In practice, there often occurs uncontrolled doping and diffusion characteristics of oxides are known to be extremely sensitive to the presents of impurities. Naturally, such materials are difficult to be analyzed. At the same time, nonisovalent doping of the cationic sublattice of $LaMnO_{3+\delta}$ does not lead to the formation of structural vacancies in the oxygen sublattice and oxygen volume diffusion coefficients change only slightly in this oxide (Fishman et al., 2003). Probably, this effect is connected to variable valence of manganese ions (Mn^{2+}, Mn^{3+}, Mn^{4+}).

The micropowder $LaMnO_{3+\delta}$ was synthesized by the standard ceramic technology from oxides La_2O_3 and Mn_3O_4. A single-phase manganite powder with the orthorhombic crystal lattice and the particle size of about several microns was obtained. A FRITSCH planetary ball monomill was used to produce a nanostructured material from the initial powder. The grinding was in air in ethyl alcohol. Zirconium dioxide grinding balls and cups were used. The time of grinding was 13 h. According to X-ray diffraction results, the ground powder had an orthorhombic modification with the average coherent scattering domains of about 15 nm.

Isothermal annealings of nano- and micropowders were carried out in oxygen enriched by 80% with ^{18}O isotope. Oxygen pressure was 0.26 atm. The change in the isotope composition of the gaseous atmosphere during annealing was negligible. Previously, stabilization annealings of the powders had been performed in air at the same temperatures as in oxygen containing tracer atoms. Their duration was approximately the same as the maximal time of annealing in $^{18}O_2$. The annealings were carried out in a quartz tube. The temperature of samples was measured with a chromel-alumel thermocouple with an accuracy of 2°C.

The concentration of ^{18}O in the samples was determined with NRA using a 2 MV Van de Graaff accelerator (reaction $^{18}O(p, \alpha)^{15}N$), the incident beam energy being 762 keV. The sample plane surface was mounted perpendicular to the incident beam axis. The acceptance angle of the nuclear reaction products was 160°. The energy spectra of the reaction products were registered with a silicon surface barrier detector of 10 mm in diameter. The diameter of the primary proton beam was from 1 to 2 mm. The number of incident beam particles reaching the sample was measured with an accuracy of ~1% using a secondary monitor. For control by means of reaction $^{16}O(d, p)^{17}O^*$ when the incident beam particle energy is 900 keV, the content of ^{16}O isotopes in samples was measured. The total oxygen concentration in all the samples was the same to within several percents.

The concentration of ^{18}O and ^{16}O isotopes was measured immediately on powders. For this purpose, the powder particles were pressed into a plate of indium. As a result, a layer containing only particles of oxide with thickness large than 2 μm was formed on the plate of indium. Nondestructive analysis by means of the NRA technique was performed to the depth of about 2 μm. In the test experiments the spectra of reaction products $^{16}O(d, p)^{17}O^*$ and $^{18}O(p, \alpha)^{15}N$ for a bulk oxide and for powders, which were not annealed in the atmosphere of tracer atoms, did not differ within experimental error. The concentration profiles were calculated using the stopping power data for the examined samples (Vykhodets et al., 1987).

Experimental and calculated $C(t)$ dependences for the ground powder $LaMnO_{3+\delta}$ are given in Fig. 2. In our model calculations we considered several OIE scenarios for systems with low oxygen volume diffusion coefficients. In particular, the experimental data were processed using relaxation (4) and diffusion (7) expressions, as well as different powder particle size distribution functions. We also considered models for simultaneous realization of relaxation and diffusion mechanisms. For numerical calculations, corrections were introduced to expressions (4) and (7), which are due to the experimental conditions: presence of ^{18}O atoms in powders before annealing and value of gas enrichment with ^{18}O atoms. A similar correction of theoretical expressions for $C(t)$ was made while processing the experimental data for other oxides.

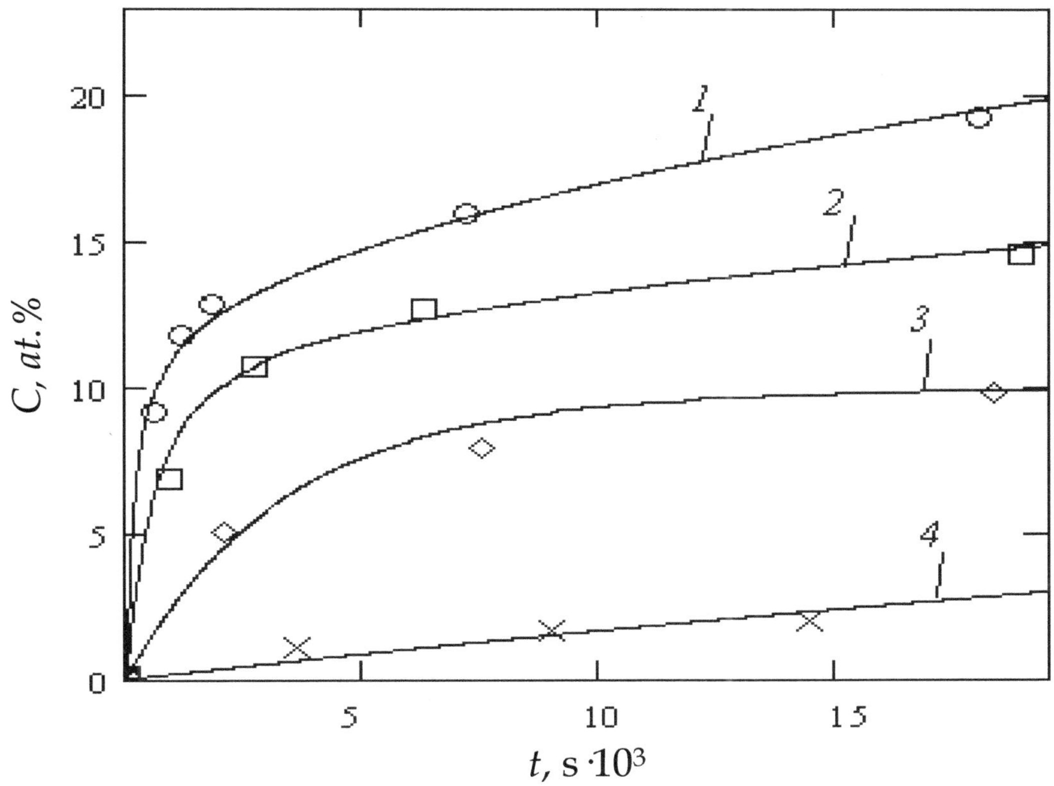

Fig. 2. $C(t)$ dependences for ground powder $LaMnO_{3+\delta}$ (Gizhevskii et al, 2008): 1 – 560°C; 2 – 500°C; 3 – 400°C; 4 – 300°C. Dots – experimental data; lines – calculation in the model for simultaneous realization of relaxation and diffusion mechanisms. Calculated curve parameters: 560°C - $\Gamma = 5.17 \cdot 10^{-3}$ s^{-1}, $D = 1.35 \cdot 10^{-23}$ m^2/s; 500°C - $\Gamma = 1.70 \cdot 10^{-3}$ s^{-1}, $D = 3.55 \cdot 10^{-24}$ m^2/s; 400°C - $\Gamma = 3.07 \cdot 10^{-4}$ s^{-1}; 300°C - $\Gamma = 1.93 \cdot 10^{-5}$ s^{-1}. At 400 and 300°C, the diffusion contribution to the calculated dependences was negligibly small. In the calculations, $r = 7.5$ nm and $\Delta = 0.5$ nm were assumed

It turned out that the experimental dependences $C(t)$ are not described in terms of diffusion theoretical models assuming no surface energy barrier during isotope exchange. In relaxation models, satisfactory fit with experimental data was shown only for low temperatures (300 and 400°C). For elevated temperatures (500 and 560°C), the best fit with experimental data was achieved in the model for simultaneous relaxation and diffusion. The caption to Fig. 2 contains the values of isotope exchange frequencies Γ and oxygen volume diffusion coefficients D, which provide satisfactory fit with experimental dependences $C(t)$. The temperature dependence of frequency Γ (see Fig. 7) can be described with the Arrhenius expression for the frequency factor $\Gamma_0 = (0.97 \cdot 10^3 \pm 0.70 \cdot 10^3)$ s^{-1} and the activation energy $E = (0.88 \pm 0.07)$ eV.

The results presented in Fig. 2 showed that the technique for OIE examination on oxide nanopowders proposed in work (Gizhevskii et al, 2008) is consistent. It furnishes information about nanopowder dimensional characteristics, reaction rates on the oxide particles surface during isotope exchange, and oxygen volume diffusion coefficients in oxides. The reliability of the technique is supported by the following facts. The best possible fit with the experimental dependences $C(t)$ was provided by the nanoparticle radius r = 7.5 nm, which is close to the experimentally determined average radius of the X-ray coherent scattering domain for the examined powder (6.5 nm). Thus, the main dimensional parameter of the powder responsible for the isotope exchange rate became apparent, namely, the size of the X-ray coherent scattering domains. The difference between 6.5 and 7.5 nm is not fundamental since the notion of the monolayer thickness is not strictly defined; in fact, it is a parameter of theory. The value Δ = 0.5 nm was assumed by the authors merely by tradition. For example, in the grain boundary diffusion studies, exactly this value was postulated for the grain boundary width. The oxygen volume diffusion coefficients (see caption to Fig. 2) obtained from $C(t)$ dependences processing also proved to be quite reasonable. For strontium-doped $LaMnO_3$-based perovskites, experimental oxygen volume diffusion coefficients are available in the literature for higher temperatures (Carter et al. 1992; De Souza et al. 2000; Fishman et al., 2003; Routbort et al., 1997). Strontium was not found to have any profound effect on the volume diffusion coefficients of tracer oxygen atoms (Fishman et al., 2003). Extrapolated diffusion coefficient values ranged from $1 \cdot 10^{-24}$ to $6 \cdot 10^{-24}$ m^2/s (for 500^0C) and from $1.6 \cdot 10^{-22}$ to $3.2 \cdot 10^{-23}$ m^2/s (for 560^0C) (see Fig.5). The values obtained in this work (see Fig. 5 and caption to Fig. 2) are within the above-mentioned intervals.

From Fig. 2 it is seen that the calculated and experimental dependences $C(t)$ agree to a high accuracy. This finding is important in two respects. First, it reveals that the size distribution function of powder particle was rather narrow when the grinding technology was used. Second, it shows that the differences in annealing conditions in $^{18}O_2$ for individual particles contained in the powder assembly were insignificant. Therefore, the same technology was applied in experiments with other powders.

From general considerations it was obvious that the value of frequency Γ should depend on the oxide type and the kind of oxygen-containing molecules in the gaseous phase. For the theoretical parameter Δ, the presence or absence of such dependence is not trivial. Therefore, we compared OIE kinetics for $LaMnO_{3+\delta}$ nanopowders annealed in $^{18}O_2$ and $C^{18}O_2$. The degree of carbon dioxide enrichment with ^{18}O isotopes was 90%; the pressures of oxygen and carbon dioxide during annealing were similar. The $C(t)$ data for 400^0C obtained during annealing in $C^{18}O_2$ were satisfactorily described by expression (4), while the concentrations of ^{18}O isotopes in the powder were from 1.5 to 2 times higher than those in oxygen. It was also found that the values of Γ and Δ increased when $^{18}O_2$ was replaced by $C^{18}O_2$. The corresponding values were $3.07 \cdot 10^{-4}$ and $6.4 \cdot 10^{-4}$ s^{-1} for Γ and 0.5 and 0.66 nm for Δ. The question why the surface layer thickness Δ varies when the kind of the oxygen-containing molecule of the gaseous phase is changed calls for further research.

Figure 3 demonstrates experimental $C(t)$ dependences for $LaMnO_{3+\delta}$ micropowder. They do not differ qualitatively from those for the nanopowder (Fig. 2) and can be also used to receive the information about surface reaction rates and oxygen diffusion coefficients in oxide. It is seen from comparison of the data in Figs. 2 and 3 that the dimensional factor showed up very vividly: the concentrations of ^{18}O isotopes in the micropowder were tens times lower. Evidently, the $C(t)$ dependences for nanopowders are more convenient for theoretical analysis than the data for mircopowders. In case of nanopowders, a better accuracy of analysis results is achieved. Besides, for microparticles, the particle spherical shape approximation and, consequently, the self-diffusion equation may turn out unsatisfactory.

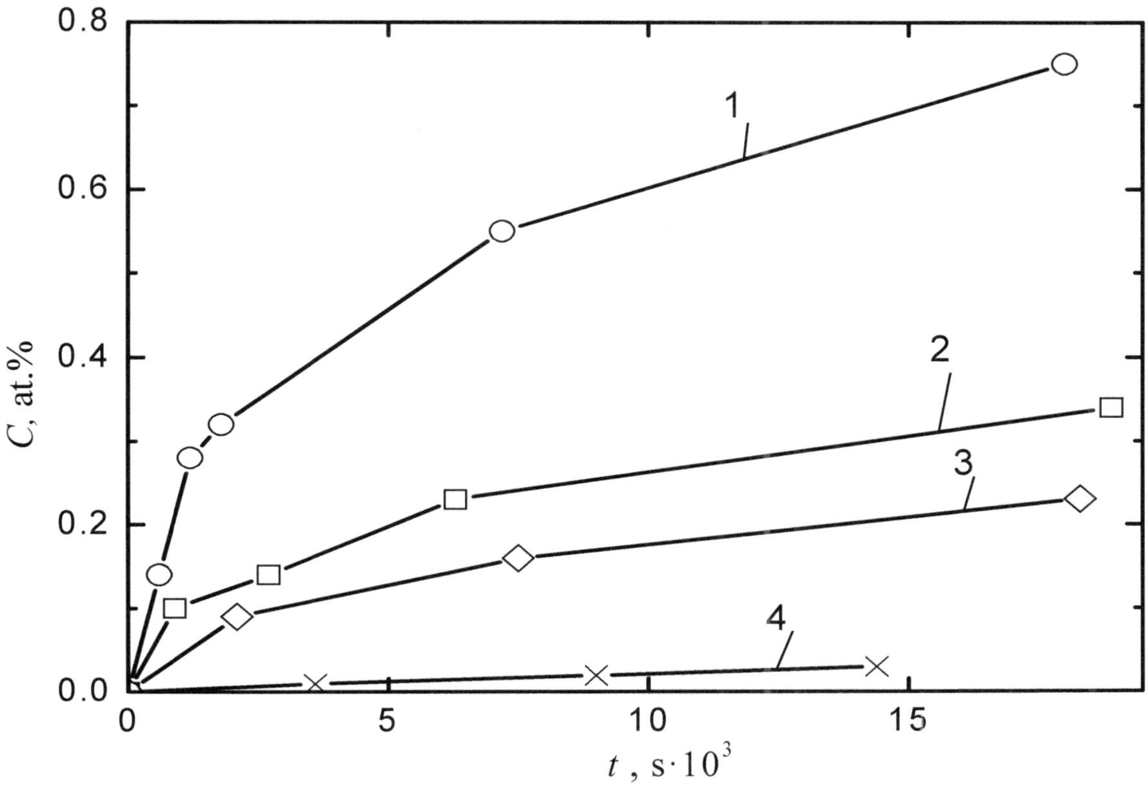

Fig. 3. Experimental dependences $C(t)$ for $LaMnO_{3+\delta}$ micropowder: 1 – 560^{0}C; 2 – 500^{0}C; 3 – 400^{0}C; 4 – 300^{0}C

4.2 Oxygen isotope exchange with α-Mn_2O_3 oxide powders

Manganese oxides are technically important and scientifically interesting materials. However, for these species, as far as we know, there is no information about OIE, oxygen diffusion coefficients, and surface reaction rates in the interaction with gaseous oxygen. Therefore, in this work we carried out appropriate studies on α-Mn_2O_3 oxide powders produced by mechanoactivation.

A planetary mill AGO-2 was used to prepare Mn_2O_3 powders. Nanoscale domains were formed for a much shorter period of time than in the FRITSCH monomill: the grinding time for the examined samples was 60 s. The average particle size during grinding decreased insignificantly: from 1026 to 344 nm. The corresponding data were obtained by method of dynamic light scattering using a laser analyzer. X-ray diffraction studies showed that the phase composition of the ground powder remained unchanged upon heating in air to 950^{0}C. The difference between $LaMnO_{3+\delta}$ and α-Mn_2O_3 powders produced by mechanoactivation consisted in a pronounced enhancement of the average size of domains with an increase in temperature. In the temperature range from 300 to 700^{0}C the radius varied from 16 to 35 nm (Petrova et al., 2010). We used these data as reference points in analysis of OIE results. At the same time, it was taken into account that sizes of coherent scattering domains during OIE investigations could differ from those established in work (Petrova et al., 2010) owing to different temperatures and duration of heating.

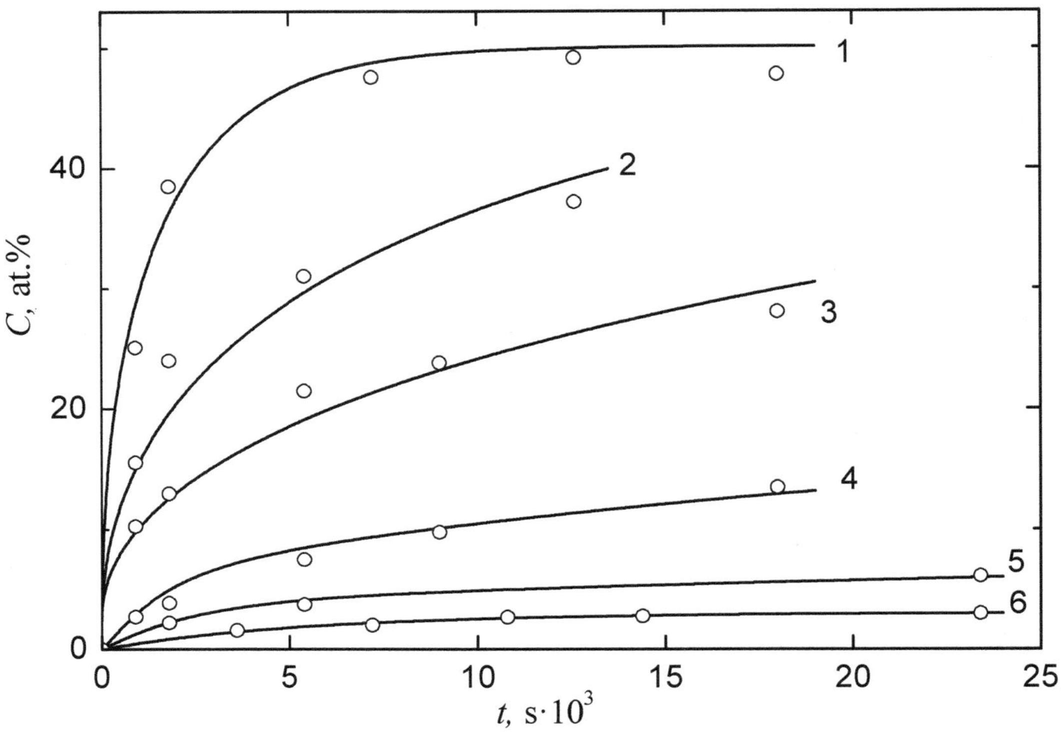

Fig. 4. $C(t)$ dependences for ground powder α-Mn_2O_3: 1, 2, 3, 4, 5, and 6 – results for 700, 550, 500, 400, 350, and 300°C, respectively. Dots – experimental data. Lines – calculation with expression (7) for curves 1, 2, and 3; with expression (4) for curve 6; and in the model for simultaneous relaxation and diffusion for curves 4 and 5. Calculated curve parameters: $D = 0.5 \cdot 10^{-19}$, $0.7 \cdot 10^{-20}$, $1.7 \cdot 10^{-21}$, $1.5 \cdot 10^{-22}$, and $1.5 \cdot 10^{-23}$ m^2/s for 700, 550, 500, 400, and 350°C, respectively; $\Gamma = 0.9 \cdot 10^{-3}$, $5.0 \cdot 10^{-4}$, and $1.8 \cdot 10^{-4}$ for 400, 350, and 300°C, respectively; nanoparticle radii r = 34, 30, 27, 26, 25, and 24 nm for 700, 550, 500, 400, 350, and 300°C, respectively. Δ = 0.5 nm was assumed in the calculations

Experimental and calculated $C(t)$ dependences for the ground powder α-Mn_2O_3 are depicted in Fig. 4. On the whole, the isotope exchange features for $LaMnO_{3+\delta}$ and α-Mn_2O_3 powders prepared by grinding were analogous. This showed up first of all in the fact that the dimensional parameter of the powder responsible for the isotope exchange rate was the X-ray coherent scattering domain radius. However, the OIE parameters for these oxide powders differed drastically. So, at similar temperatures the frequencies Γ (see captions to Figs. 2 and 4) for α-Mn_2O_3 were three to ten times higher than for $LaMnO_{3+\delta}$. Even more dramatic differences were observed for the oxygen volume diffusion coefficients D (see Fig. 5). The data on oxygen diffusion coefficients in manganese oxides were obtained for the first time and can be commented in the following way. The measurement error shown in Fig. 5 was ~50%; it was due mainly to the uncertainty in the powder nanoparticle size in our experiments. We believe that the diffusion coefficient measurement error can be reduced several times if the nanoparticle radius is determined upon each isothermal annealing in gaseous oxygen. Thus, the technique employed can probably provide precision measurements of low values of oxygen volume diffusion in oxides. Of importance is the following finding. The least typical diffusion length $(Dt)^{1/2}$ in our experiments was about 0.1 nm. This level of measurements can hardly be achieved with other techniques based on concentration profile measurements of light element atoms in solids. Based on the obtained results, we suppose that the use of oxide nanopowders in diffusion experiments will allow one to make a qualitative leap in oxygen diffusion coefficient measurements in oxides.

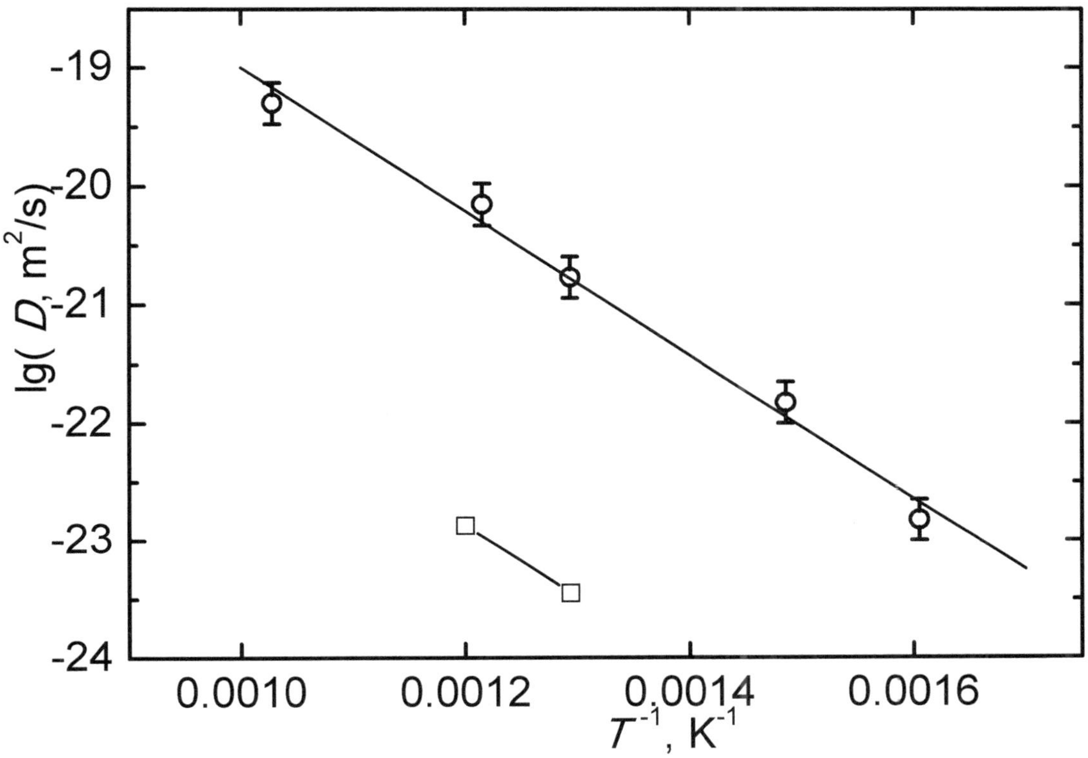

Fig. 5. Temperature dependence of volume diffusion coefficients D in α-Mn_2O_3 (circles) and $LaMnO_{3+\delta}$ (squares)

The oxygen diffusion activation energy in α-Mn_2O_3 was (1.20 ± 0.08) eV. For oxygen diffusion in oxides, this value is typical of oxygen ion migration energy. When oxygen diffusion was due to structural vacancies in the oxygen sublattice of the oxide, approximately the same value was observed also for the volume oxygen diffusion activation energy. For example, in yttrium-stabilized cubic zirconium oxides, oxygen diffusion activation energies were found (Solmon et al., 1995; Brossmann et al., 2004) to be 1.23 and 1.11 eV, respectively. The pre-exponential factor in the temperature versus diffusion coefficient dependence for α-Mn_2O_3 turned out to be very low, namely, $D_0 \approx 10^{-13}$ m²/s. On the basis of this result, the structural oxygen vacancy concentration c_v in the examined manganese oxide can be estimated (Shewmon, 1989) as $c_v \approx D_0/a^2\nu \approx 10^{-6}$ (a - oxide crystal lattice parameter, ν - oxygen ion vibration frequency in the lattice).

In the temperature range 350 – 700⁰C in α-Mn_2O_3 oxide containing no structural vacancies in the oxygen sublattice, the oxygen diffusion coefficients will be evidently much smaller than those established by the authors, since in oxides without structural vacancies the diffusion activation energy is approximately equal to the sum of migration energy and vacancy formation energy. We don't know any literature data on oxygen vacancy formation energies in manganese oxides.

4.3 Oxygen isotope exchange with cubic zirconium oxide powders

In this section, as distinct from 4.1 and 4.2, we shall consider OIE for gaseous oxygen with an oxide characterized by very fast oxygen volume diffusion coefficients. The condition $(Dt)^{1/2} \gg d$ (d is the particle linear size) corresponding to such species is fulfilled for many oxides containing structural vacancies in the oxygen sublattice.

OIE studies (Fishman et al., 2009) were performed on yttrium-stabilized cubic zirconium oxide YSZ containing 9 mol. % of Y_2O_3. It can be shown that the condition $(Dt)^{1/2} \gg d$ at the minimal isothermal annealing experiment time ($t \approx 10^3$ s) is met not only for nano-, but also for micropowders of this oxide in a wide range of experimental conditions. For example, at 500^{0}C the oxygen diffusion coefficient value is $5.9 \cdot 10^{-13}$ m^2/c (Solmon et al., 1995) and the typical diffusion length $(Dt)^{1/2}$ is 24 μm. Since at $(Dt)^{1/2} \gg d$ all oxygen ions of the oxide particle take part in isotope exchange, the concentrations of ^{18}O isotopes in the YSZ oxide upon annealing were higher than for $LaMnO_{3+\delta}$. This allowed us to carry out high-accuracy OIE studies on YSZ oxide powders.

YSZ micropowder was produced by coprecipitation of components. For better particle size homogeneity, it was calcined at 1100^{0}C in air for 3 h. The powder specific surface was 2.15 and 0.64 m^2/g before and after calcination, respectively. Figure 6 displays the experimental and calculated with eq. (9) $C(t)$ dependences for YSZ micropowder. In the calculations we postulated that $\Delta = 0.5$ nm, as in the case of $LaMnO_{3+\delta}$. The experimental and theoretical $C(t)$ dependences are seen to be in reasonable agreement.

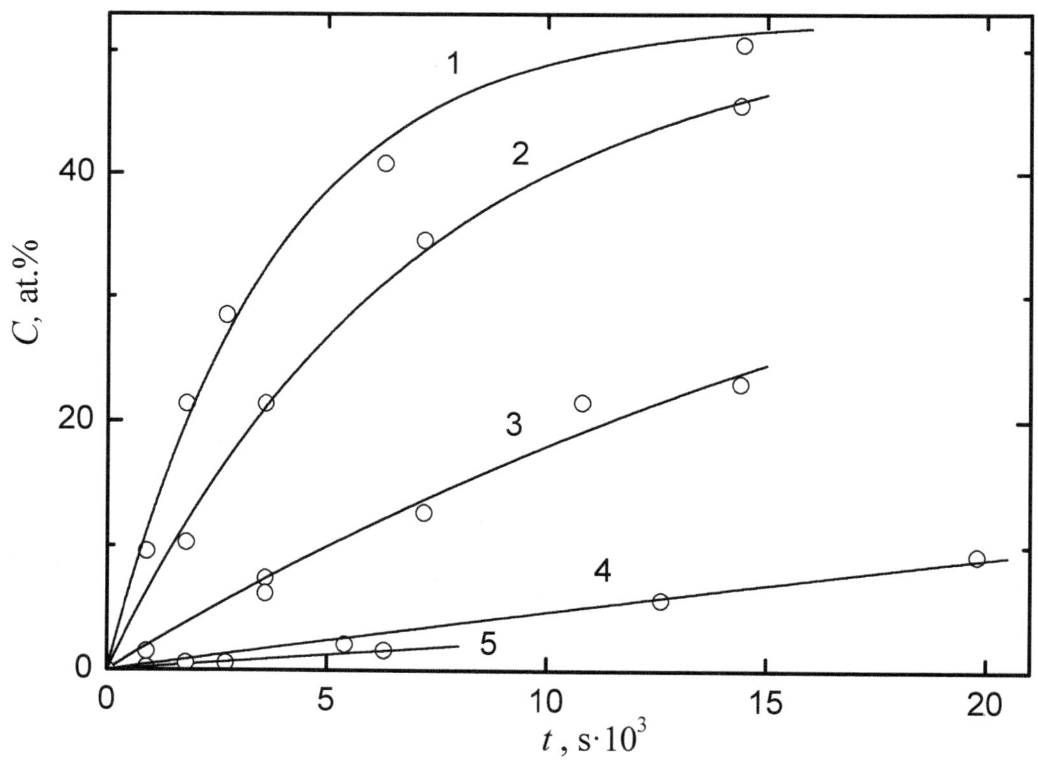

Fig. 6. $C(t)$ dependences for calcined YSZ micropowder. Dots – experiment. Curves – calculation with equation (9). 1, 2, 3, 4, 5 – results at 800, 700, 600, 500, and 450^{0}C, respectively

The temperature versus frequency Γ dependence is presented in Fig. 7. The frequency factor Γ_0 = 0.84·10^3 s^{-1} and the activation energy E = 0.79 eV correspond to this dependence Γ(T). The mean-square error in their determination was 60 and 4%, respectively. These results for the YSZ powder can be compared with the literature data (Manning et al., 1997) obtained in concentration profile measurements for ^{18}O tracer atoms in bulk samples (see eqs. (1)-(2)). These data are also given in Fig. 7 (triangles). In study (Manning et al., 1997), the surface reaction rate was described using the parameter k. Γ was calculated on the basis of k values with eq. (6) at l = 2·10^{-10} m. The good agreement of Γ data for the micropowder and the bulk sample of YSZ is indicative of the reliability of the results obtained in works (Fishman et al. 2009; Manning et al., 1997).

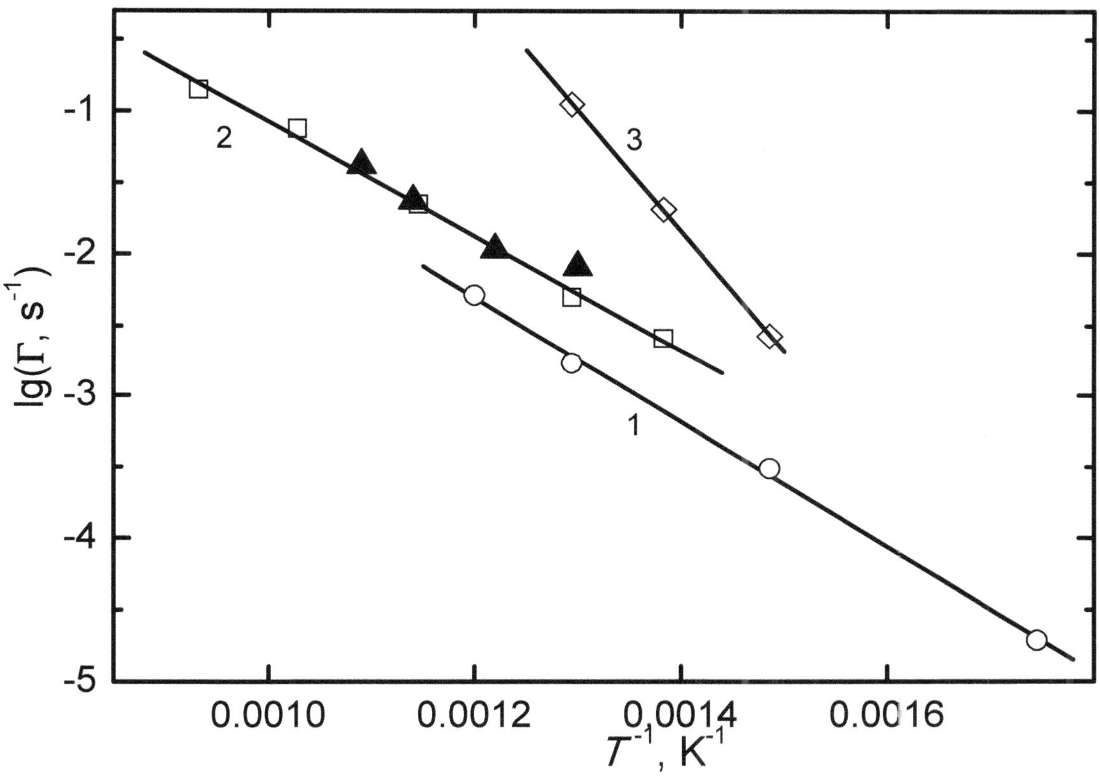

Fig. 7. Temperature dependence of exchange frequency Γ for $LaMnO_{3+\delta}$ nanopowder (line 1), YSZ micropowder (line 2), and nanoboundaries of $LaMnO_{3+\delta}$ (line 3). Triangles – data of work (Manning et al., 1997) for a bulk YSZ sample

Attention is drawn to the very low values of the frequency factor Γ_0 in the temperature versus Γ dependence, which are about 10^3 s^{-1} for $LaMnO_{3+\delta}$ and YSZ oxides. Almost the same frequency factor values were obtained also for bulk samples of other oxides: $La_{0.8}Sr_{0.2}MnO_{3+\delta}$ (De Souza et al., 1999), $La_{0.8}Sr_{0.2}Mn_{1-y}Co_yO_{3\pm\delta}$ (De Souza & Kilner, 1998), $La_{1-x}Sr_xYMnO_{3-\delta}$ (Ruiz-Trejo & Kilner, 1997). These values are by several orders of magnitude smaller than other frequencies characterizing the examined systems. So, the vibration frequency of light atoms in solids is ≥ 10^{12} s^{-1}, whereas the frequency of gaseous molecule collisions with solid particle atoms during isotope exchange is about 10^9 s^{-1}. Such a considerable discrepancy in the characteristic frequency values points to a very low concentration of active centers on the surface of oxides participating in isotope exchange.

Their identification is of current importance. Moreover, such discrepancies in the characteristic frequency values may be due to the existence of several isotope exchange mechanisms, instead of only one mechanism of dissociation adsorption - desorption (Odzaki, 1979).

The YSZ nanopowder containing 9.5 mol. % Y_2O_3 was produced by laser sputtering of a ceramic target (Ivanov et al., 2006). It was subjected to sedimentation in isopropyl alcohol to remove particles with diameter of 200 nm or larger from the powder. The specific surface of the nanopowder was (58.6 ± 0.4) m^2/g. Upon annealings at 500^{0}C, it decreased to (53.3 ± 0.8) m^2/g, which is likely to be connected with agglomeration of smaller particles.

The experimental and calculated $C(t)$ dependences for the YSZ nanopowder are exhibited in Fig. 8 (Fishman et al., 2009). Comparing the experimental data in Figs. 6 and 8 we see that the nanopowder is characterized by higher ^{18}O isotope concentrations, which were also observed for other oxides. The calculated curve 2 in Fig. 8 was plotted with expression (11). In the calculations we used powder particle size distribution established by transmission electron microscopy (Ivanov et al., 2006); it was narrow with the width of about 13 nm and had a maximum at the particle diameter of 11 nm. It was also assumed that the isotope exchange frequencies Γ are not different for nano- and micropowders. From Fig. 8 it is seen that the calculated (curve 2) and experimental $C(t)$ dependences do not agree with each other. At the same time, the powder specific surface calculated with the above nanopowder particle size distribution (58.1 m^2/g) practically coincided with its experimental value (58.6 m^2/g). The agreement between theoretical and experimental $C(t)$ results was not improved when possible changes in the frequency Γ in going from micro- to nanopowder were taken into account. The experimental $C(t)$ dependence was not at all described by theoretical expressions supposing a narrow powder particle size distribution.

It was suggested on the basis of these results that the discrepancy in the calculated and experimental $C(t)$ data is due to the existence of large particles in the nanopowder, which were not presented in the size distribution function of powder particles obtained with transmission electron microscopy. The corresponding calculation had an evaluating character, where all large particles were postulated to have a spherical shape and similar size. The calculations showed that a satisfactory fit between calculated and experimental $C(t)$ dependences can be achieved within this model. Figure 8 (curve 1) demonstrates one of the model versions postulating the presence of particles with radius 130 nm in the nanopowder. Their part in the total number of particles was estimated to be 0.0003. A similar satisfactory agreement between calculated and experimental $C(t)$ data was achieved for large particle radii ranging from 100 to 130 nm. However, this interpretation of the experimental $C(t)$ data was in conflict with the specific surface experimental data. It was calculated that the nanopowder containing particles with a 130 nm radius would have the specific surface 20.4 m^2/g if the part of large particles is 0.0003 (the experimental value being 53.3 m^2/g).

The problems of interpretation of experimental data on specific surface for nanopowders produced by laser sputtering of a ceramic target were also known earlier. They manifested themselves in the following result: the specific surface of sedimentated and non-sedimentated powders were virtually the same according to numerous studies. For example, in work (Kotov et. al., 2004) the specific surface of sedimentated and non-sedimentated $Ce_{0.78}Gd_{0.22}O_{2-\delta}$ nanopowder was found to be 57 and 56 m^2/g, respectively, though the non-sedimentated powder contained to 8 mass % of large particles. Such results are usually explained in the following way: during sedimentation both large and finest particles are removed. However, this explanation seems doubtful. Some estimations show that small differences in the specific surface for sedimentated and non-sedimentated powders can occur only in the case of multi-layer coating of large particles with nanoparticles. According to the available data, less than one nanoparticle layer on large particles was usually registed on photographs.

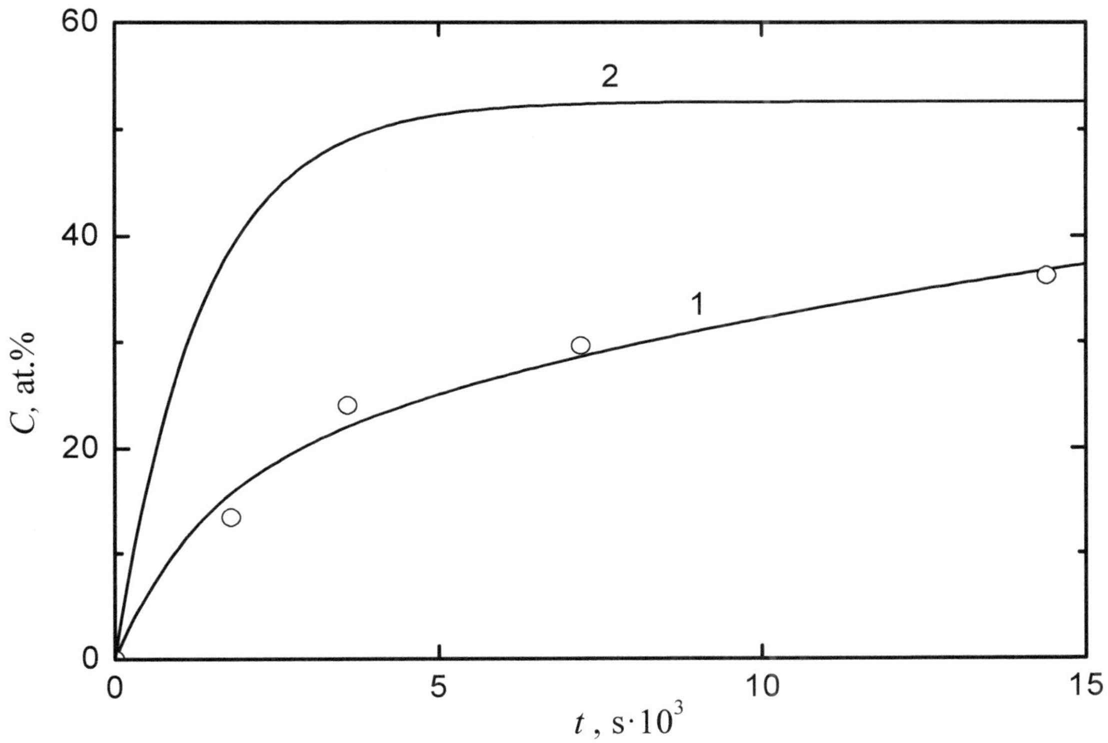

Fig. 8. $C(t)$ dependences for the YSZ nanopowder at 500^{0}C. Dots – experiment; curves 1 and 2 – calculation with equation (11). Curve 2 – calculation with the particle size distribution function obtained by transmission electron microscopy. Curve 1 – calculation with the particle size distribution function obtained by transmission electron microscopy and supplemented by the presence of particles with the radius of 130 nm

It can be supposed as an alternative that large YSZ particles appearing during laser sputtering of a ceramic target are conglomerates of usual nanoparticles. When the BET technique is used, nitrogen molecules penetrate into the space between nanoparticles contained in large particles. That is why the specific surface values for non-sedimentated and sedimentated nanoparticles are almost the same. At the same time, the presence of large particles in nanopowder was reliably established with the OIE technique. These findings show that the OIE technique can yield supplementary (as compared with the BET method) information about the dimensional characteristics of powders and is very sensitive to the presence of small amounts of large particles in nanopowder, which lie beyond the main peak of the size distribution function of powder particles. According to these results, the expected sensitivity of the isotope exchange technique is about 0.01% of the number of particles.

4.4 Oxygen isotope exchange with aluminum oxide nanopowders

In this section we consider OIE in nanopowders produced by electrical explosion of wire. The powders were sedimentated in isopropyl alcohol. The sedimentation regime provided removal of particles with diameter greater than 200 nm from the powder. The main characteristics of the powders are listed in the Table 1.

S, m^2/g	d_{BET}, nm	γ-Al_2O_3/δ-Al_2O_3	d_x, nm	d_p, nm	Θ, nm
23.2	73	30/70	18.5	16	19
38.3	45	40/60	17	17	24
41.7	40	15/85	21	-	-
84	20.5	53/47	21	13	13

Table 1. Characteristics of aluminum oxide powders

The values of specific surface S were determined with the BET method; the nanoparticle diameter d_{BET} was calculated with the expression $d_{BET} = 6/S\rho$, where ρ is the calculated density for the crystalline component of the powder. X-ray phase analysis data showed that all the powders contained two crystalline phases: γ-Al_2O_3 and δ-Al_2O_3. The ratios of their mass fractions are presented in the Table 1. X-ray powder diffraction patterns were also indicative of an amorphous component; however, its amount in the powders could not be determined. The tabulated sizes d_x of X-ray coherent scattering domains characterize the δ-Al_2O_3 phase. The size distribution functions of powder particle were determined using transmission electron microscopy. All the distributions featured a narrow peak; its widths Θ and the most probable particle diameters d_p are also listed in the Table 1.

As is seen from the Table 1, the aluminum oxide nanopowders are much more complicated objects than those examined in sections 4.1, 4.2, and 4.3. Since the OIE technique for nanopowders made a good showing in the case of relatively simple oxide systems (see the above sections), it is interesting to apply this method to more complicated species. The major task in this section was to estimate the sensitivity of the OIE technique to the multi-phase state of nanopowders and, in particular, to the presence of an amorphous phase.

$C(t)$ dependences at 500^{0}C for OIE of aluminum oxide nanopowders with $^{18}O_2$ are depicted in Fig. 9. For one of the oxides (with specific surface of 41.7 m^2/g), investigations were also performed for a powder, which was not exposed to preliminary stabilization annealing in air. In this case, the concentrations $C(t)$ were appreciably higher than for the stabilized powder. This finding points to a nonstoichiometric chemical composition of the aluminum oxide powders produced by electric explosion of wire. Such peculiarities of nanooxides can be easily revealed with the NRA method.

The $C(t)$ dependences in Fig. 9 differ from those for other nanopowders in appreciably lower ^{18}O concentrations. This hampered experimental studies of IOE kinetics at temperatures, for which the nanoparticle agglomeration effect can be disregard. For this reason, the main body of OIE data for aluminum oxides was obtained with the use of carbon dioxide CO_2 enriched with ^{18}O isotopes, rather than with gaseous oxygen. As was shown in section 4.1, replacement of $^{18}O_2$ by $C^{18}O_2$ led to an increase in the ^{18}O isotope concentration in $LaMnO_{3+\delta}$ powders. The same effect is also observed for aluminum oxide nanopowders, which permitted us to perform $C(t)$ dependence measurements in the temperature range from 500 to 100^{0}C. Figure 10 shows $C(t)$ dependences for the powder with specific surface 38.3 m^2/g. For other powders, $C(t)$ dependences were analogous, namely, the concentration of ^{18}O isotopes in the powders increased with the time and temperature of annealing and powder specific surface.

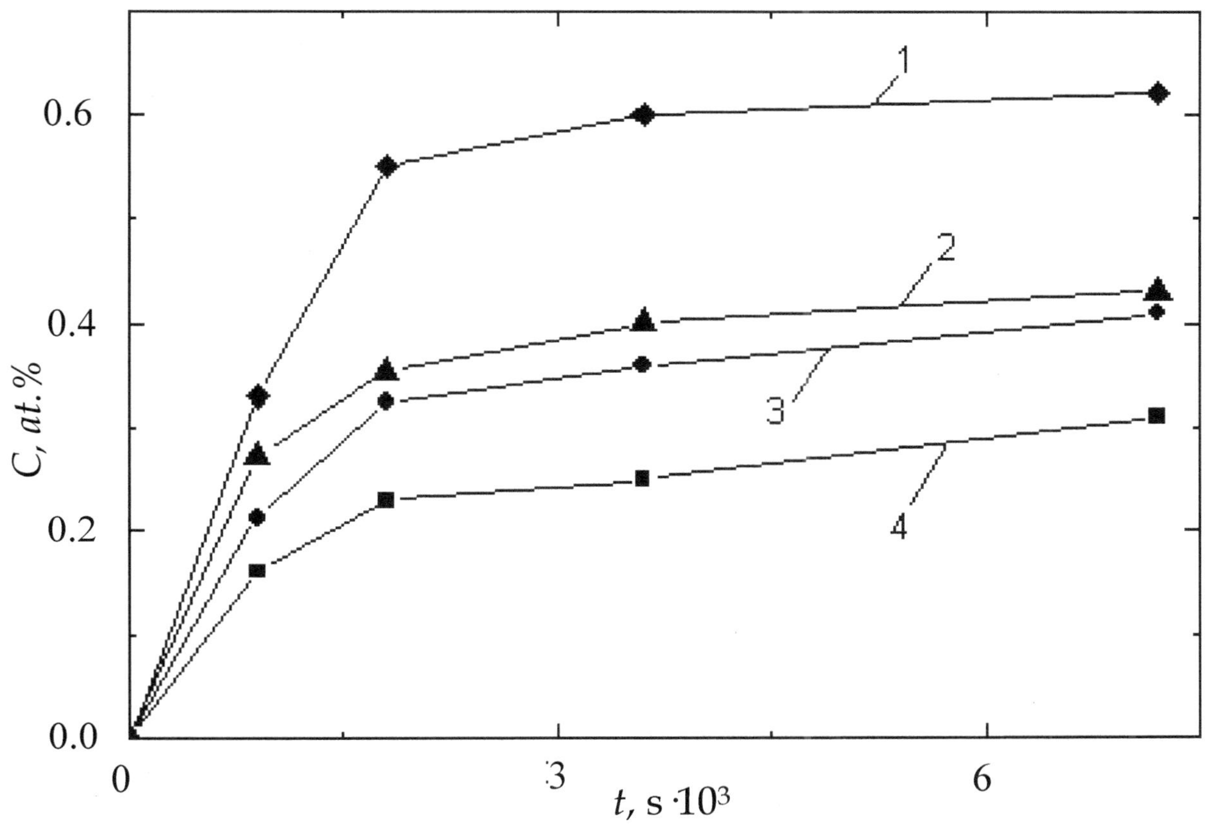

Fig. 9. Experimental $C(t)$ dependences for aluminum oxide nanopowders at 500^{0}C. 1,3,4 – results for powders with specific surface 84, 41.7, and 23.2 m^2/g, respectively. 2 – results for the powder, which was not exposed to preliminary stabilization annealing in air, its specific surface is 41.7 m^2/g

Figure 11 demonstrates several $C(S)$ dependences of ^{18}O isotope concentration versus the powder specific surface S. For each dependence, the time and temperature of annealing in the atmosphere of tracer atoms were constant. All the $C(S)$ dependences, including those not given in Fig. 11, were linear. As is seen in Fig. 11, for the systems powder - $^{18}O_2$ they are fulfilled to a very high accuracy, while for the systems powder - $C^{18}O_2$ a more considerable deviation form the linear dependences is observed. Thus, it can be stated quite definitely that for multiphase oxide powders the concentration of ^{18}O isotopes during isotope exchange is also a linear function of the powder specific surface. Note that at 500^{0}C the extrapolation of $C(S)$ dependences to the region $S \to 0$ gives values $C(S \to 0) \neq 0$. This is likely to be due to oxygen diffusion in nanopowders. This peculiarity of $C(S)$ dependences can be used for the estimation of oxygen diffusion coefficients in aluminum oxides at 500^{0}C.

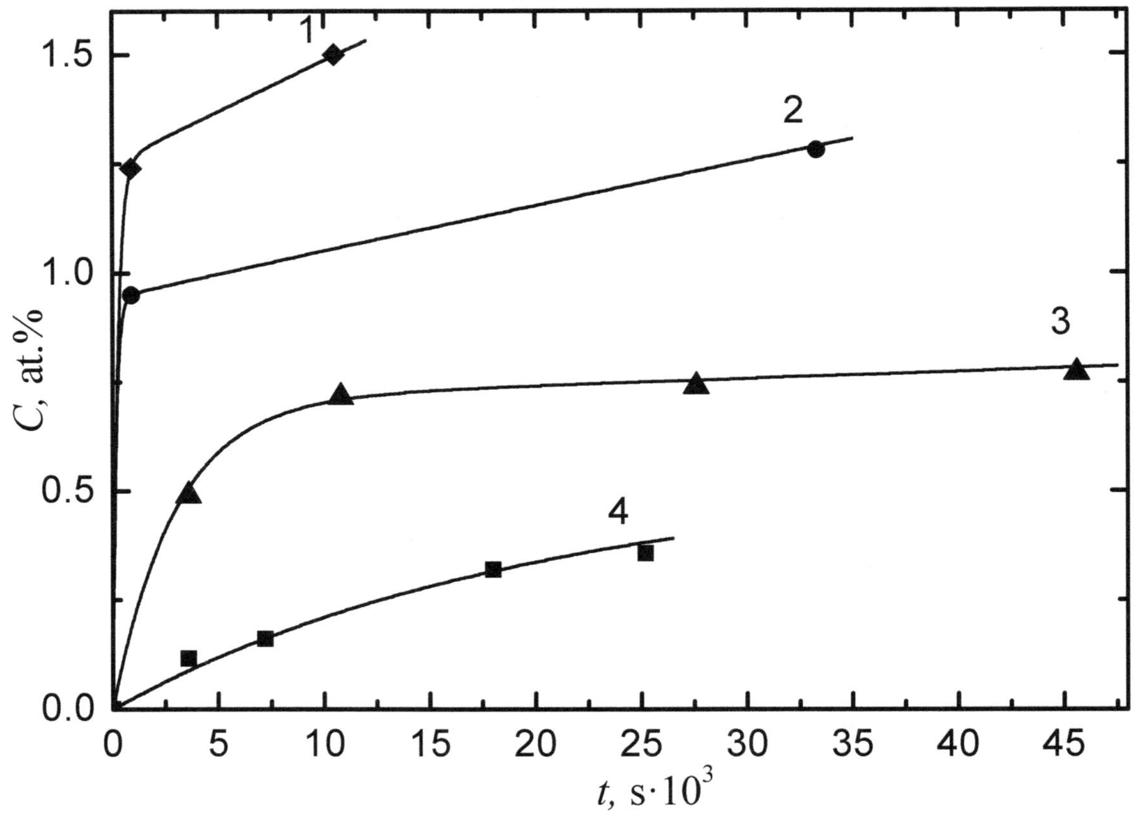

Fig. 10. $C(t)$ dependences for aluminum oxide nanopowder during annealing in carbon dioxide enriched with ^{18}O isotopes. Dots – experimental data. 1,2,3,4 – results for 500, 400, 200, and 100°C, respectively. Curves – calculation with equation (14). Calculated curve parameters: $\Delta_2 = 0.66$ nm, $\alpha_1 = 0.13$, $\Gamma_1 = 5.0 \cdot 10^{-5}$, $3.4 \cdot 10^{-4}$, $6.0 \cdot 10^{-3}$, and $1.0 \cdot 10^{-2}$ s^{-1}, $\Gamma_2 = 2.5 \cdot 10^{-8}$, $3 \cdot 10^{-7}$, $2.6 \cdot 10^{-6}$, and $0.57 \cdot 10^{-5}$ s^{-1}, $\Delta_1 = 0.66$, 0.88, 1.17, and 1.55 nm at 100, 200, 400, and 500°C, respectively

Thus, some OIE regularities for single- and multi-phase oxide nanopowders were found to be of the same type. This is true for linear $C(S)$ dependences and the effect of the oxygen-containing gas type on ^{18}O isotope concentrations. At the same time, multi-phase aluminum oxide powders exhibited some features that distinguished them from other examined systems, namely, the aforementioned low level of ^{18}O isotope concentrations in powders upon annealing in $^{18}O_2$. When the gaseous atmosphere was changed to $C^{18}O_2$, the concentrations of ^{18}O isotopes increased, but still remained low as compared with those for oxide powders earlier considered. Moreover, the $C(t)$ concentrations for aluminum oxides powders were also unusual (see Fig. 10). They were described neither by relaxation-type expressions (4)-(5), nor by the model for simultaneous relaxation and diffusion. The same is true of the attempts to explain the experimental $C(t)$ data by the presence of larger particles in aluminum oxide powders, which are not present in the main peak of size distribution function of powder particles.

No strict theoretical description of $C(t)$ data could be given for aluminum oxides since IOE in amorphous oxide phases has not been studied at all. Therefore let us analyze $C(t)$ dependences in a crude model, which allows for the multi-phase character of aluminum oxide powders and, especially, for the presence of an amorphous phase in them. For this phase we postulate that the number of oxygen ions participating in isotope exchange between the amorphous phase and the gaseous atmosphere increases with temperature. It means that parameter Δ in eqs. (4)-(5) for the amorphous phase increases with temperature too. We shall also restrict our analysis to the model considering only two - amorphous and crystalline - phases. True, X-ray structural analysis revealed the existence of two crystalline phases: γ-Al_2O_3 and δ-Al_2O_3. However, the approximation proposed for aluminum oxide powders may be acceptable since no noticeable distinctions were observed in the $C(t)$ dependences for powders with close specific surface values (38.3 and 41.7 m^2/g) and with drastically different volumes of γ-Al_2O_3 and δ-Al_2O_3 phases (see the Table 1). Assume that the oxygen diffusion rate in oxide particles is negligibly small (at least for $T < 500^0C$).

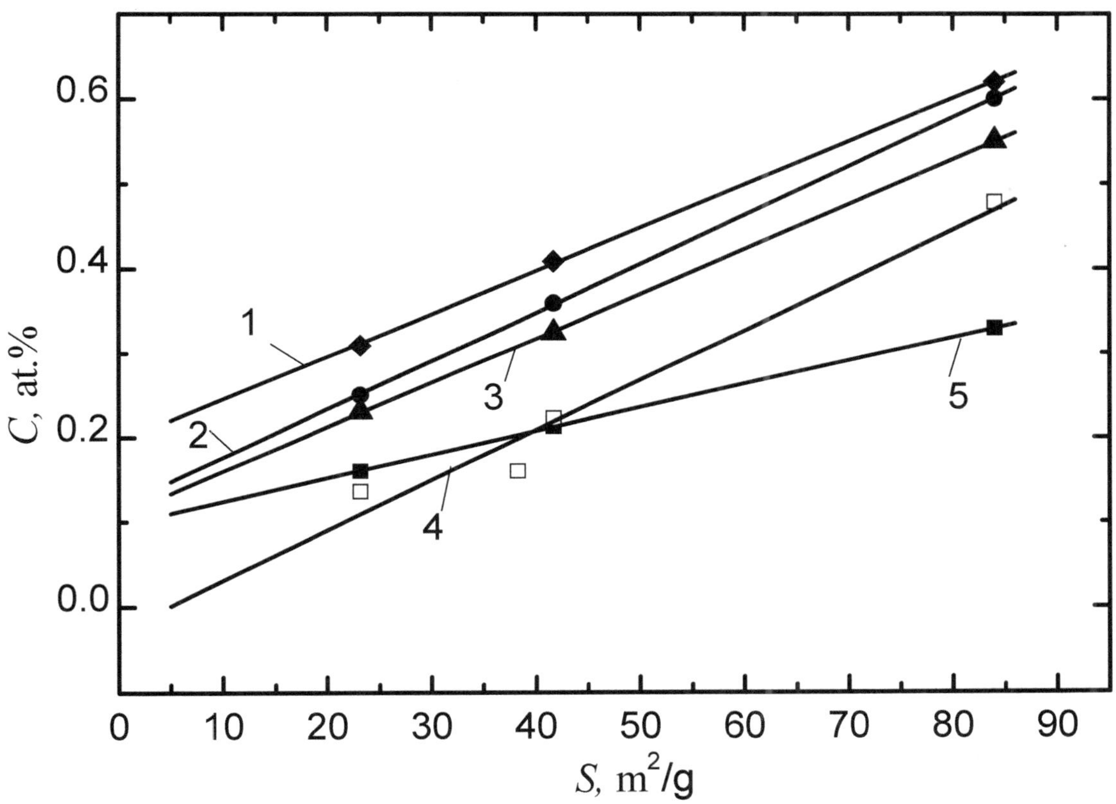

Fig. 11. Average ^{18}O isotope concentration C in aluminum oxide nanopowders as a function of the powders specific surface S during annealing in oxygen and carbon dioxide enriched with ^{18}O isotopes. Annealing conditions: 1,2,3,4 - annealing in $^{18}O_2$ at 500^0C for 120, 60, 30 and 15 min., respectively; 5 - annealing in $C^{18}O_2$ at 100^0C for 120 min

In terms of the above approximations, the expression for $C(t)$ is a sum of relaxation-type contributions (see eq. (5)) from amorphous and crystalline powder components

$$C(t) = C_0 S\left[\alpha_1\rho_1\Delta_1(1-\exp(-\Gamma_1 t)) + \alpha_2\rho_2\Delta_2(1-\exp(-\Gamma_2 t))\right] , \quad (12)$$

where indexes 1 and 2 relate to the oxide powder amorphous and crystalline phases, respectively; $\alpha_{1,2}$ - parts of these phases in the total surface area of the powder ($\alpha_1 + \alpha_2 = 1$). The results of $C(t)$ calculations with eq. (14) are given as solid lines in Fig. 10. The values of parameters α_1, Δ_1, Δ_2, Γ_1, and Γ_2 are presented in the caption. In the calculations we also took into consideration that the densities of the aluminum oxide amorphous phase, the γ-Al_2O_3 phase and the δ-Al_2O_3 phase are 3.0, 3.35 and 3.62 g/cm^3, respectively.

It is seen that the "two-phase" model easily provides a satisfactory description of experimental $C(t)$ dependences during isotope exchange for a powder with specific surface 38.3 m^2/g at all temperatures. Such a satisfactory description should be obtained also for powders with other specific surface values (all model parameters being the same) owing to the linear dependence $C(S)$. No special problems are expected also in the description of $C(t)$ dependences for isotope exchange in aluminum oxides - $^{18}O_2$ systems by means of eq. (12). However, the hypothesis that the number of oxygen ions in the amorphous phase surface layer participating in isotope exchange depends on the annealing temperature requires special experimental verification.

Thus, the studies of aluminum oxide powders with complex phase compositions showed that the OIE technique is very sensitive not only to the particle size of nanopowders, but probably to the presence of an amorphous phase in them.

5. Oxygen grain boundary diffusion and oxygen isotope exchange in nanocrystalline oxide $LaMnO_{3+\delta}$

Grain boundary diffusion in metal oxides is a poorly studied phenomenon and the data from different authors are controversial. In some of the most reliable works, the presence of enhanced diffusion at the grain boundaries is questioned (Kaur et al., 1995). At the same time, this fact has been reliably determined for metallic systems. The question of the possible common properties of the grain boundary diffusion for oxides and metals is still open. New experimental data on the coefficients of grain boundary diffusion in oxides attract the interest of theoreticians due to the specific properties of the grain boundaries in ionic compounds; in particular, these boundaries are charged. The investigations under consideration are of an applied interest, because metal oxides are widely used as functional materials in applications. The coefficients of the grain boundary diffusion determine the stability of the properties of polycrystalline materials. Presently, most researchers explain this inadequate situation with the diffusion data for oxides citing methodological reasons. It is difficult to obtain correct results primarily because of the strong influence of impurities in the oxides (Kaur et al., 1995). Control of the purity and stoichiometry of a material, especially near grain boundaries, is a complicated problem. In this connection, the study (Vykhodets et al., 2008) dealt with the problem of obtaining correct data on grain boundary diffusion in oxides.

The measurements were carried out for $LaMnO_{3+\delta}$ oxide. This is a suitable model object, because it does not have structural vacancies in the oxygen sublattice. Similar systems are characterized by low values of the volume diffusion coefficients of oxygen D. This facilitates the choice of the annealing regimes for which the contribution of the grain boundary diffusion to the formation of the concentration profiles for the tracer atoms will be determinative. The research was focused on the realization of two kinetic regimes: the B type, when the grain boundary diffusion and volume diffusion take place simultaneously; and the C type, when the diffusion of the tracer atoms into the bulk of grains is negligible. Such an approach allows one to determine two parameters characterizing grain boundary diffusion: the diffusion coefficient for grain boundaries D_{GB} and the boundary width δ (Kaur et al., 1995).

The research was carried out for nanocrystalline samples. For them, the conditions to obtain reliable results are more favorable due to the presence of the branching net of the boundaries, which enhances the contribution of the grain boundary diffusion to the diffusion flux. Besides, the nanocrystalline samples were prepared with the method in which the time of the nanostructure formation is very short. This requirement is fulfilled by the shockwave loading method (Kozlov et al., 1997) using explosives. In this case, it is expected that the doping of the nanograin boundaries is weak or absent. The initial material for the shockwave method is a coarse-grained powder of $LaMnO_{3+\delta}$ with 15–30-μm grains. According to the results of x-ray diffraction analysis, the material obtained after an explosion was nanocrystalline with an average particle size of 41 nm. These characteristics did not change after further diffraction annealing. Samples 3-mm-thick with an area of 1 cm^2 were cut out of this material. Mechanical operations of cutting, polishing, etc., were not performed to avoid the surface contamination of the samples. The scanning electron microscopy investigation of the fractured surface of the samples revealed that the main element of the microstructure of the samples under consideration is a crystallite with an average size of 3μm. The distance between the crystallites was very large (submicron); as a result, $^{18}O_2$ molecules penetrate freely into the space between the crystallites during annealing. This feature was taken into account in mathematical treatment of $C(x)$ profiles.

The results presented in Fig. 2 allow one to choose the optimal conditions for the diffusion annealing of compact samples. At 500^{0}C, the C and B kinetic types are expected for $t \le 1$ h and for a few hours, respectively. The concentration of ^{18}O atoms near the surface of the samples should be less than $(3/2)C_0\Delta/r$ in the C regime and about or larger than this value in the B regime. This estimate was obtained by taking into account that, in contrast to an insulated nanoparticle, one boundary inside the massive sample simultaneously belongs to two particles. Similarly, the optimal conditions of annealing were also determined for other temperatures. Diffusion annealings were performed at 500, 450, and 400^{0}C for ~ 1, 4, and 26 h, respectively, to realize the C type kinetics. At 500^{0}C, diffusion annealings for ~4 and 8 h oriented to the B type kinetics were carried out.

The concentration profiles $C(x,t)$ of ^{18}O atoms for bulk samples (some of them are shown in Fig. 12) were determined with the NRA method, the measurement conditions being the same as for powders (see section 4.1). The concentration profiles were characterized by the presence of background concentration $C_{PH}(t)$. It appears due to the free penetration of ^{18}O molecules into the space between crystallites during annealing. It is easy to show that the following relation should be fulfilled in this case: $N(t)/C_{PH}(t) = n/4$, where $N(t) = \int_0^\infty dx(C(x,t) - C_{PH}(t))$, n is the linear size of a crystallite. The experimental value of $N(t)/C_{PH}(t)$ for ~1-, 4-, and 8-h annealing did not differ within the statistical error (~6%) and the average value of the parameter n was approximately 3 μm. This value is in agreement with the results of the investigations of the microstructure of the samples. Thus, there is good reason to believe that the difference profiles ($C(x,t)$ - C_{PH}) were due to diffusion into crystallites of size ~3 μm, which, in their turn, have a nanocrystalline structure.

In $C(x,t)$ profiles processing, expression (1) was used for the C regime. Here, the volume diffusion coefficient D was replaced by the grain boundary diffusion coefficient D_{GB} and it was assumed that $C_S^{max} = 3C_0\Delta/2r = 1.75$ at. %. For the B regime, the following expression (Kaur et al., 1995) was used for $C(x,t)$ profiles processing:

$$D_{GB}\delta = 1.332\left(D/t\right)^{1/2}\left(\partial \ln C(x)/\partial x\right)^{-2} \quad , \tag{13}$$

where δ is the grain boundary width. Expression (13) satisfactorily describes the corresponding experimental data within measurement error.

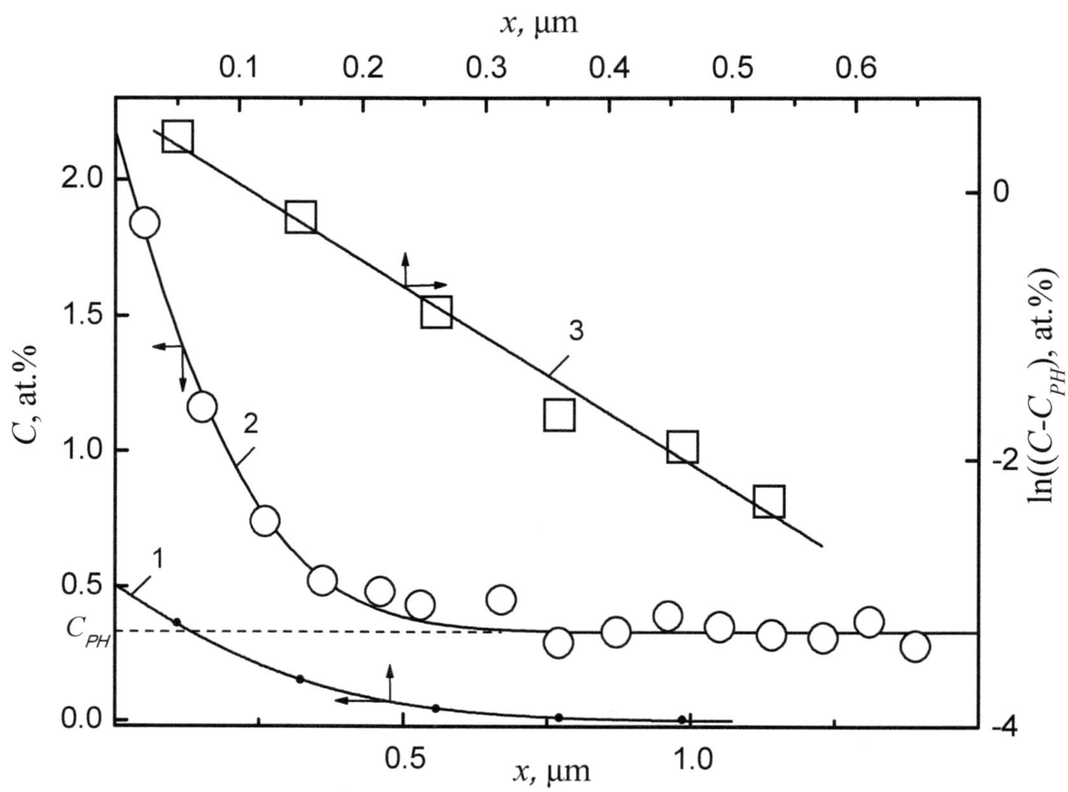

Fig. 12. Concentration profiles of ^{18}O atoms in $LaMnO_{3+\delta}$ nanocrystal: 1 - 400^{0}C, t = 26 h, curve – calculation by eq. (3), background C_{PH} is deducted; 2 - 500^{0}C, t = 4 h 11 min., curve – experimental dependence; 3 - 500^{0}C, t = 8 h 12 min.; dots – experiment

Figure 13 shows the results of temperature versus grain boundary diffusion coefficients studies. The grain boundary diffusion activation energy Q_{GB} was found to be 1.95 eV. Using the data of study (De Souza et al., 2000) for volume diffusion activation energy Q_V we find $Q_{GB} \approx 0.7Q_V$. Such a relationship is typical of metallic systems for vacancy-type diffusion (Kaur et al., 1995). Thus, no qualitative differences in the diffusion behavior for oxide and metallic systems were established.

A value of 0.044 nm was found for boundary width δ. It is an order of magnitude larger than the value δ ≈ 0.5 nm obtained in all of the correct experiments on grain boundary diffusion. Note that using eq. (13) one can only determine an effective thickness of the boundary, δ, disregarding its structure and, correspondingly, diffusion inhomogeneity. For an inhomogeneous boundary with the mixing of tracer atoms between intervals with strong disordering (dislocation cores) and with a weakly distorted lattice, the ln $C(x)$ dependence is linear (Kondrat'ev & Trakhtenberg, 1986), as in the case of the homogeneous boundary. The left-hand side of Eq. (13) is reduced to the form $D_{GB}\delta p$, where p is the volume fraction of the trajectories with a high diffusion permeability. It is seen that the δ value for $p \ll 1$ is strongly underestimated. We can point out some reasons why the structural inhomogeneity of the boundaries was manifested in our experiments. This can be associated with a specific type of nanoboundaries formed in a brittle material after shock loading and with a short diffusion length $(Dt)^{1/2}$, which was on the order of the atomic size in our experiments.

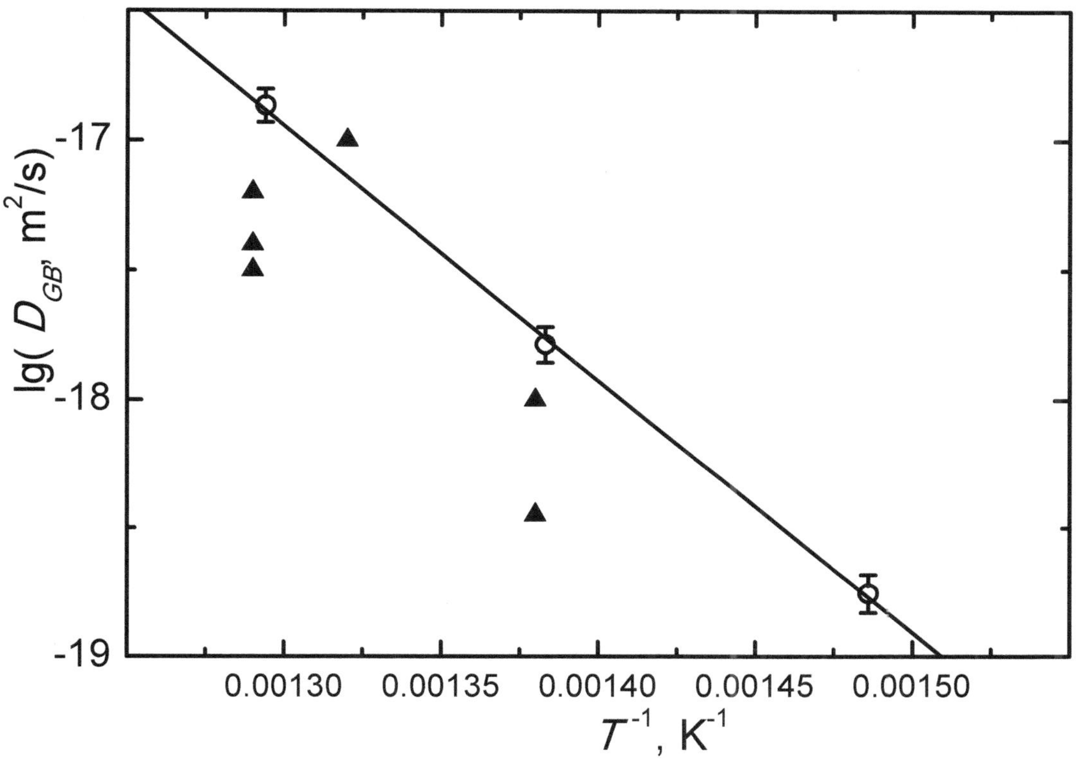

Fig. 13. Temperature dependence of grain boundary diffusion coefficients D_{GB} in $LaMnO_{3+\delta}$ nanocrystal. Triangles – results of work (Brossmann et al., 2004) for ZrO_2 nanocrystal

We know only one work (Brossmann et al., 2004) devoted to oxygen grain boundary diffusion in oxide nanocrystalline. This study was performed on a high-purity oxide ZrO_2. The ZrO_2 nanocrystalline was produced by sintering of oxide nanoparticles with the average nanograin size 100 nm. Diffusion was examined with the use of the B type kinetics; the grain boundary width value δ = 0.5 nm was postulated for determination of grain boundary diffusion coefficients D_{GB}. The value of Q_{GB} = 1.95 eV obtained in work (Brossmann et al., 2004) coincided with the result of study (Vykhodets et al., 2008). In both works, enhanced grain boundary diffusion was registered. The D_{GB} values from work (Brossmann et al., 2004) are given in Fig. 13 for comparison. The oxygen grain boundary diffusion coefficient in ZrO_2 and $LaMnO_{3+\delta}$ nanocrystallines are seen to be very similar. This coincidence of the results reported in the two works (Brossmann et al., 2004; Vykhodets et al., 2008) is considered to be accidental; nevertheless, we draw attention to these facts. Thus, an important finding for the diffusion theory has been obtained in the above studies, namely, no qualitative differences in volume and grain boundary diffusion were found for oxide and metal systems. Note that both studies were performed on nanocrystallines; besides, systematic measurement errors due to uncontrolled doping of oxide grain boundaries with impurities were probably excluded.

As was mentioned above, the grain boundary diffusion coefficients D_{GB} in study (Vykhodets et al., 2008) at 500, 450, and 400^{0}C were determined with the use of the C type kinetics, i.e. in the absence of oxygen volume diffusion. By means of expressions (1), (2), and (6) we calculated for the same temperature the parameters *H*, *k*, and Γ characterizing the surface energy barrier during isotope exchange of gaseous oxygen with $LaMnO_{3+\delta}$ oxide nanoboundaries (Fishman et al., 2010). The results for Γ are given in Fig. 7. As in section 4.3, the *k* values were recalculated in Γ using expression (6) at $l = 2 \cdot 10^{-10}$ m. Figure 7 shows a great difference in the frequency factor Γ_0 and activation energy *E* values in the temperature versus Γ dependence, which were obtained, respectively, for the nanograin boundary and the regular lattice of $LaMnO_{3+\delta}$. The difference is almost 7 orders of magnitude for Γ_0 ($0.9 \cdot 10^{10}$ and $0.97 \cdot 10^{3}$ s^{-1}) and 2 times for *E* (1.67 and 0.88 eV). In our opinion, such a dramatic difference in Γ_0 and *E* parameters for the nanograin boundary and the regular lattice of one and the same oxide is difficult to understand from the isotope exchange mechanism only. Competing mechanisms can take place simultaneously, and, depending on the relationship between the system parameters, one or another mechanism will prevail. A similar situation happens, for example, in studies of diffusion in solids, when volume and grain boundary diffusion occur simultaneously. For such objects, the contributions from volume and grain boundary diffusion to the mass transfer rate will depend on the experiment conditions; as a result, drastically different values of effective diffusion coefficients, activation energies, and frequency factors in the temperature dependence *D* can be obtained.

6. Conclusion

In the studies presented here it was shown that the OIE parameters are sensitive to the dimensional characteristics of oxide powders. Therefore, the approach based on OIE examination shows promise for receiving the information about the size of nanopowder particles. Similar data are obtained with other techniques (electron microscopy, X-ray diffraction, specific surface measurement, dynamic light scattering by means of a laser analyzer etc.); in this respect, the OIE technique has an important specific feature. In addition to the conventional methods, it makes it possible to determine the stoichiometry of oxide powders, is sensitive to chemical activity of powders and, probably, to the presence of amorphous phases and very small amounts of larger particles in nanopowders. These peculiarities of the considered technique are of great interest both for theory and practice, since all existing technologies for nanopowder synthesis lead to the formation of non-equilibrium materials and assemblies consisting of particles with different sizes and shapes. Moreover, when the OIE technique is used on nanopowders, very low oxygen volume diffusion coefficient values can be measured as compared with traditional approaches. The application of the OIE technique for bulk nanocrystalline samples also holds much promise. Already in several works it was shown that considerable progress can be achieved with the use of this technique in the examination of oxygen grain boundary diffusion in oxides.

7. Acknowledgment

This work was supported by RFBR (grant 10-03-96016-p_ural_a and grant 09-03-00335_a), the Program of fundamental research of Presidium of Russian Academy of Sciences N 27

"Foundations of fundamental research of nanotechnology and nanomaterials" and the Federal Target Program "Scientific and scientific-pedagogical staff of innovation Russia (contract 02,740. 11.0641).

8. References

Brossmann, U. et al. (2004), Oxygen diffusion in nanocrystalline ZrO_2, Rev. Adv. Mater. Sci., Vol. 6, p. 7-11, ISSN 1605-8127

Carter, S. et al. (1992), Oxygen transport in selected nonstoichiometric perovskite structure oxides, Solid State Ionics, Vol. 53-56, p. 597-605, ISSN 0167-2738

De Souza, R.A. & Kilner, J.A. (1998), Oxygen transport in $La_{1-x}Sr_xMn_{1-y}Co_yO_{3\pm\delta}$ perovskites - Part I. Oxygen tracer diffusion, Solid State Ionics, Vol. 106, p.175-187, ISSN 0167-2738

De Souza, R.A. et al. (2000), A SIMS study of oxygen tracer diffusion and surface exchange in $La_{0.8}Sr_{0.2}MnO_{3+\delta}$, Materials Letters, Vol. 43, p. 43-52, ISSN 0167-577X

Fishman, A.Ya. et al. (2003), Diffusion in concentrated Jahn-Teller systems, Advances in quantum chemistry, Vol. 44, p.497- 509, ISBN: 0-12-034844-6, ISSN 0065-3276

Fishman, A.Ya. et al. (2009), Oxygen isotope exchange in nanocrystal oxides Journal of NanoResearch, Vol.7, p. 33-41, ISSN 1662-5250

Fishman, A.Ya. et al. (2010), Isotope Exchange between Gaseous Oxygen and Nano-Grain Boundary of $LaMnO_{3+\delta}$ Oxide, Defect and Diffusion Forum, Vol. 297-301,p. 1301-1305, ISSN 1012-0386.

Gizhevskii, B.A. et al, (2008), Oxygen Isotope Exchange Between the Gaseous Phase Enriched with the ^{18}O Isotope and Nanocrystal Oxides $LaMnO_{3+\delta}$ Obtained By Severe Plastic Deformations, Defect and Diffusion Forum, Vol.273-276, p.233-238, ISSN 1012-0386

Ivanov, M.A. et al. (2006), Formation of a thin-layer electrolyte for SOFC by magnetic pulse compaction of tapes cast of nanopowders, Journal of Power Sources, Vol. 159, p. 605-612, ISSN 0378-7753

Kaur, Y et al. (1995), *Fundamentals of Grain and Interphase Boundary Diffusion* (3rd ed.) Wiley, ISBN 978-0-471-93819-4, New York, USA

Kondrat'ev, V.V. & Trakhtenberg, I. Sh. (1986), Grain-boundary diffusion of atoms in a model of structurally heterogeneous boundary, The Phys. Metal & Metallogr., Vol. 2, p. 434-441, ISSN 0031-918X

Kotov, Yu.A. et. al. (2004), Characteristics of nanopowders produced by vaporation CeO_2/Gd_2O_3 emission targets repetitively pulsed CO_2 laser, Journal of Technical Physics, Vol. 74, N3, p.72-77, ISSN 0044-4510

Kozlov, E. A. et al. (1997), Patent of Russian Federation No. 2124716, December 24, 1997

Manning, P.S. et al. (1997), The kinetics of oxygen transport in 9.5 mol % single crystal yttria stabilized zirconia, Solid State Ionics, Vol. 100, p. 1-10, ISSN 0167-2738

Odzaki, A. (1978), *Isotopic studies of heterogeneous catalysis,* Acad.Press, ISBN 0125319509, New York, USA

Routbort, J.L. et al. (1997), Diffusion-controlled creep in mixed-conducting oxides, Def. and Dif. Forum, Vol. 143-147, p.1201-1205, ISSN 1012-0386

Ruiz-Trejo, E. & Kilner, J.A. (1997), Oxygen diffusion and proton conduction in $La_{1-x}Sr_xYO_{3-\delta}$, Solid State Ionics, Vol. 97, N 1-4. - P. 529-534, ISSN 0167-2738

Petrova, S.A. et al. (2010), Phase states of mechanoactivated manganese oxides, *Proceedings of the Ninthight Israeli - Russia Bi-National Workshop 2010 "The optimization of composition, structure and properties of metals, oxides, composites, nano and amorphous materials"*, p. 138-152, Belokurikha, Russia, July 25-30, 2010

Shewmon, P. (1989), *Diffusion in solids* (2nd ed.) The Minerals, Metals & Materials Society, ISBN 0-87339-105-5, Warrendale, PA, USA

Solmon, H. et al. (1995), Ionic transport properties of yttria-doped zirconia, Solid State Phenomena, Vol. 41, p.103 – 112, ISSN 1012-0394

Vykhodets, V.B. et al. (1987), Oxygen diffusion in α-Ti. II. The calculation of the concentration profile of impurities in the nuclear microanalysis, The Phys. Metal & Metallogr., Vol.64, p.920-923, ISSN 0031-918X

Vykhodets, V.B. et al. (1994), Potential barriers in the migration of tracer oxygen atoms in the $YBa_2Cu_3O_{7-\delta}$ lattice, JETP, Vol.106, N2(8), p.648-662, ISSN 0044-4510

Vykhodets, V. B. et al. (2000), Atomic structure and diffusion properties superanizotropnyh systems, Journal of Solid State Physics, Vol.42, N4, p.595-601, ISSN 0367-3294

Vykhodets, V.B. et al. (2008), Grain Boundary Self-Diffusion of Tracer 18O Atoms in Nanocrystalline Oxide $LaMnO_{3+\delta}$, JETP Letters, Vol. 87, No. 2, pp. 115–119, ISSN 0370-274X

6

Considerations about Degradation of the Red Ceramic Material Manufactured with Granite Waste

Xavier Gustavo de Castro, Saboya Fernando,
Maia Paulo Cesar de Almeida and Alexandre Jonas
State University of Norte Fluminense Darcy Ribeiro – UENF, Department of Civil Engineering, Campos, Rio de Janeiro
Brazil

1. Introduction

Ceramics are used as construction materials since the earliest days of civilization, around 5,000 B.C. Manufactured fired earth bricks had firstly been used as protection walls against invaders and fierce animals. However, during the industrial revolution on century XIX, the steam machines allowed a marked development of the ceramic industry.

The term "Ceramic" is originated from *Keramikos*", which, in Greek, means fired raw material. The most important materials used in the composition of the red ceramic bodies are clayey soils. Chemically the clay mineral is composed basically by hydrous aluminum, iron and magnesium silicates, which can be dispersed in fine particles. Presence of impurities such as ferric oxides, quartz among others plays an important role on the characteristics of the final product.

Despite its great technological advances, some industries in the developing countries have not invested in high technology in order to manufacture bricks with high durability and acceptable mechanical properties.

Therefore, for the achievement of the better quality of the ceramic bricks and in accordance with the environmental appeal, some industrial by-products have been used in the raw material in ceramic industries.

In the case presented herein, by-products from decorative stone industry have been used aiming to improve the durability of the ceramic bricks as well as their mechanical properties.

In this particular perspective, the ceramic industry emerges as a great attractive to promote the use of industrial by-products. Moreover, the use of industrial by-products, as being granulated materials, does not affect the path normally followed during ceramic processing. As a secondary effect, but not less important, the insertion of industrial by-products can also help to save noble materials in the pit borrow, avoiding, thus, the increasing on the rate of exploration works.

2. Red ceramic materials

Red ceramic raw materials are generally taken from natural deposits of clayey soils composed by clay minerals and others organic and/or inorganic particles that are very "reactive" in the presence of water. This property is known as plasticity, which is very dependent on chemical interaction among particles and the liquid, which percolates through the pores.

Most of ceramic materials are fragile and stiff, with very low tenacity. They have high fusion/sintering temperature and good chemical stability specially, in hostile environment, due to their strong chemical bonding between structural units.

2.1 Durability of the red ceramic materials

Despite its stable mechanical behavior, it is not rare to observe problems in ceramic bodies regarding its durability, i.e., strength against weathering.

This is commonly observed in artifacts that are fired at temperature bellow its sintering point. Cracks and weathering signs are well developed inside the ceramic body, with time, due to work done by external agents like temperature, humidity and insulation. The main problem found in construction materials regarding their durability are:

1. Deterioration in brick wall and tiles resulting in cracks and infiltration (Fig. 1);
2. Degradation of the first lines of masonry due to capillarity effect of sulphate water (NA_2SO_4) (Fig. 1)
3. Rapid degradation due sudden changes in temperature (Fig. 2)
4. Quickly degradation in coastal environment resulting in complete deterioration of the ceramic brick (Fig. 4)
5. Mechanical and abrasive deterioration of the floor (Fig. 4)

Fig. 1. Fast degradation of masonry and tiles

Fig. 2. Fast degradation of masonry due to sulfate water and temperature change

White stain on masonry surface (effluorescence) appears as shown in Figure 3. This is consequential from a chemical process where the cement reacts with water resulting, thus, in a soluble binder, calcium hidroxile ($Ca(OH)_2$) that adheres on the ceramic brick. As long as the mortar layer contains cement and this layer has high porosity, the free calcium hydroxile found in its composition reacts with air (carbonic anidre). This reaction between these two components results in calcium carbonate, non-soluble white salt, which leads to the expansion of the brick (Fig. 3).

Fig. 3. Detachment of the floor brick and walls

Fig. 4. Fast degradation in masonry due to saline spray and abrasion

The process of degradation takes place when the body gets in contact with atmospheric conditions. Therefore, degradation can be understood as the effect of atmospheric agents on the ceramic bodies without crystallographic modification.

Alterability is defined as the susceptibility of a ceramic body to suffer degradation. Durability, on the other hand, means the capacity in keeping its structural stability, as well as the appearance with time. Therefore, alterability and durability will be closely dependent on the material and environmental characteristics and the exposure time, as well.

In tropical regions a wide range of temperature and humidity variation is commonly observed. The weathering agents that act directly on ceramic bodies are:

- Rain and air humidity
- Periodic insulation and drop of temperature in a short period of time, causing contraction and dilation, respectively.
- Air temperature favoring chemical reactions
- Wind and kinetic energy favoring abrasive deterioration.

Mechanisms that act on degradation process are the same as those that act on weathering. Dilation and contraction resulting from temperature and humidity changes can give rise to cracks that commonly leading the ceramic body to undermine.

Chemical agents can flow into these cracks and pores, reducing, thus, the bond between grains, making easier the erosion process and leaching the constituent materials. These transported constituents can induce the formation of new crystals resulting from chemical reactions, mainly when soluble salts are present in the mortar. When the air humidity

penetrates into the structure, these crystals are generally dissolved and carried to the surface, provoking thus, stains on the ceramic artifact. If these salts remain inside the ceramic body, they will re-crystallize causing expansion, favoring thus, degradation of the material.
These alterations can be very harmful in particular in coastal zones and industrial areas due to the saline sprays and acid rain, respectively.
The intensity of the effects of these degradation processes is closely dependent on the environmental conditions and also on the level of the thermal treatment specified to the ceramic material.

3. Characterization of the materials and evaluation of the alterability and durability of ceramic mass with addition of granite waste

This chapter presents the adopted methodology in order to evaluate the durability of ceramic bricks when granite waste by-product is incorporated in the raw material. Figure 5 shows the flow chart for complete characterization of the materials used in the confection of brick ceramic.
It is presented here the test sequence for the identification of the studied raw material: the clay mass and the granite by-product.

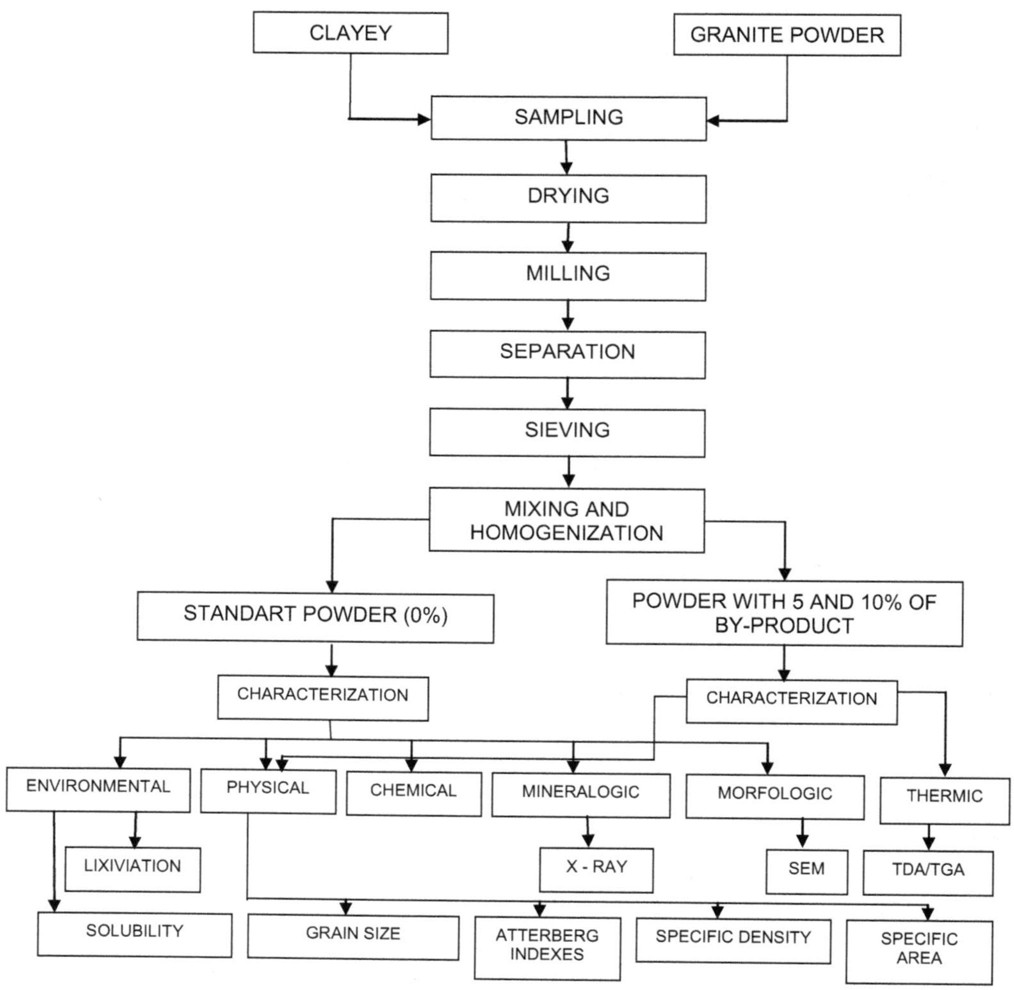

Fig. 5. Flow chart for complete characterization of the raw material with different content of granite powder

Further results and information can be reached in Saboya *et al* (2006), Xavier (2006), Xavier (2001) and Alexandre (2000).

3.1 Environmental characterization of the granite waste

The description of the results obtained from leaching test is shown in Table 1 and Table 2 where some results obtained from literature are shown in order to make the reader familiar with characteristic values and allow a direct comparison. Maximum allowed values for each chemical element are also presented aiming the classification of the material harmfulness.

Chemical Element	Waste Granite (mg/l)	Waste Silva (1998) (mg/l)	Waste Neves (2002) (mg/l)	Maximum limits (leaching test lab) NBR- 10005 (mg/l)
Ag	<0.005	< 0.010	<0.010	5.000
Cd	<0.002	< 0.010	<0.047	0.500
Cr	0.020	< 0.020	<0.010	5.000
Pb	0.030	< 0.050	2.120	5.000
Ba	0.110	1.160	N.S.	100.000

Legend: N.E.: Not Specified.

Table 1. Results of the leaching test and presented in literature

Condition test lab	Granite Waste (mg/l)	Waste (Silva,1998) (mg/l)	Waste (Neves, 2002) (mg/l)	Maximum limits (Leaching test in laboratory) NBR- 10005 (mg/l)
pH Begin	9.400	10.300	9.400	N.S.
pH final	5.150	6.500	5.500	N.S.
Vol. Acid (ml)	80	130	400	N.S.
Time (h)	28	28	28	N.S.

Legend: N.S.: Not Specified.

Table 2. Conditions of the leaching test

Evaluating the results obtained in the leaching test and comparing them with the maximum limits established by the norm ABNT 10004 (1987), it is observed that none of its chemical elements of the gross mass presented superior concentrations than permitted, being the sample considered non toxic.

The pH of the leached extract from the gross mass (Table 2) falls in the interval between 5.15 and 9.4, being, thus, characterized as not corrosive (ABNT 10004, 1987).

The results of the leaching tests together with the results obtained in the literature are shown in the Table 3.

Evaluating the results obtained in Table 3 and comparing them with the maximum limits established by the norm ABNT 10004 (1987), the chemical element Pb (lead) of the mass gross has presented concentration slightly superior to the established as the maximum limit. Therefore, the residue is classified as class II, not inert.

Chemical Element	Granite Waste (mg/l)	Granite Waste (Silva, 1998) (mg/l)	Granite Waste (Neves, 2002) (mg/l)	Maximum limits solubilized NBR - 10.004 (mg/l)
Ag	<0.005	<0.010	0.070	0.050
Cd	<0.002	<0.001	<0.001	0.005
Cr	<0.020	<0.020	<0.001	0.050
Pb	<0.060	<0.050	0.001	0.050
Ba	<0.020	0.120	0.230	1.000
Al	0.150	1.080	2.990	0.200
Cu	<0.010	<0.010	<0.040	1.000
Fe	0.030	0.4000	0.010	0.300
Mn	<0.010	<0.010	<0.010	0.100
Zn	<0.003	<0.010	0.005	5.000
Na	33.800	134.000	24.000	200.000
NaCl	9.440	78.000	35.500	250.00
Ca^{+}	19.900	3.700	19.000	500.000
Sulfate	11.170	<10.000	44.000	400.000

Table 3. Results of the leaching test and some values reported in literature

The results obtained in this work are similar to those obtained by Silva (1998) and Neves (2002), which can be conclude that the granite waste is a non toxicant material, not corrosive, not inert of class II.

3.2 Chemical characterization

Following, the Table 4 depicts semi-quantitative chemical compositions of the standard clayey raw material granite waste.

Raw Materials	L.O.I. (%)	SiO_2 (%)	Al_2O_3 (%)	Fe_2O_3 (%)	CaO (%)	MgO (%)	Na_2O (%)	K_2O (%)	TiO_2 (%)
Clay	15.40	42.30	32.00	6.87	0.20	0.70	0.04	0.94	1.24
Granite Waste	1.55	69.20	14.60	3.49	3.21	0.032	3.13	4.49	0.16

Table 4. Chemical Components of the clayey and granite waste materials

Observing the values obtained in Table 4, it is verified that for the standard clay sample, the amount of SiO_2 (42.3%) indicates the probable presence of kaolinite (Al_2O_3. $2SiO_2$. $2H_2O$) and ilite, as well as the probable presence of free quartz in the total sample. The amount of Al_2O_3 (32.0%) almost totally is responsible for forming clay minerals, but it can also be credited to the presence of hydroxides as gibsite ($Al(OH)_3$). The total value of 64.3% (SiO_2 + Al_2O_3) indicates the refractory character of the raw material.

The high amount of coloring oxide Fe_2O_3 (6.87%) characterizes it as fondant agent and also indicates the prone for the red color achieved after burning. It can also indicate the goethite presence (FeO.OH). The oxides of calcium and magnesium (CaO + MgO) corresponds a total of 1%. They are fondant agents and, as consequence, they lower the refractory properties of the clay. The presence of alkalis (Na_2O + K_2O) in the amount of 0.98% indicates also fondants forming liquid phase during firing, reducing, thus, the porosities of the material.

The coloring oxide Ti_2O in the amount of 1.24% can be credited due to the presence of rutile or anatase (Toledo, 2003) and it contributes to the yellow color of the raw material. The loss of ignition (LOI) of 15.40% indicates the loss of free, adsorbed and constitution water of the raw material and, also, the degradation of the organic content of the sample.

It is noticed in the chemical analysis carried out on granite waste, that the amount of silica is higher than 69% and the amount of Al_2O_3 can reach values higher than 14%, indicating chemical compositions of primary minerals (quartz, feldspar and minerals of the group of the mica). The traces of CaO (3.21%) and Fe_2O_3 (3.49%) come mainly from the whitewash and from the metallic particle used as lubricant and abrasive, respectively.

The oxides of sodium and of potassium (Na_2O + K_2O) in the amount of 7,62% presented by the by-product, are almost totally originated from the feldspar and mica and they are considered as fondant agents that could fill the voids. The loss of ignition (LOI) of 1.55% of the by-product is considered small indicating structural stability of the waste. However, this loss can also be credited to the degradation of the mica.

Comparison the results of the chemical analysis shown on Table 4 between the standard clay mass and that of red ceramic reported by Souza Santos (1989), Alexandre (2000) and Vieira (2001), there is a clear indication that this material is quite suitable to be used in the Brazilian industry of red ceramic.

In the same way, comparison the results of the chemical analysis shown in Table 4 for the granite waste with the chemical analyses for granite waste carried out by Neves (2002), Mothé Filho (2003) and Vieira *et al.* (2003), it can be shown that the incorporation of the by-product in ceramic mass can be used in the process of production of red ceramic goods without any special measure.

3.3 Thermal characterization from Differential Thermal Analysis

Observing the thermal curves of the ceramic mass and the mass with addition of waste with 5R (5%) and 10R (10%) in Figure 6, it can be detached, for each curve, two endothermic events of medium intensity very close to each other.

In this figure it is verified that the clay samples 0R, 5R and 10R present endothermic events of medium intensity between 258,50°C and 260,90°C, indicating a shift to the left of the peak values (in relation to the sample 0R) due to the presence of waste in the ceramic mass. These peaks are mainly due to the loss of OH^- of the hydroxides of aluminum ($Al(OH)_3$) and iron ($FeO(OH)$) probably credited by the presence of gibbsite and goethite.

There are also endothermic events of medium intensity between 448.50°C and 499.84°C due to the beginning of the reaction of loss of hydroxyls of the kaolinite creating, thus, an amorphous phase (metakaolinite). Endothermic bands are observed starting from 900°C. It could indicate the transformation of the metakaolinite ($Al_2O_3.2SiO_2.H_2O$) amorphous in spinally ($2Al_2O_3.3SiO_2$) forming a stable crystalline structure.

The corresponding curve of the granite waste in the Figure 6 shows an endodermic event of small intensity for 570°C, indicating the transformation of the quartz for generating expansion. At 663.18°C an endothermic peak of small intensity corresponding to the beginning of loss of OH^- of the mica is verified in the sample.

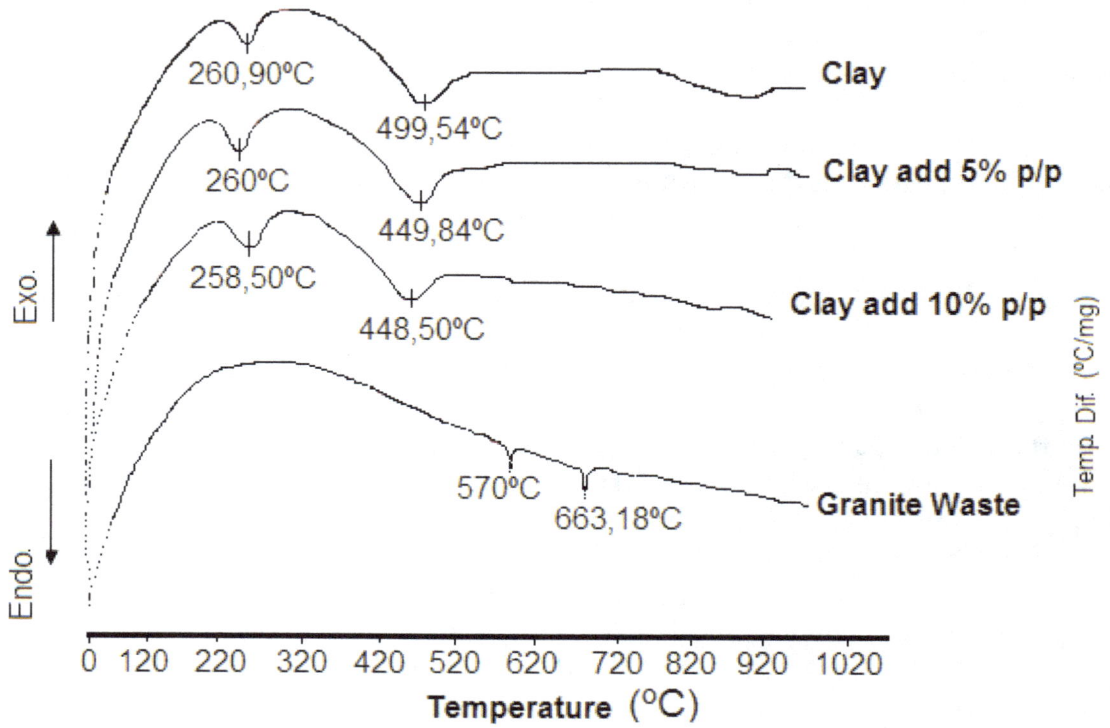

Fig. 6. Curves of the Thermal Analysis Differential (ATD) of the ceramic samples 0R, 5R and 10R

4. Experimental program for natural and accelerated degradation of samples 0R, 5R and 10R

The condition of alteration of a ceramic piece depends primarily on their physical-chemical, mechanic properties and also on the characteristics that are correlated with the time of alteration. Another important features is the degradation potential caused by the aggression of the environment, affecting the durability of the material, especially, the red ceramic in tropical countries.

For the evaluation of durability of the red ceramic materials incorporated with granite waste in the proportions of 0%, 5% and 10% known herein as 0R, 5R and 10R respectively, fired at temperatures of 500°C, 700°C and 900°C, it was used the methodology presented in the Figure 7, in the form of illustrative organization chart.

For the evaluation of the changes experienced by the red ceramic materials, it should be considered the following aspects (adapted from Maia, 2001):

- The characteristic of the change processes of the ceramic material used in the civil construction;
- The methodology adopted for the evaluation of the technological properties after alteration tests carried out in laboratory and in the field;
- The representatively of the considered intact and naturally altered samples,
- Correlation of the properties with the time of natural alteration, for the prediction of the potential of the damage caused by this alteration, i.e., long time prediction.

To evaluate the changes in the red ceramic materials it should identify the mechanisms that influence the natural alteration, whose processes of the evolution of the degradation are visually accompanied.

For the determination of the characteristics of the alteration of the red ceramic pieces incorporated with granite residue in the field and in the laboratory in an induced and accelerated way, it was used some procedures considered essential, as follows:

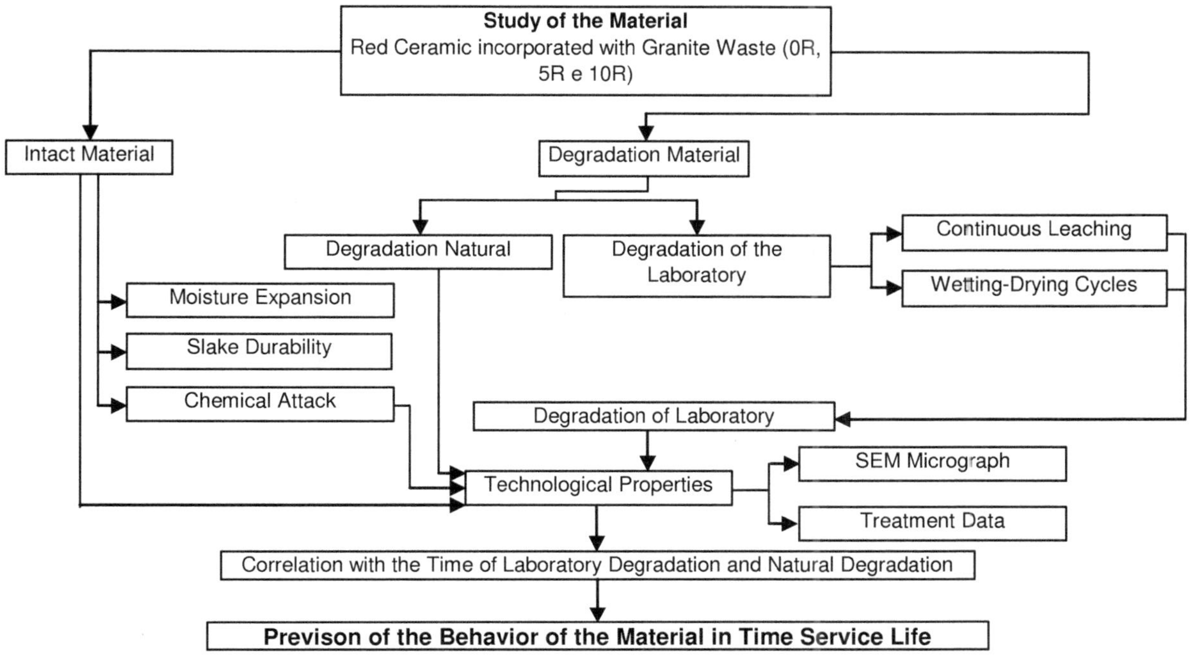

Fig. 7. Methodology for evaluation of the degradation of red ceramic samples incorporated with 0R, 5R and 10R of granite waste

4.1 Test of continuous leaching in laboratory

This particular laboratory test is considered to mimic the climatic conditions that degrade the red ceramic artifacts. When it is intended to determine the state of alteration of a rock or other natural material, the extractor Soxhlet is most suitable because the samples are submitted to leaching in distilled hot and cold water. The extractor Soxhlet allows the samples to be submitted to temperature variation in controlled periods of precipitation and water level variation.

Farjallat (1971) used the extractor Soxhlet to evaluate changes in rocks specimens using correlation with Los Angeles abrasion test.

After associating the characteristics of the rock with the quality indexes, this author compared these characteristics in different alteration stages. The quality indexes are applied in the specific quantification of measurable properties as strength, porosity, water absorption and others (Maia, 2001).

Maia (2001) has used an extractor Soxhlet of great dimensions to evaluate the alteration of rocks used in rockfill. Based on this author's work, it was manufactured in LECIV/CCT/UENF in Campos/RJ a degradation equipment (Figure 8) similar to that extractor Soxhlet.

The Equipment of Degradation developed at LECIV was designed to make possible to control periods of temperature variation and of precipitation of the solution. Distilled water is used as leaching solution. During the leaching, this solution can be monitored and, if necessary, changed.

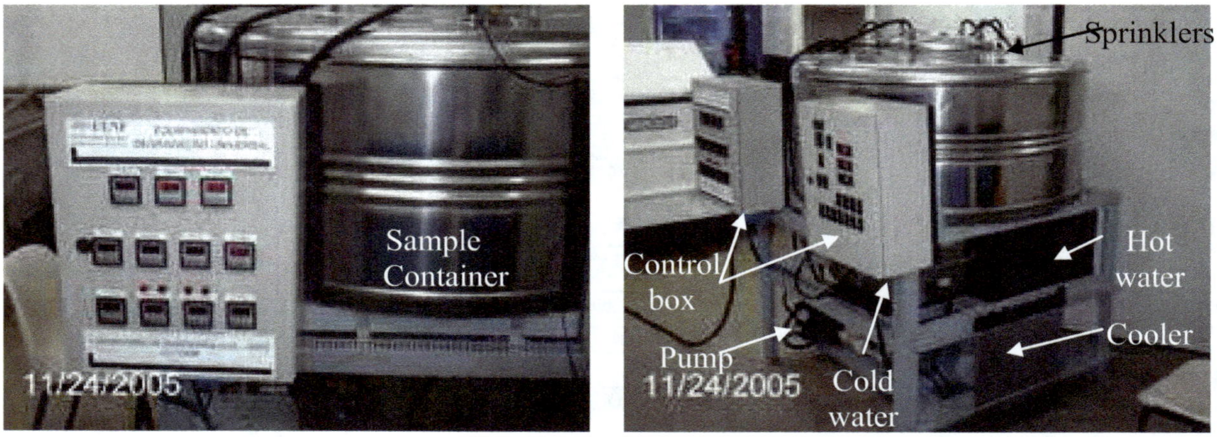

Fig. 8. LECIV/CCT/UENF Laboratory Degradation Equipment

Inside the tank, samples are subjected to environment similar to atmospheric conditions, where the material is placed on shelves of stainless-steel and submitted the temperature variations. At the same time, the samples are washed periodically with hot and cold water, simulating the natural precipitation. The wash is made by the leaching solution, coming from the pumps of hot or cold water that goes through the sprinkles placed at the top of the tank creating a saturated vapor atmosphere (this phase is only when there is the circulation of hot water).

The vapor condensation also takes place when the cycle of hot water is in operation. The equipment was programmed to work for 1 hour for each cycle of hot or cold water, meaning that when the leaching for hot water is in operation, the cold water system keeps turned off. During the cycle (hot or cold water), the water is pumped every each 15 minutes.

The hot water is controlled to be kept at approximately 70ºC and the cold water around 35ºC. The time periods of 300 hours, 658 hours and 1500 hours were used to submit to the samples to this complete process. The equipment possesses an electronic system of operation (I-square of command) which ignores manual control of the time and water cycles.

It also allows the verification of the temperature of the hot and cold water and the tank of samples as well (Figure 9). To avoiding shortage of water in the system, a device of production of distilled water is connected directly to a column of water. It is linked to the equipment and it monitors the level of the water tanks by means a level buoy, renewing the solution constantly allowing monitoring the pH of the solution.

To guarantee the required temperature of the cold water, it was installed a cooling system that starts operation when it reaches a pre-established temperature. The pumping system is constituted of three pumps. The 1st and 2nd are responsible for the suction and hot and cold water pressure toward the sample tank. The third pump, besides the cold water suction, direct it to the coolers and is also responsible to guide the water back to the cold water tank.

The pump of hot water possesses a system to support high temperatures. The tanks of hot and cold water possess a capacity of approximately 200 liters (each) and the container of samples possess 500 liters of volumetric capacity. The equipment is constantly monitored by a computer and by a camera installed in the laboratory to notice possible defects during the night and weekends when the same is in operation.

Curves shown in Figure 9 depict the temperature balance inside the sample tank (500 liters) and also the temperature balance inside the hot and cold water tanks. It can be noticed that at around 100 minutes from the beginning of the operation the balance is reached. This means that the electronic circuits work with synchrony and precision and that the samples are always under the same conditions inside the equipment.

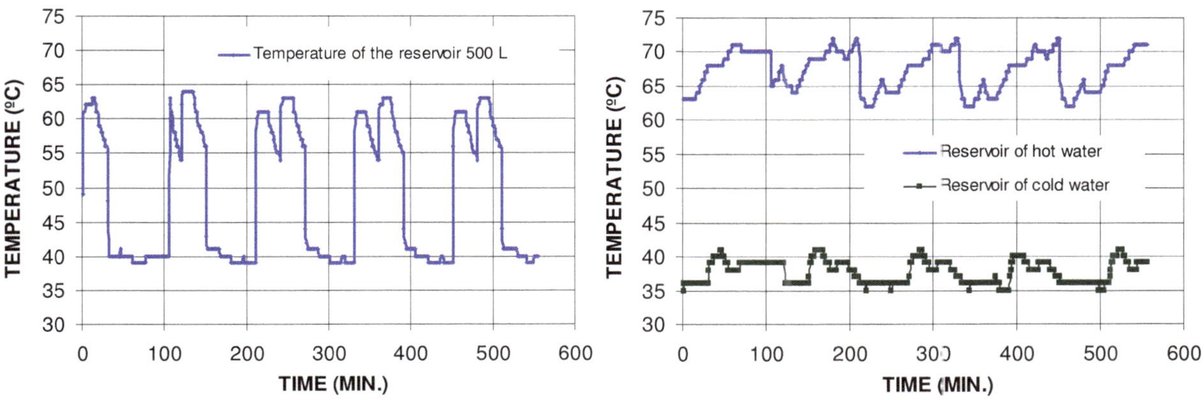

Fig. 9. Temperature of the sample container, hot and cold water tanks with time

In the change of cycle of hot and cold water inside of the samples tank, it is noticed the continuous fall in temperature, simulating in an accelerated way the climatic conditions of temperature variation between the days and the nights, provoking dilation and retraction in the samples due to thermal gradients.

4.2 Laboratory test of wetting and drying cycles

This particular test intends to simulate, in laboratory, the aggression suffered by the ceramic when subject to environmental natural conditions. It is known that the laboratory conditions are not the same that of the field. However, the results can be somewhat representative (Maia, 2001; Cunha, 2006).

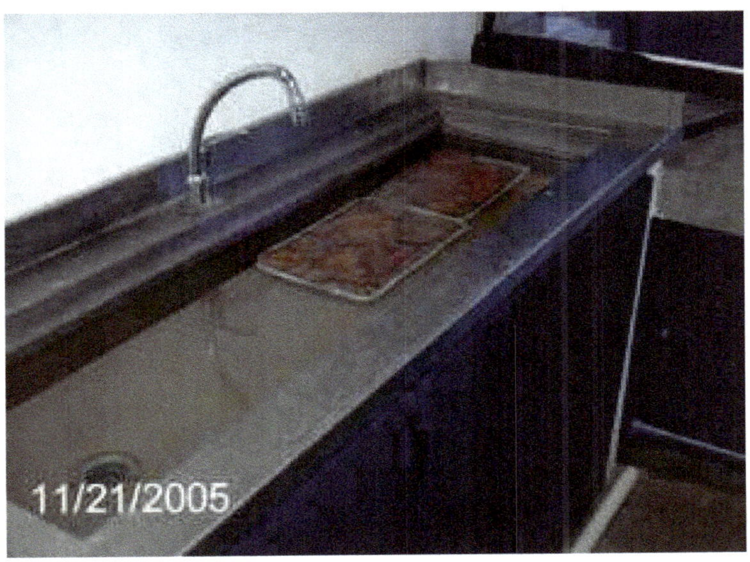

Fig. 10. Laboratory tank for wetting and drying cycling test

The washing water was monitored periodically every 15 days through pH meter aiming to assure keeping the solution as basic acidy. The environment temperature inside the room was controlled to keep it about 21±1°C and relative humidity of the about 64±5%. Cycling was composed of 50 and 150 cycles, representing respectively 1.200 hours and 3.600 hours of soaking and drying steps.

4.2.1 Field natural degradation

This procedure has as the objective comparison of the degradation accelerated in laboratory to the alteration of the material under local environmental conditions. Samples were exposed to periods of 6, 8 and 10 months where variables like air temperature, relative humidity, wind speed, precipitation and the solar radiation, were recorded.

The natural degradation of the samples was carried out at the Campus of the State University of the Norte Fluminense Darcy Ribeiro - UENF, Campos of Goytacazes/RJ, with latitude of 21st 45 ' South, longitude of 41st 18 ' West and altitude of 11 m. This kind of degradation is used for the long time prediction which will serve as a base to obtain the information about the effects of the degradation of a specific material (Cunha, 2006).

The samples were fastened in a special support made of wood placed on the roof of a UENF building, according to the Illustration 11. It was placed in such way to guarantee ideal incidence and absorption of solar radiation, facilitating, thus, the effective degradation process during the research period. The supports for samples were designed to guarantee free air circulation.

Fig. 11. Samples on terrace exposed on the roof directed to the best insulation direction according to the local coordinates

The climatic conditions define the processes of degradation of ceramic materials exposed to the environment. It is naturally believed that those ceramics material that have been under more severe conditions of temperature variation and humidity are more susceptible to damage.

5. Evaluation of the alterability and durability of incorporated red ceramic pieces with granite waste

5.1 Technological properties of the intact material

The curves related to the technological properties of the standard clay for 0R, 5R and 10R after firing are presented. These curves were obtained from laboratory tests carried out on the so-called intact samples. These curves considered as standard are compared to the results obtained from laboratory and natural degradation process.

Figure 12 presents the curves corresponding of the mean results of five determinations of the technological properties after firing. Results of the diametrical compression test are presented in Table 1.

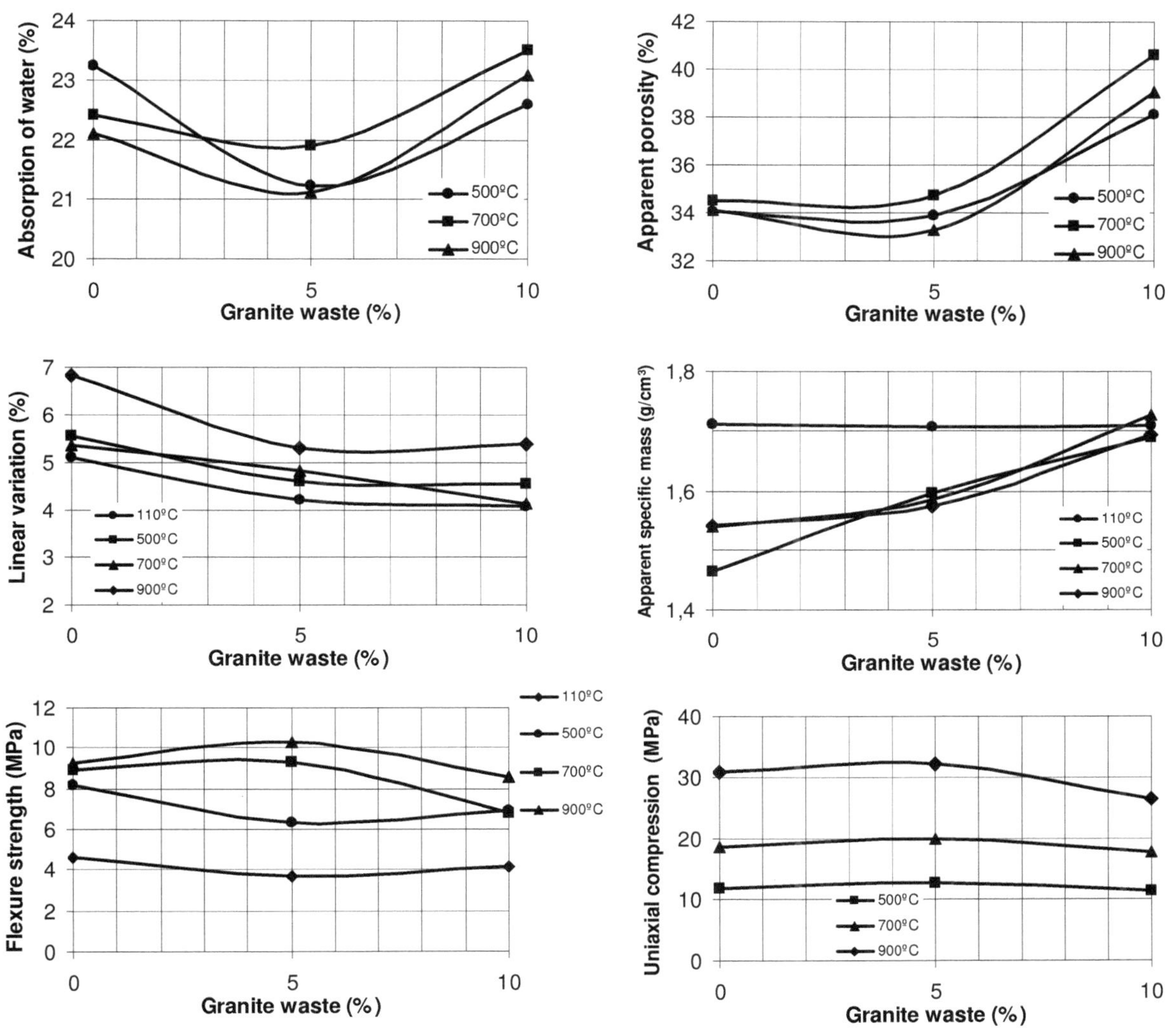

Fig. 12. Technological Properties of the samples 0R, 5R and 10R after firing

Values of the Diametrical compression (MPa) of the intact samples						
	Mix (%)					
Temp. (°C)	0R	σ (SD)	5R	σ (SD)	10R	σ (SD.)
500	1.06	±0.5	1.99	±0.03	0.60	±0.09
700	1.69	±0.01	0.85	±0.10	1.97	±0.62
900	1.74	±0.16	1.21	±0.20	3.02	±0.12

Table 5. Values of the diametrical test on so-called intact samples

The Figure 12 also shows the tendency of the absorption of water and apparent porosities of the samples 0R, 5R and 10R after firing. It is observed that the results indicate reductions in the mean values when added 5R in the sample for all firing temperatures, however, for 700ºC and 900ºC this effect is accentuated indicated by smaller peak.

This behavior is closely linked to the formation of liquid phase during the thermal treatment due to presence of alkaline ($Na_2O+K_2O>7\%$) and earthy alkaline ($CaO+MgO>3\%$), according to the Table 4, provoking, thus, a reduction in the mean values of the absorption of water and apparent porosities, as expected.

In this case, the amount of 5% of granite waste can be used in massive bricks and ceramic blocks because of the obtained absorption lower than 25% (ABC, 2000). This represents the formation of rigid vitreous phase during cooling filling voids of the ceramic pieces.

In general, it is observed a new increase in the curves starting from the addition of 10R, revealed by the higher values of absorption of water and porosities. Extrapolating the mean values, it seems that when the amount of waste is increased, the water absorption is also increased, which means that it may not be attractive for practical use.

However, the curves of the open porosities of the green samples converge for very close values (between 33 and 35%), showing that when rising the temperature, it does not provoke significant densification of the pieces, being close to that range shown for values of open porosities.

It is observed in the Figure 12(c), that the linear retraction of the ceramic mass with 5R at 900ºC, presents tendency of stabilization, being the largest values found for the samples with 0R. As expected, the higher the firing temperature, the higher the sample retraction is.

The Figure 12(d) indicates a tendency of increase of the apparent specific mass with the increase of the waste percentage, independently of the firing temperature. That increasing tendency can be justified for the granolas presence in the waste.

The convergence of the mean points for the samples is verified with 5R of addition, characterizing the densification of the ceramic pieces (Figure 12d). It is observed in the Figure 12(b) the same convergence tendency in the curves for open porosities. The tendency of these curves is to show opposite behavior, because when the open porosities are reduced the volumetric density is increased.

In Figure 12(e) and (f) it is noticed higher values for the curves of simple compression and flexural strength for sample 5R, regardless the firing temperature. The differences obtained from both tests are solely in the failure stress level. Higher mechanical strengths are achieved for sample 5R fired at 900ºC.

When the thermal curves (Figure 6) are observed, it can be observed transformations in the solid state of the particles increasing, thus, stiffness and strength of ceramic bodies. Consequently, the increase of the mechanical strength with the increase of the firing temperature is independent of the addition waste amount (Figure 12e).

It is noticed that when passing from 0R to 5R of addition of granite waste for samples fired above 700ºC, the strength increases considerably, which indicates that the addition of 5R for process of ceramic production can be a very acceptable value.

Observing the Table 5, it is noticed that the results of the diametrical compression strength for standard clay, show themselves varied and dispersed when compared to those obtained for incorporated samples which also depend on firing temperature, indicating that this test cannot bet conclusive, not being indicated to evaluate the material behavior. This can be explained by the presence of cracks randomly distributed in all samples provoked by air drying process.

5.2 Technological properties of the natural and accelerated altered material

5.2.1 Physical properties: Water Absorption (WA) and Apparent Porosity (AP)

Figures 13 and 14 present the variations of the absorption of water and apparent porosities with the time for both natural and laboratory degradation of the samples fired at temperatures of 500°C, 700°C and 900°C and with 0, 5 and 10% of waste content in the ceramic mass. It should be stood out that the absorption of water and apparent porosities are inter-correlated properties, presenting the same pattern of variation of the degradation.

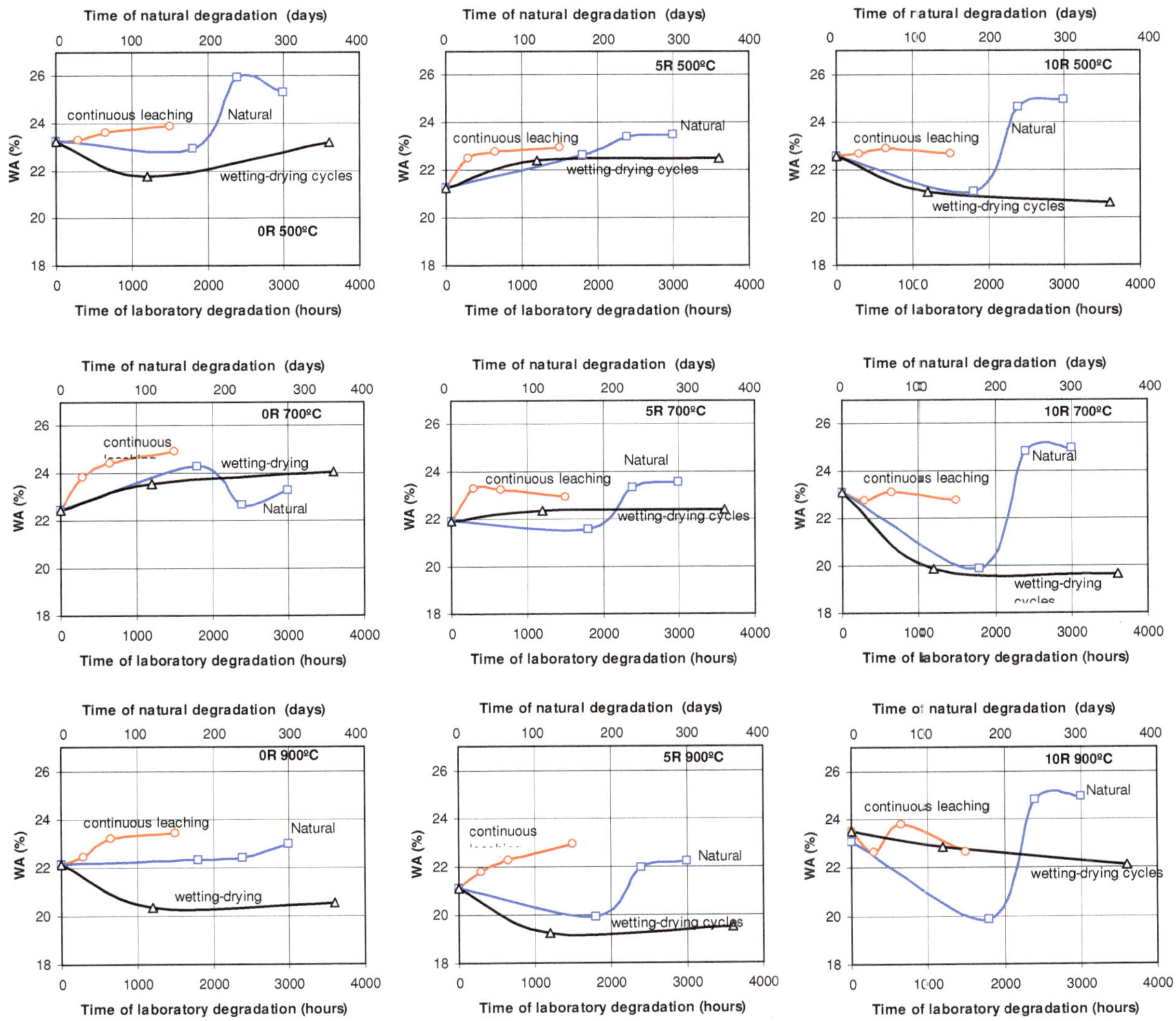

Fig. 13. Effect of the degradation type in the absorption of water for the samples with addition of granite waste for firing temperatures of 500°C, 700°C and 900°C

The variation in the magnitude of the absorption of water with the degradation time observed in the Figure 13 is, in general, not much affected by the firing temperature. It is noticed in the last alteration stages a variation of the absorption of water about 2% in relation to that of intact material. Variations of the order of 3% were noticed in the naturally degraded material with 10R.

It is observed in Figure 14 that the variation of the range of the apparent porosities with the time of degradation was not very much affected by the firing temperature. It is also observed in the last alteration stages, a variation of the open porosities about 4% in relation to that of intact material. Variations of the order of 5% were noticed in the naturally degraded material with 10R.

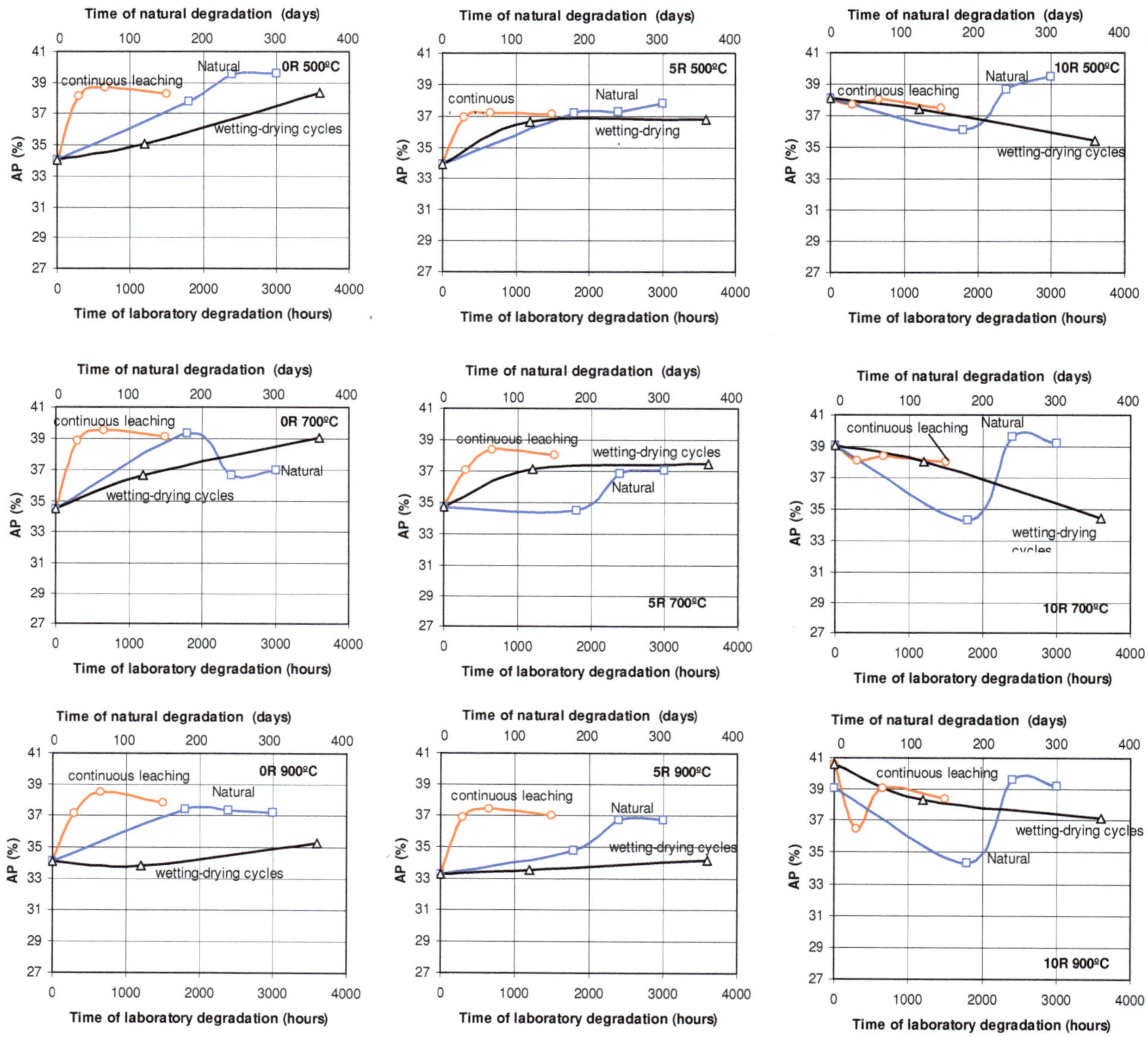

Fig. 14. Effect of the degradation type in the apparent porosities for samples with granite waste for firing temperatures of 500ºC, 700ºC and 900ºC

5.2.1.1 Continuous leaching

For the leaching conditions also shown in Figures 13 and 14, the largest effect of the degradation was noticed in the samples with 0R at 700ºC, which can evidence the loss of cohesion among particles due to the increase of the water absorption.

In the samples fired at 900ºC under continuous leaching, the variation of the absorption of water is small due to the great influence of vitreous phase and larger changes in the present crystalline phases of the samples maintaining it, therefore, more resistant against leaching effects. It should be stood out that the elevation of the firing temperature is also responsible for the densification of the ceramic material due to the fondant properties of some existing oxides.

The vitreous phase is characterized by the presence of alkaline (Na_2O + K_2O) and earthy alkaline (Dog + MgO), that are fondant oxides, mainly when granite waste is added, as observed in the Table 4. When the temperature increases, a eutectic formation takes place. During cooling a plug partially or totally seals the pores, making difficult the effect of the continuous percolation of distilled water through voids of the ceramic sample.

The results of the apparent porosities are similar to the absorption of water for the inter-correlation among these properties. In general, the increments of open porosities observed in all cases provoked by the field and laboratory degradation facilitates the displacement of the particles welded by firing before submitting them to the continuous leaching. It has been observed reductions on the values obtained in the samples with 10R.

Open porosity decreasing in these samples reflects on the reduction of the absorption of water of the ceramic piece, when compared to the intact samples. This is accredited to the presence of the CaO and MgO (Table 4) in the granite waste, corroborating, thus, for plugging openings.

The largest increments of the porosities opened after the continuous leaching, were observed in the samples with 0R in all range of the thermal treatment. That indicates the susceptibilities of the samples to the degradation conditions imposed by the laboratory tests and calls for the attention for the suitability of waste use, aiming to assess more durability of red ceramic goods.

5.2.1.2 Cycles of drying and wetting

It is verified in the Figures of 13 and 14 that the values of the water absorption are similar to those of open porosity for 0R, 5R and 10R, as expected. It can be noticed in the curves of 500ºC after 50 cycles a quite similar behavior of that observed for 150 cycles with reductions of the water absorption. The samples 5R at 500ºC depict significant increase in water absorption for 50 (and after) cycles.

On the other hand, the curves of the samples fired at 900ºC have shown reduction in water absorption after 50 and 150 cycles when compared to those values of the intact samples. This can be credited to the severity of the tests in laboratory, mainly if the stove is used to dry samples causing increase in volumetric density, as shown later.

The largest effect of the degradation given by larger water absorption was noticed for the samples with 0R at 700ºC in all conditions of wetting and drying cycles due to the loss of mass induced by these cycles which favors the cohesion reduction among particles.

The samples fired at 900ºC show the reduction of the water absorption after cycling, due to the great influence of the vitreous phase and also due to the crystalline phases presented in the sample at this temperature, increasing, thus, the volumetric density as a result of pore plugging.

The tendency of increasing the open porosity is verified for 50 wetting-drying cycles for sample 5R. On the other hand, small reductions are observed for the samples 10R under same conditions. This applies for all firing temperature.

After 150 wetting-drying cycles, the sample 10R has its porosity reduced for all tested firing temperature.

The open porosity values after degradation tests for each waste addition (0R, 5R and 10R), suffered alteration in the wetting-drying cycles. The samples 0R and 10R presented higher increase in porosity, and the samples 5R did not show any change in their initial porosity.

5.2.1.3 Natural alteration (Field)

The Figures 13 and 14 show that after 8 and 10 months of field alteration the values of the absorption of water and apparent porosities are quite similar. The values for the samples molded with standard clay at 500°C after 6 months of natural degradation, are kept unchanged. For samples 5R Increasing in water absorption can be observed. On the other hand, for samples 10R this values decreases.

The largest variation took place for samples 0R after 8 months of natural degradation. However, the samples 10R tend to converge to the same point of water absorption, slightly below 25% after 8 months. For these same samples, after 10 months of natural degradation, the mean values of water absorption rise above 25%.

The smallest values are still observed for samples 5R for all firing temperatures, revealing, thus, their stability against natural degradation. The samples 0R are more inclined to suffer natural degradation as a consequence of high values of water absorption.

It is also noticed the similarity between natural degradation and wetting-drying cycles test regarding water absorption values obtained for the two groups. These results indicate that 6 months of natural degradation can be replicated by around 50 to 150 wetting and drying cycles, revealing that such laboratory test may, somehow, represent field effects acting on degradation process.

The largest variation of open porosity is shown for samples 0R after 8 months of natural degradation. Samples 10R, after 8 months of natural degradation, tend to converge to the same values of open porosity at values slightly below 40%, staying stable after further 10 months of natural degradation.

The smallest value of apparent porosities is observed in the samples with 10R to 900°C, revealing the stability to the natural degradation of the samples with addition of waste.

5.2.2 Mechanical properties

5.2.2.1 Index of Flexural Strength (FS)

Variation of the index of the Flexural strength (I_{FS}) with time for natural and laboratory degradation is shown in Figure 15.

This index expresses a correlation between a given material property for both intact and degraded states. The index I_{FS} which uses the flexural strength (FS) to define de degradation index is defined as follows:

$$I_{FS} = \left(\frac{ES_D - ES_I}{ES_D} \right) \times 100$$

Where, subscripts "D" and "I" mean "degraded" and intact, respectively.

It can be observed that the tendency of the curve of the index Flexural strength ($I_{\text{Flexure Strength}}$ (%)) with firing temperature follows the same pattern regardless the kind of degradation. These curves present similar pattern of those obtained for water absorption and open porosity, as expected. This indicates that the firing temperature has little or no influence on I_{FS}.

The variation of the I_{FS} with degradation time is significantly affected by the degradation process regardless the firing temperature and waste content, as well.

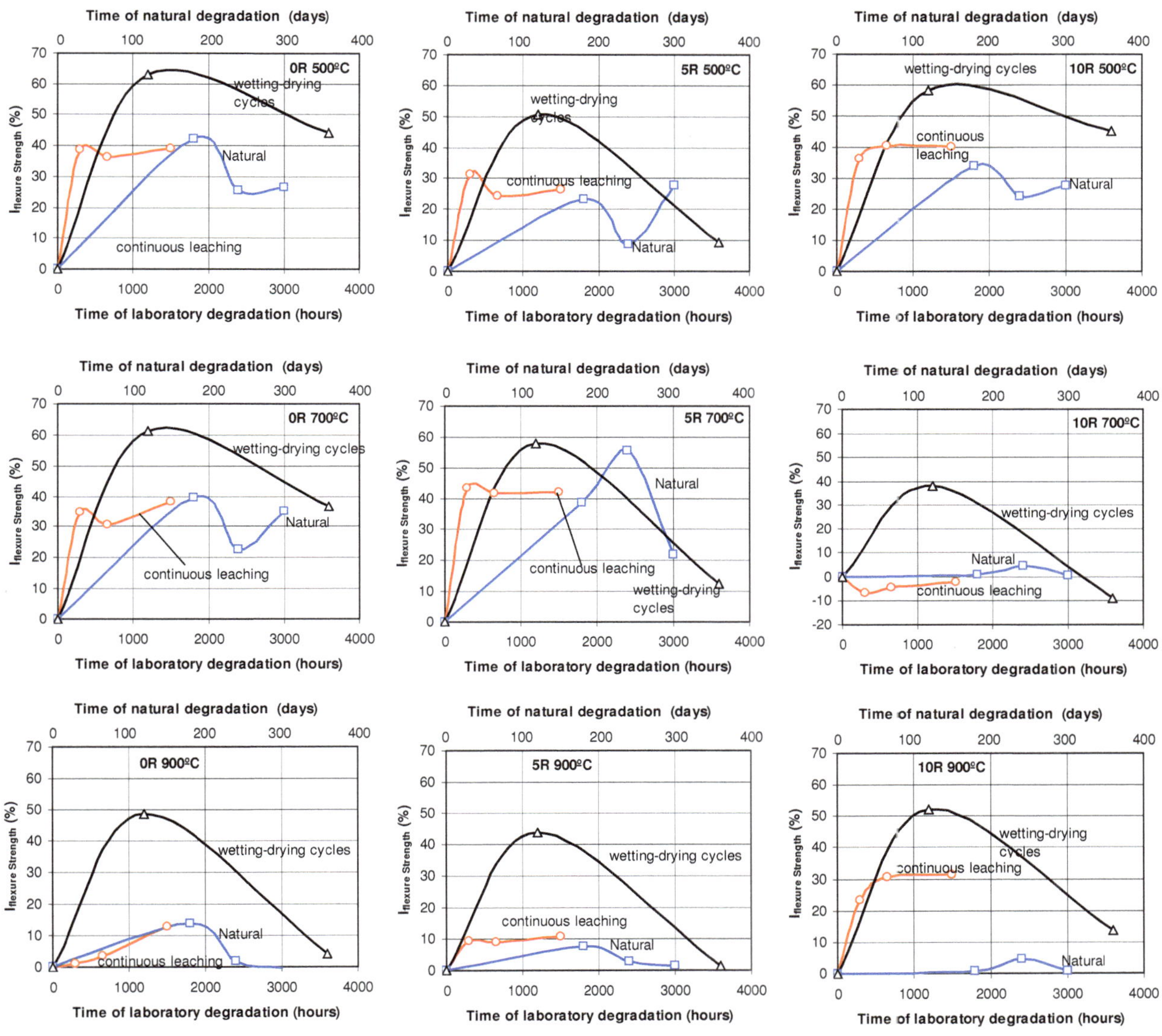

Fig. 15. Effects of the degradation process in the index of FS for samples with addition of granite waste fired at temperatures of 500°C, 700°C and 900°C

5.2.2.2 Continuous leaching

The effect of the reduction of mechanical strength of the ceramic pieces after the continuous leaching is shown in Figure 15. It can be observed that this reduction on mechanical strength is accentuated at 500°C and 700°C, which can be explained by the low firing temperatures that are suitable for promoting coalescences among particles.

Flexural strength is reduced with leaching time for those standard clay samples (without waste content) when fired at 900ºC, (300 hours). After 1500 hours of leaching, samples with waste content (5R and 10R) did not show any loss of strength which indicates that the granite waste content is a very suitable mechanical stabilizer.

In the leaching test, the samples are submitted to the cycles of hot (≈70°C) and cold water (≈35°C). These conditions intend to simulate the temperature variations and humidity of the environment and can be efficient when it is compared to those results from natural degradation.

5.2.2.3 Wetting - drying cycles

The loss of mechanical strength is verified for the samples after 50 cycles. For 150 cycles there is a significant gain of strength of I_{FS}. The samples 5R show higher mechanical strength, once they present smaller strength indexes for all firing temperature.

Some degradation mechanisms may occur contributing for the increase of the I_{FS}: Formation of new clay minerals, restructuring, suction, arrangement of glass phase and pores clogging. It is believed that, amongst these mechanisms involved in this process, suction and pores clogging are most presented.

5.2.2.4 Field or natural degradation

The Figure 15 depicts the values of I_{FS} for the samples 0R, 5R and 10R for firing temperatures of 500°C to 900°C. It is observed that the samples, in these conditions, show higher increase in I_{FS} after 6 months of natural degradation. After 8 months of natural degradation, the I_{FS} decreases slightly and are kept stabilized thereafter, except sample 5R which shows increase in I_{FS} after 10 months of leaching.

No significant degradation is verified for samples fired at 900°C for all waste content. This suggests high cohesion amongst the flat particles guaranteeing, thus, larger mechanical strength under the natural degradation. For samples 5R higher strength are obtained, showing, thus, that the firing temperature of 900°C or higher is suitable when waste is used.

6. Scanning electronic microscopy of the intact and degraded samples

For the evaluation of the surface morphology and texture of the intact and degraded samples, the Scanning Electronic Microscopy (SEM) was used. Images of the ceramic samples 0R, 5R and 10R, fired at 700°C in intact and degrades samples (wetting-drying cycling) were obtained and presented in Figure 16.

The Figure 16 (a, b and c) shows the evolution of the densification of the ceramic mass with and without addition of waste. It can be noticed that the absence of granite waste makes easy micro-cracks to occur and also to avoid small diameter pores clogging, characterizing the coarse texture of the sample

The Figure 16 (b) also shows some defects associated to the pores of wide range of magnitude for samples 5R. It is observed that the texture is finer than of 0R and 10R, for the same firing temperature, indicating that 5R may be considered an optimum waste content for achieving higher densities.

The Illustration 16 (c) presents the image of the intact sample 10R, evidencing a coarse texture with pores of large diameters. It is noticed an intrusion of waste particles into the larger pores which can be attributed to waste that did not reacted to firing.

Figure 17 (a, b and c) shows SEM images from samples 0R, 5R and 10R fired at 700°C after 150 wetting - drying cycles.

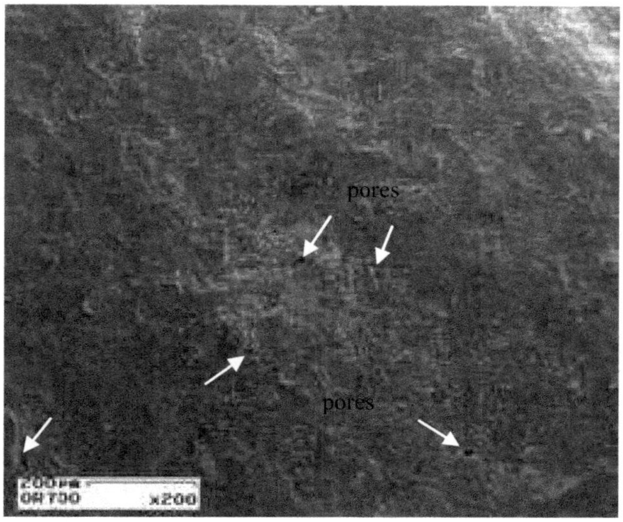

(a) - 700°C 0R

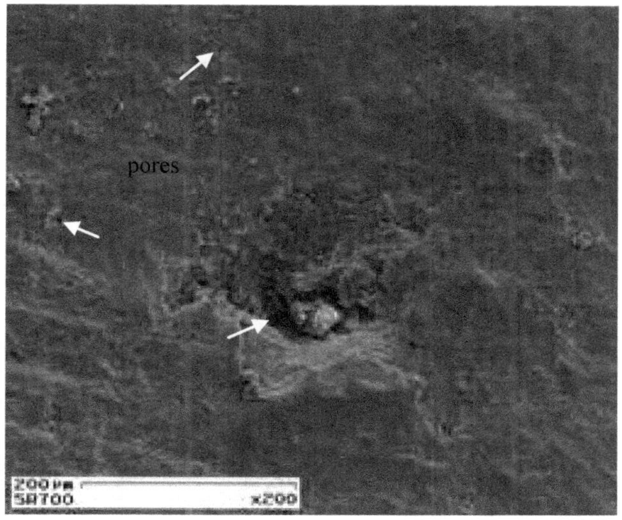

(b) - 700°C 5R

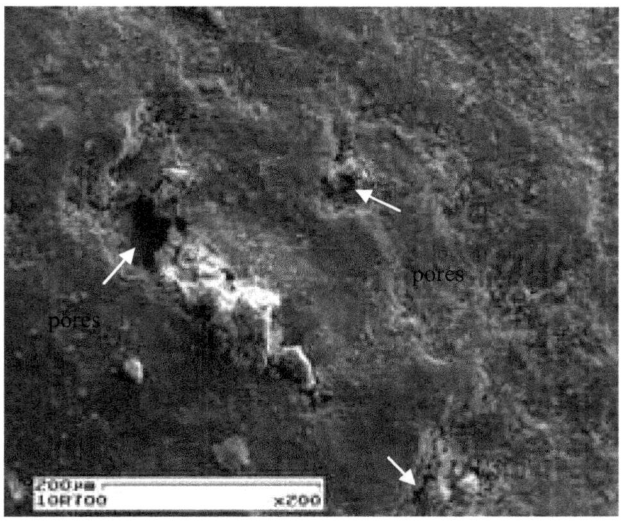

(c) - 700°C 10R

Fig. 16. SEM micrograph of the intact samples (0R, 5R, 10R) fired at 700°C

It is observed in the Illustration 17 (a) an agglomeration of free flat particles on the surface along with a micro-crack and a considerable concentration of shallow pores which are believed to be created by the wetting-drying cycles. This pore size distribution and their characteristics are strongly linked to high values of water absorption shown by these samples.

The effect of the wetting-drying cycles was more intense than that of samples 5R and 10R, causing higher strength decrease (Figure 15).

In the Figure 17 (b) sample 5R presents a smother surface texture and very few loose particles. This sample has, as expected, less porosity and less surface degradation when compared to sample 0R. This can be a clear indicative of pores clogging caused by thermal movement hysteresis and humidity variation imposed during the test.

It is also verified in the Illustration 17 (c), free particles, pores and defects on the whole surface area, aggravated by the degradation after 150 cycles.

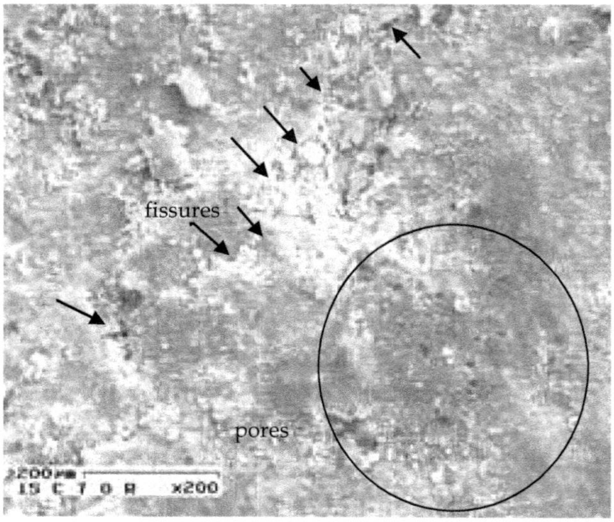

(a) - 700°C 0R

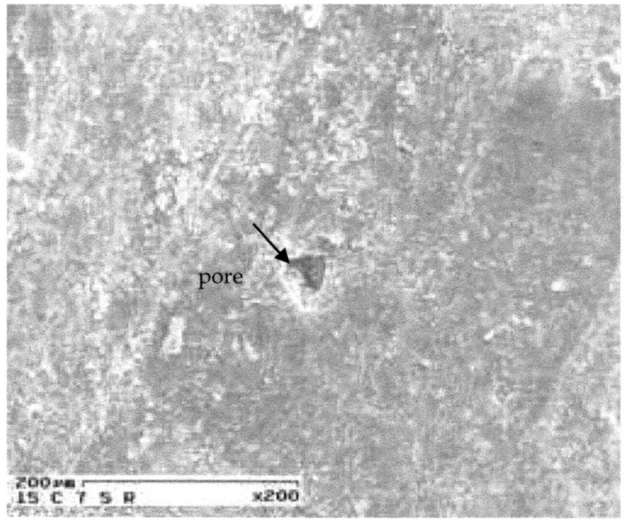

(b) - 700°C 5R

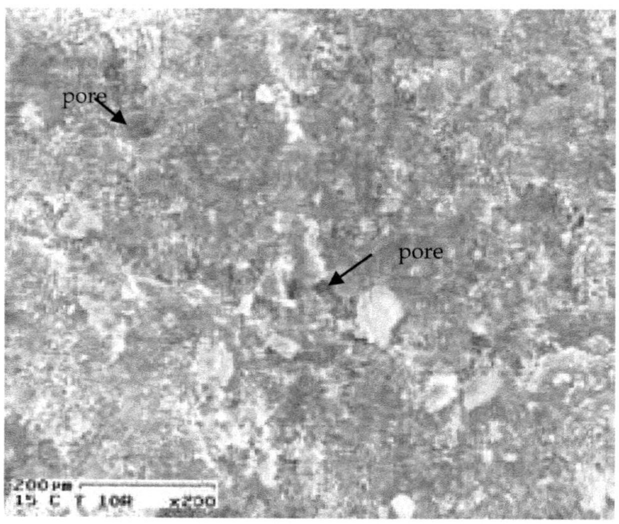

(c) - 700°C 10R

Fig. 17. (a, b e c) SEM micrograph of the samples 0R, 5R and 10R fired at 700°C degraded after 150 wetting - drying cycles

7. Remarks

In this chapter, it was studied the behavior of the ceramic material with additions of granite waste (0R, 5R and 10R) fired at 500°C, 700°C and 900°C. The samples were submitted to artificial and natural degradation. Through these tests it was possible to obtain some response that can help the understanding the physical mechanisms associated to the mechanical behavior of the ceramic-based material used in civil construction industry.

The laboratory tests, in special that of wetting and drying cycles tests showed themselves suitable for testing samples fired at 500°C and 700°C. However, for 900°C the results did not matched well with those obtained from natural degradation test, indicating that, for this temperature, the samples seemed to be somewhat stronger than those submitted to natural degradation process.

When technological properties of intact and degraded ceramic are evaluated in laboratory, it is possible to assess the mechanisms involved in this process based on changes in porosity, volumetric density, shrinkage and dilation amongst other parameters.
The images obtained in SEM allowed the analysis of the so called intact and degraded ceramic material surface, reflecting the effects suffered by the ceramic pieces, in accordance with the evolution of the degradation process imposed in laboratory and natural environment.
It was also observed that the natural and laboratory degradation causes heterogeneity on the mechanical behavior of the materials without following a pre-defined pattern.
It is verified that the granite waste addition influences in the degradation of the red ceramic materials, for propitiating them better durability characteristics, mainly for high firing temperature that, based on the results obtained herein, is the most important variable governing the ceramic materials behavior against degradation

8. References

ABNT (Associação Brasileira de Normas Técnicas) (1987). Resíduos Sólidos - Classificação. NBR 10004.

ABNT (Associação Brasileira de Normas Técnicas) (2004). Procedimentos para Obtenção de Extratos Solubilizados de Resíduos Sólidos. MB 2616 (1987_Lixiviação de Resíduos - Procedimento. NBR 10005).

ABNT (Associação Brasileira de Normas Técnicas) (2004). Procedimentos para Obtenção de Extratos Solubilizados de Resíduos Sólidos. NB 1067 (Solubilização de resíduos - Procedimento: antiga NBR 10006_1987).

ABNT (Associação Brasileira de Normas Técnicas) (2004). Amostragem de Resíduos Sólidos. NB 1068 (1987 NBR 10007).

ALEXANDRE, J. (1997). Caracterização das Argilas do Município de Campos dos Goytacazes para Utilização em Cerâmica Vermelha. Tese (Mestrado em Ciências de Engenharia com ênfase em Geotecnia) - Campos dos Goytacazes - RJ. Universidade Estadual do Norte Fluminense - UENF. 164p.

ALEXANDRE, J. (2000). Análise de Matéria-Prima e Composições de Massa Utilizada em Cerâmica Vermelha. Tese de Doutorado em Ciências de Engenharia - Campos dos Goytacazes - RJ. Universidade Estadual do Norte Fluminense - UENF. 174p.

ALGRANTI, E.; HANDAR, Z.; RIBEIRO, F.S.N.; BOM, A.M.T.; SANTOS, A.M. BEDRIKOW, B. (2004). Exposición a Sílice y Silicosis em Brasil (PNES). *Ciência do Trabalho.* Capítulo 6 (11) págs 1-13.

ASSOCIAÇÃO BRASILEIRA DE CERÂMICA (2005). Panorama Nacional da Indústria Cerâmica. Disponível: www.abceram.org.br/asp/abc_21.asp [capturado em 05 de novembro de 2005].

ASTM C 373 (American Society Tecnology Materials) (1977a). Standard Test Method for Water Absorption, Bulk Density, Apparent Porosity and Apparent Specific Gravity of Fired Whiteware Products.

ASTM C 674 (American Society Tecnology Materials) (1977b). Standard Test Method for Flexural Properties of Ceramic Whiteware Materials.

ÁVILA JÚNIOR, M. M.; Ferreira, H. S.; Menezes, R. R.; Neves, G. A.; Lira, H. L.; Ferreira, H. C. (2002). Expansão por Umidade (EPU) de Blocos Alternativos Utilizando Rejeitos de Granito. *Anais do 46º Congresso Brasileiro de Cerâmica.* São Paulo - SP. 12p.

BERNETT, F.E. (1976). Effects of Moisture of Installed Quarry Tile. *Ceramic Bulletin*. Vol.55, nº12.

BIZERRIL, C.R.S.F.; ARAÚJO, L.M.N.; TOSIN, P.C. (1998). Contribuição ao Conhecimento do Rio Paraíba do Sul - Coletânea de Estudos. *ANEEL/CPRM*. Rio de Janeiro-RJ. 128p.

BUSTAMANTE, G.M.; BRESSIANI, J.C. (2000). A Indústria Cerâmica Brasileira. *Revista Cerâmica Industrial*. Vol. 5 (3). Maio/junho.

CABRAL, G. L.L. (2005). Metodologia de Produção e Emprego de Agregados de Argila Calcinada para Pavimentação. Dissertação de Mestrado. Rio de Janeiro-RJ. IME - Instituto Militar de Engenharia. 358p.

CASTRO, E.M. (2005a). O que esperar para o setor em 2006? Editorial. *Revista Rochas de Qualidade*. Edição 185: pág. 06.

CASTRO, E.M. (2005b). Um ano bom que poderia ser melhor. Editorial. *Revista Rochas de Qualidade. Edição* 180: pág. 07.

CALLISTER, W. D. J. (2000). Materials Science and Engineering, 5ª Edição, New York-USA, Editora John Wiley & Sons, Inc. 852p.

CEN (Comité Européen de Normalisation). (1995) - Methods of Testing Cement - Part X: Determination of the Resistance of Cements to Attack by Sulphate Solution and by Seawater, PR ENV 196-X.

CHIODI FILHO, C. (2001). Situação brasileira do setor de rochas ornamentais. *Revista Rochas de Qualidade*, Edição 158: págs. 87-106.

CUNHA, V. P. (2006). Estudo da Degradação de Geotêxteis Tecidos. Dissertação de Mestrado em Engenharia Civil - Campos dos Goytacazes - RJ. Universidade Estadual do Norte Fluminense - UENF. 151p.

FALCÃO BAUER, R.; RAGO, F. (2000). Expansão por Umidade em Placas Cerâmicas para Revestimento. *Revista Cerâmica Industrial*. São Paulo-SP. 5(3), maio/junho. 5p.

FARJALLAT, J. E. S.; DE BARROS, F.P. YOSHIDA, R.; DE OLIVEIRA, J. A. N. 1972. Alguns Problemas de Desagregação dos Basaltos da Barragem de Capivara, Rio Paranapanema. *4ª Semana Paulista de Geologia Aplicada*, São Paulo, vol. II A, pp. 73-102.

FRASCA, M.H.B.O. (2005). Durabilidade e Alterabilidade em Rochas Ornamentais e para Revestimento. *Revista Rochas de Qualidade*. Edição 180: pág. 178-188.

GUIMARÃES, J. E. P. (1998). A Cal. São Paulo. Editora PINI. 285p.

ISRM - International Society for Rock Mechanics. (1981). Rock Characterization Testing and Monitoring. - ISRM Suggested Methods, Ed. E. T. Brow, Pergamon Press, Oxford, pp. 51-60.

JOHN, V.M. (2000). Reciclagem de Resíduos na Construção Civil: Contribuição à metodologia de Pesquisa e Desenvolvimento. Tese (Livre Docente) de Doutorado - São Paulo - SP. Universidade de São Paulo. 102p.

JOHN, V.M.; Zordan, S.E. (2001). Waste Management 21. 213-219.

LUZ, A. B. (2005). O setor de rochas ornamentais. Revista Rochas de Qualidade, Edição 180: pág. 129.

MACHADO, M. A. (2002). "O Setor de Rochas Ornamentais". "Rejeitos". Palestra apresentada no CREA/ES sobre rejeitos da Indústria de Rochas Ornamentais. Cachoeiro de Itapemirim-ES.

MAIA, P. C. A. (2001). Avaliação do Comportamento Geomecânico e de Alterabilidade de

Enrocamentos. Tese de Doutorado - Rio de Janeiro-RJ. Pontifícia Universidade Católica do Rio de Janeiro. 351p.

MOTHÉ FILHO, H.F.; POLIVANOV, H.; MOTHÉ, C.G. (2002). O Uso do Resíduo do Mármore e Granito para Indústria da Cerâmica Vermelha. *Anais do 46º Congresso Brasileiro de Cerâmica.* São Paulo - SP. 8p.

MOTHÉ FILHO, H.F. (2003). Caracterização e Aplicação Tecnológica do Rejeito da Indústria do Mármore e Granito. Tese de Doutorado em Geologia - Rio de Janeiro - RJ. Universidade Federal do Rio de Janeiro - UFRJ. 137p.

MOTTA, J.F.M.; ZANARDO, A.; CABRAL, M.J. (2001). As Matérias-primas Cerâmicas. Parte I: O Perfil das principais Indústrias Cerâmicas e seus Produtos. *Revista Cerâmica Industrial.* Vol. 6 (2). Março/Abril.

NEVES, G.A. (2002). Reciclagem de Resíduos da Serragem de Granitos para Uso como Matéria-Prima Cerâmica. Tese de Doutorado em Engenharia de Processos - Campina Grande - PB. Universidade Federal de Campina Grande - UFCG. 242p.

NITES - Núcleo Regional de Informação Tecnológica do Espírito Santo. (1994). Desperdício na Indústria de Mármore e Granito. *Rochas de Qualidade.* São Paulo. Ed. 118, p. 27-32.

PEITER, C.C.; CARANASSOIS, A.; VIDAL, F.W.H.; ALBUQUERQUE, G.S.C.; FERREIRA, G.E.; FILHO, C.C.; CALAES, G.; VALE, E. (2001). Rochas ornamentais no século XXI. CETEM, ABIROCHAS. (Relatório Técnico sobre o Setor de Rochas Ornamentais do Brasil, TR - 30/07). 113p.

PESAGRO/RIO (2005). Boletim de Dados Climatológicos da Região de Campos dos Goytacazes-RJ. *Estação Automática Thies-Clima.*

ROC MAQUINA. (1989). Referências arqueológicas sobre la explotación de piedras naturales em época romana de España. Cuarto trimenstre.

SABOYA, F.A.J., XAVIER, G. C., ALEXANDRE, J. (2006).The Use of the Powder Marble By-Product to Enhance the Properties of Brick Ceramic. In: *Construction and Building Materials,* Vol. 20, pág. 1-8.

SILVA, S.A.C. (1998). Caracterização de Resíduo da Serragem de Blocos de Granito. Estudo Potencial de Aplicação na fabricação de Argamassas de Assentamento e de Tijolos de Solo-Cimento. Tese (Mestrado em Engenharia Ambiental) - Vitória - ES. Universidade Federal do Espírito Santo - UFES. 159p.

SMITH, W.F. (1998). *Princípios de Ciência e Engenharia dos Materiais.* 3ª Edição, Lisboa-Pt, Ed. Mc GRAW-HILL, 892p.

SOUZA SANTOS, P. (1989). Ciência e Tecnologia das Argilas. 2ª Ed. São Paulo: Editora Edgard Blucher Ltda. Vol. 1, 499p.

TOLEDO, R. (2003). Técnicas Fototérmicas e de Raios-X para o Monitoramento de Sólidos e Gases em Materiais Cerâmicos. Tese (Doutorado em Engenharia e Ciência dos Materiais) - Campos dos Goytacazes - RJ. Universidade Estadual do Norte Fluminense - UENF. 315p.

VIEIRA, C.M.F. (2001). Caracterização de Argilas de Campos dos Goytacazes Visando à Fabricação de Revestimento Cerâmico Semiporoso. Tese de Doutorado em Engenharia de Materiais - Campos dos Goytacazes - RJ. Universidade Estadual do Norte Fluminense - UENF. 118p.

VIEIRA, C.M.F.; SOARES, T.M.; SÁMCHEZ, R.; MONTEIRO, S.N. (2003). Incorporation of Granite Waste in Red Ceramics. Materials Science & Engineering. Campos dos Goytacazes - Brazil. 7p.

WARREN, J. (1999). Conservation of Brick. New York, Oxford University.

WIKPEDIA, (2005). História da Cerâmica. Disponível: http://www.en.wikpedia.org/wiki/brick. [capturado em 20 de março de 2005].

WINGER, R.J.J. (1965). In place permeability tests use for subsurface Drainage investigation. Division of Drainage and Groundwater Egineering. Denve, Colorado-USA. 9p.

XAVIER, G.C. (2001). Utilização de Resíduos de Mármore e Granito na Massa de Conformação de Cerâmica Vermelha. Dissertação de Mestrado em Ciências de Engenharia - Campos dos Goytacazes - RJ. Universidade Estadual do Norte Fluminense - UENF. 285 p.

XAVIER, G.C. (2006). Resistência, Alterabilidade e Durabilidade de Peças Cerâmicas Vermelhas Incorporadas com Resíduo de Granito. Tese de Doutorado em Engenharia Civil - Campos dos Goytacazes - RJ. Universidade Estadual do Norte Fluminense - UENF. 201 p.

7

Physical and Metallurgical Characteristics of Fiber Reinforced Ceramic Matrix Composites

Zdeněk Jonšta, Evelyn A. Bolaňos C., Monika Hrabalová and Petr Jonšta
VŠB-Technical University of Ostrava
Czech Republic

1. Introduction

Ceramics represent a very wide range of materials, which differ from each other by their chemical composition, structure and properties. These materials can be defined as predominantly crystalline materials, formed mainly by inorganic compounds of non-metallic character, which are bound particularly by ionic and ion-covalent bonds.

Ceramics are most frequently divided by the purpose of its use into **traditional** and **structural ceramics**. Traditional ceramics are produced entirely from natural raw materials (such as kaolin, clays, quartz, feldspar), which are formed and fired.

These ceramic materials can also be designated as vitreous or porous due to the fact that they contain a considerable portion of vitreous phase, pores and crystals of various compositions. Traditional ceramics can be further divided into constructional, artistic, utility and refractory.

Structural ceramics are produced only from synthetic raw materials. It is a newer type of ceramic, which has a considerable development from the seventies of the 20th century. It is used in numerous technical applications, especially under extreme conditions where, thanks to its properties, completely replaces the traditional ceramics. These ceramic materials include also functional and bio-ceramic materials.

Structural ceramics represent a fine-grained, poly-crystalline material, usually with grains smaller than 1 μm. It contains random technological defects and micro-structural heterogeneities. The structure of ceramic materials can be characterized as **heterogeneous, poly-crystalline and poly-phased.**

2. Classification of structural ceramics

2.1 Oxidic ceramics

The ceramics based on aluminium oxide (Al_2O_3) called also sintered alumina (Al_2O_3 with addition of ZrO_2) has the biggest significance for structural applications. Thanks to the zirconium oxide this ceramics has better mechanical properties. Increasing the Al_2O_3 content increases its mechanical strength and hardness, as well as thermal expansion coefficient.

Oxidic ceramics have high strength, hardness, corrosion resistance, and they are very good electric insulators. They are used mainly at the machining of cast iron, as a structural material for components of high-temperature devices, melting crucibles, cutting plates for metals machining, spark plug insulators for engines, etc.

2.2 Non-oxidic ceramics

This group includes mainly nitrides, carbides and borides. Non-oxidic ceramic materials have higher strength at high temperatures in comparison with oxidic ceramics. They have higher heat conductivity and lower thermal expansion coefficient than oxides, which means they have a better resistance to thermal shocks. For this reason they are in practice used mainly as high-temperature structural materials. Their disadvantage consists in lower resistance to oxidation. In non-oxidic compounds covalent bond prevails (for example SiC, B_4C, Si_3N_4, BN) or metallic bond (for example TiC, UC, WC, TaC, TiN). Carbides and nitrides have the broadest use of non-oxidic ceramics, mainly in special fields of mechanical and electrical engineering.

2.3 Mixed ceramics

It is formed by a combination of structural oxidic and non-oxidic ceramics. Composite cutting tools based on Al_2O_3+ZrO_2+SiC are used for the lathe work of quenched steels and hardened cast irons, at the end of the machining process, such as milling. Except SiC, the most frequently used non-oxide additives are TiC or TiN.

3. Ceramic properties

The physical, mechanical and chemical properties of ceramic materials depend on their chemical composition and micro-structure.

3.1 The positive properties of ceramic materials are the following:

- low density
- low thermal expansion coefficient
- high hardness
- high temperature and corrosion resistance
- high wear resistance
- high melting temperature
- stability of mechanical properties in a wide temperature interval

3.2 Main disadvantages of ceramic materials are the following:

- brittleness
- difficult reproducibility of complex shapes
- highly demanding processing
- difficult joining of different ceramic components or between ceramic components and other materials

3.3 Causes of brittleness

Brittleness is the main problem of structural ceramics. It expresses itself by:

- the creation of brittle failure
- static fatigue
- the degradation of properties at mechanical and thermal shocks

It results from the constitution of ceramic materials, that it is suppressed in them the development of plastic deformation under conventional conditions.

The causes of brittleness are divided into:

- Fundamental
- Secondary

FUNDAMENTAL CAUSES result from the type of inter-atomic bond (physical cause), but also from the type of crystalline lattice (geometric cause).

Fundamental causes of brittleness are essentially caused by **low mobility of dislocations** and by **bad manoeuvrability**.

Low mobility of dislocations is related to the high level of the Peierls–Nabarro stress in materials with an ionic and covalent bond. Bad manoeuvrability is connected with non-fulfilment of the requirement of five independent slip systems in single crystals - so called von Mises criteria.

The fundamental causes of ceramic brittleness as a result of ionic and covalent bonds comprise also a less perfect connection of the lattices along the grain boundaries, than with metallic bonds which manifest itself by a lower cohesion strength on the grain boundaries.

Brittleness transition temperature T_P exists in ceramic materials. Below this temperature the materials suffer brittle failure - they are unable to deform plastically. The brittleness is caused by the fact that a plastic zone is not formed in front of initiated cracks and the crack propagates by cleavage along the grain boundaries or through the extend of the grains.

Fundamental causes do not lead to absolute suppression of macroscopic plastic deformation below the transition temperature T_P. For example, BeO, AlN, Al_2O_3 in a poly-crystalline state at superimposed multiple pressures are capable even at room temperature of a plastic deformation up to 10% - easier formability of ceramics while loading with a big hydrostatic pressure. The strength of the boundaries is sufficient for the movement of dislocations and the activation of additional slip planes while realizing shear stress.

SECONDARY CAUSES include the heterogeneity of the structure and properties, as well as the significant weakening of the grain boundaries due to an imperfect sintering and to the presence of pores, cracks and vitreous phases along grain boundaries. The characteristic defects of ceramics include cracks and other defects caused by incorrect technological process at forming, drying or sintering.

The level of mutual contact of crystalline grains or of their contact with vitreous or binding phase is of great importance. The pores significantly influence the strength of ceramics by reducing the carrying section, as well as by the fact that they act as stress concentrators.

The mentioned secondary causes of brittleness are usual sources of failure of traditional ceramics already at low stresses. They are partly suppressed at structural ceramics.

These secondary causes are strongly applied at high temperatures, when the softening of the vitreous phase occurs and there is an easy slip along the grain boundaries, which cannot be compensated by the plastic deformation of the grain. It leads to the initiation of cracks and to their brittle inter-crystalline propagation above the transition temperature T_P.

4. Possibilities to enhance the mechanical properties by toughening the ceramics

Minimum capability of plastic deformation has an essential influence on the achieved level of mechanical properties of structural ceramics. The enhancement of the level of the mechanical properties can be therefore achieved by increasing the toughness of the ceramic materials, i.e. by the achievement of a certain level of development of the process zone for the cracks under limited conditions. The increase of the fracture toughness can be achieved by several ways as follows.

4.1 Structure optimization

In this possibility to increase the toughness we talk about the so-called engineering of the grain boundaries. It consists in the reduction of porosity, the removal or reduction of the volumetric part of the vitreous phase on grain boundaries. This reduction requires highly pure initial powders and the perfection of all of the production of ceramics in sintering and pre-sintering processes.

If the ceramic contains a large portion of the vitreous phase, the crack propagates mainly through this phase and other phases have very little influence on this propagation. Ceramics with increased fracture toughness should contain only a very small part of the vitreous phase.

Another condition is a homogenous fine-grained material, a grain size from tenths of micrometers to units of micrometers. The pores of spherical shape, occurring for example in the vitreous phase, can even have a favourable influence, since they obstruct the front of the crack.

The shape of the grains and their way of strengthening is of great importance. For example Si_3N_4 **with acicular grains** has even **4-times higher** fracture toughness - K_{IC} – than in the case of **uniform equi-axial grains** (Jonšta, 1998).

Increase the strength (toughening), i.e. to increase of resistance to crack propagation, can be achieved with the effect of dispersive particles. At the use of these particles the solid grains of the dispersed phase cause a deviation of the crack from its direct course, and this results in an extension of the fracture area. Except that the particles take over a greater part of the load and they thus reduce the stress on the matrix. This final result leads to a reduction of the stress factor on the front of the crack.

4.2 Phase transformation

It is based on an induced phase transformation of the dispersed particles in the matrix of a ceramic material. It is used mainly in the area in front of the crack, where the effectuation of a **martensitic phase transformation** contributes to energy absorption. A pre-requisite for using this type of strengthening is the requirement that the transforming material must have a greater volume of its low-temperature modification than it has in the high-temperature structural modification.

The considered process must therefore lead to the creation of a phase with bigger volume, where the phase transformation should be made with a martensitic mechanism. It may also be assumed that the strains are not relaxed, since the relaxation process proceeds very slowly. Among the materials that have this behaviour, zirconium dioxide **(ZrO_2)** seems to be the most advantageous technical ceramic.

ZrO_2 crystallizes in three modifications:

- * ***cubic*** - stable up to 2680°C
- * ***tetragonal*** - stable up to 2370°C
- * ***monoclinic*** - stable up to 950°C

A big volumetric change at the transition of tetragonal modification to a monoclinic one, when relaxation by diffusion or relaxation mechanisms is impossible, is the cause of the fact that pure ZrO_2 cannot be used as ceramic. That is because at cooling from the sintering temperature it could crack along the grain boundaries, at the modification change temperature, due to big internal strains.

If we stabilize this modification till room temperature (so called complete stabilization) by the addition of various oxides that are soluble in cubic modification of ZrO_2, such as for example Y_2O_3, in such case the detrimental transition from tetragonal to monoclinic modification cannot occur in whole volume.

The transformation of tetragonal modification to a monoclinic one occurs only at the places of high tensile stresses, i.e. in the front of the crack. Due to the fact that the monoclinic phase has a volume approximately by 4% bigger than the tetragonal phase, the development of pressure strains occurs at these places, which modify the state of stress on the crack and thus slows a further development of cracks.

The following three mechanisms participate in a decisive way on energy dissipation at a material failure, initiated by strengthening at phase transformation:

- stress induced transformation in the course of load
- initiation of fine cracks
- deviation of cracks from their initial direction of growth

The stress induced martensitic phase transformation of tetragonal ZrO_2 to monoclinic is significantly manifested by a *protective effect of the crack*, which is accompanied by the above mentioned *volume changes.* The dissipation of fracture energy occurs by the energy consumed for the nucleation of phase transformation and the energy consumed for the creation of a new volume originated by transformation. As a result of the already mentioned increase of volume, the tensile stresses are also eliminated (Pokluda et al., 1994).

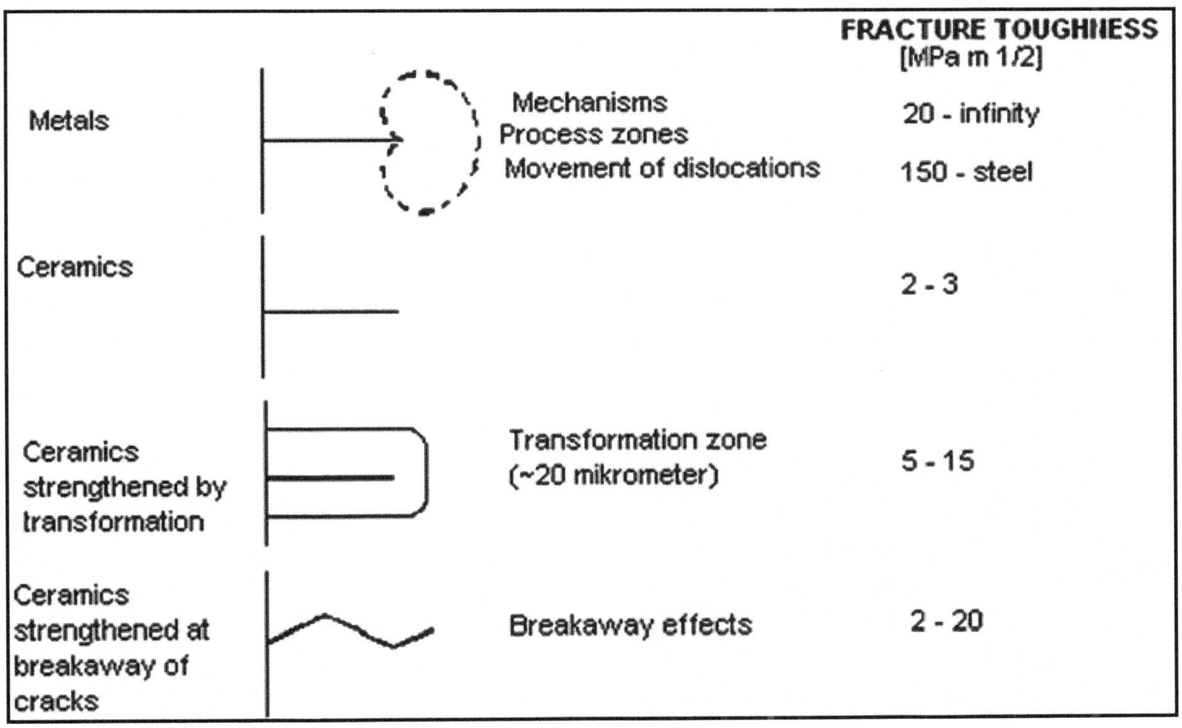

Fig. 1. Schematic representation of the effects related to the dissipation of energy at crack growth in metallic and ceramic matrices

The creation of fine micro-cracks in the regions of dispersed particles, in which the tetragonal phase has transformed into a monoclinic phase, has a favourable effect on the increase of the fracture toughness. These cracks have usually lengths of 10 to 100 nm. They are initiated around the front of the crack in the process zones. Their frequency in the matrix

can be quite high and at the increase of the load there can be appropriately oriented micro-cracks expanded by a certain part of their original length, which is also connected with their ramification.
As a result we have the reduction or elimination of local stress peaks. It is possible to imagine this process as the distribution of stresses induced on the front of the crack onto a large number of small cracks. The material behaves under these conditions in an elastic-plastic manner, because the opening of micro-cracks is at the unstressing, partly reversible.

4.3 Formation of composite materials

The biggest increase of the resistance to crack propagation, i.e. the increase of the level of ceramic´s fracture toughness, promises the use of compound ceramic materials - composites. This means using the strengthening effect of ceramics thin fibres or whiskers in the ceramic matrix. The diameter of the used fibres is usually several μm and their length is of several millimetres, while the whiskers´ dimensions are smaller almost by one order. Optimum properties are achieved, if the cracks do not propagate through the fibres, and a considerable part of energy is consumed on extraction of the fibres and at the failure of the fibre interface – matrix.

5. Evaluation of mechanical properties of structural ceramics

The basic characteristics of structural ceramics are the fracture toughness, tensile and bending strength, compressive strength, elastic constants, the parameters of the statistical distribution of strength and parameters of growth curves of the statistical or dynamic fatigue. The evaluation of these parameters is often difficult, because there are a number of empirical relationships for the characterization of ceramic materials.
Right now, the attention is devoted mainly to the determination of the fracture toughness by the use of the indentation technique. This technique, usually used to measure hardness, finds its application in the evaluation of these important material characteristics. In our case it was used the Vickers indenter. In the use of these techniques there are measured the length of the cracks initiated at the corners of the indentation print when it is exceeded the critical value of the loading force, which involves the initiation of cracks and their subsequent development as shown in Fig. 2.

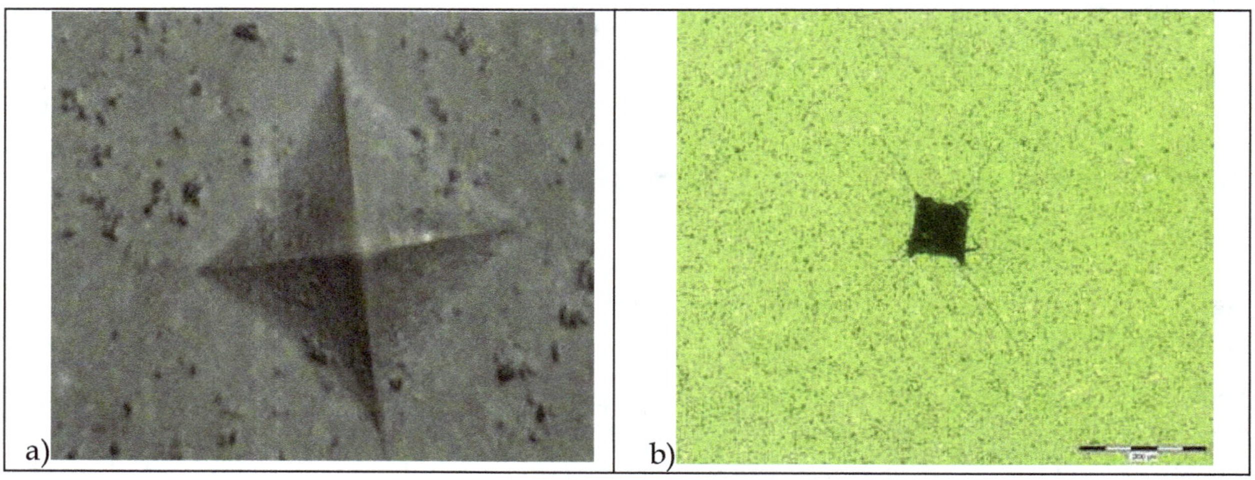

Fig. 2. Vickers´ print with a load of 10 N, for a sample material a) zoom 1000x b) zoom 100x

This approach, however, requires overcoming some limiting characteristics which are related to the type of the initiated cracks. Preferential incidence of one type or another is related to the type of materials and the intensity of load on the print. On Fig. 3 are schematically illustrated the cross-section variations of the initiated cracks by Vickers indentation.
This is known as radial / median crack (see Fig. 3a), where m_t corresponds to the median crack and l_t to the side (lateral) crack. The figure 3b shows the radial or Palmquist cracks r_t and lt shows the side (lateral) cracks. The Palmquist cracks type can be observed at lower levels of indenter loads and on the contrary the radial / median cracks are initiated at higher levels.

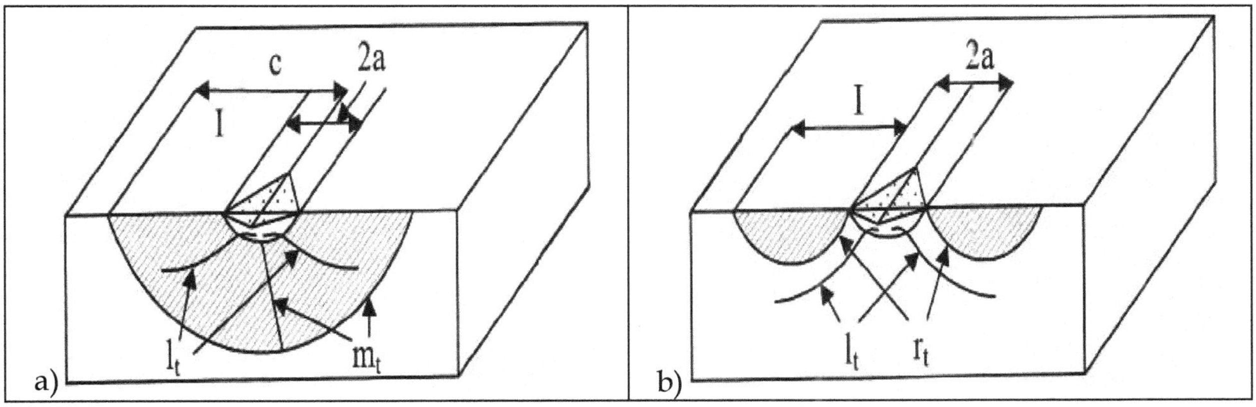

Fig. 3. a) median crack, b) lateral crack

This evaluation technique requires the preparation of a relatively small area of the material sample for the indentation prints. Of course, there a number of restrictive conditions in its application related both to the level of surface preparation, and to the conditions for the initiation and propagation of the cracks. We could also consider a risk on the evaluation associated with the use of small areas and the fact that the measurement is performed on the material surface (assuming that the homogeneity and the volume of the tested material is maintained).
Matzke pointed out these limiting factors, on behalf of his special attention to the study of this kind of situation (Matzke, 1980, 1990, 1991). It should also be taken into account the possible presence of discontinuities by major or minor deviations in the width of cracks. It cannot be excluded in some cases the possibility to initiate more than one indentation crack in the corners of the print.
According to Matzke this leads to a possibility of underestimation of the crack length, but in macroscopic terms the fracture toughness values would be realistic data (Matzke, 1991). The processes of uneven development of cracks are an integral part of a complete evaluation of the parameter of resistance to brittle fracture.
In our works we have used, for determining the values of fracture toughness, a relationship that is independent of the crack profile and allows you to evaluate different types of materials or a type of material in a wide range of applied loads. This relationship was presented by Liang et al. (in 1990) and is expressed as follows:

$$K_{IC} = \frac{\left[\frac{2l}{u}\right]^{\left(\frac{l}{9u}-1,51\right)} \cdot HV\left(\frac{u}{2}\right)^{\frac{1}{2}}}{\left(\frac{HV}{E\cdot\Phi}\right)^{0,4} \cdot \alpha \cdot \Phi} \quad [\mathrm{MPa \cdot m^{1/2}}] \qquad (1)$$

with:

$$\alpha = 14\left[1 - 8\left(\frac{4\nu - 0{,}5}{1+\nu}\right)^4\right] \quad (2)$$

and the length of the crack is:

$$l = \frac{\sum_{i=1}^{n} l_i}{4} \text{ [m]} \quad (3)$$

Where:
HV - hardness by Vickers,
u - diagonal length,
l - crack length,
α - dimensionless constant dependent on the order of the Poisson´s number of a particular type of ceramic material - 9.3,
Φ - limiting factor - 3,
E - Young´s module of elasticity.
With the indentation technique we can make an evaluation of structural ceramics response to repeated loadings. Here, we also have similar limitations as in the previous case. Informatively it may be noted that in the evaluation of the kinetics the advances of the fatigue cracks in structural ceramics the obtained and modified characteristics are compared with the results obtained at structural steels. Although this technique does not allow to establish an analogical relation with Paris' kinetic equation, it allows to determine the response of a certain type of ceramic to repeated loading based on the dependence between the number of burdened cycles and the length of the induced cracks, or in that case based on a defined number of burdened cycles leading to the chip of the evaluated material around the indentation print.

6. Indentation model of the origin of cracks

At the formation of a Vickers´ indentation print an elastic-plastic stress fields in the material are originated, which affects the subsequent expansion of cracks. The period of the cracks derivation is an important factor that determines the fracture threshold value and could be the control parameter for determining the dominant influence of the surface radial cracks or of the subsurface median cracks in the first stage of propagation.
According to the model of Marshall and Lawn, the fully loaded indenter with an elastic / plastic field is considered as a superposition of a residual field of an unloaded material and as an ideal plastic contact field. The essence of the mentioned model is the assumption that during the unloading operation of the indenter does not appears the action of a reverse plasticity. (Marshall & Lawn, 1979)
During the indenter loading, in the material the origin of elastic deformation zones occur with the subsequent development of the radial / median cracks. The growing tensions cause the subsequent stable growth of median cracks. During the subsequent unloading the closure of median cracks occurs, while the lateral cracks grow until the indenter load is completely removed, and there is when the subsequent chipping can occur (Cook & Pharr,

1990). The initiation and propagation of the radial/median and lateral cracks leads, in brittle materials, to a subsequent chipping formation. On Fig.4 a schematic representation of the cracks system under Vickers´ indentation print is shown.

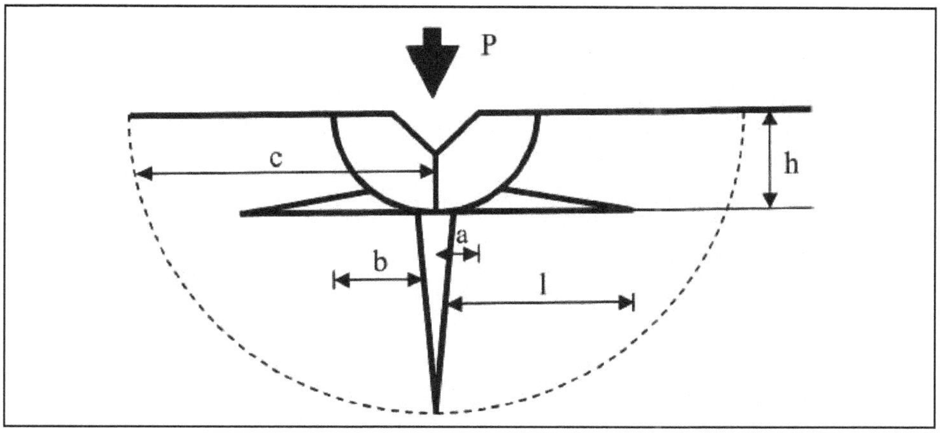

Fig. 4. Schematic representation of the crack system under Vickers´ print

On the presented picture - P- is the index load at which the origin of median cracks with a length of – c - occurs. The plastic zone is formed during the application of the Vickers´ load of the indenter with an appropriate angle 2 ψ, where - a – is half the diagonal of the Vickers´ print and – b – is the plastic deformation zone accommodated under the print in the plastic zones. The critical stress for the formation of radial/median crack is given by (Evans & Marshall, 1981; Kumar & Khamba, 2006):

$$P = \alpha \cdot \frac{K_{IC}^4}{HV^3} \quad [\mathrm{N}] \tag{4}$$

where:
α - dimensionless constant
K_{IC} - fracture toughness
HV - hardness by Vickers
At the Vickers indenter loading, just below the indentation print in the material, the formation of irreversible strains occurs. From the schematic view of the cracks resulting in a material when exposed to Vickers indentation, we have that δV is the printed volume, V is the volume of the deformed zone and ψ is half the size of the Vickers´ indentation print. Then the next equation applies:

$$\frac{\delta V}{V} = \left(\frac{a}{b}\right)^3 \cdot \cot g\psi \tag{5}$$

This zone is to maintain its constant volume elastically compressed with the surrounding material. The result is the formation of a residual force F_r on the surface of the indentation print, which then acts on the axis of the indenter in the opposite load direction. At the same time the formation of the residual forces F_m occurs, which are located in close proximity to the plastic zone. These residual forces cause on their surroundings the creation of tensile stresses and with the absence of a reverse plasticity they achieve, at full loading of the indenter, the maximum levels of the force F (F_{r0} and F_{m0}).

This level is maintained even after the subsequent unload of the indenter, if during the loading there are no material cracks. At a sufficient level of strength of F_{r0} occurs after unloading the indenter the formation of lateral cracks at the depth h, while the power F_{m0} is associated with the formation of radial/median cracks. The lateral crack grows up to the equilibrium length l and the residual strength, leading to the formation of these cracks, is reduced from the maximum level F_{r0} to the level equilibrium value F_r (Evans & Marshall, 1981).
Marshall et al. (in 1982) derived an equation from an elastic-plastic analysis that expresses the relationship between the so-called equilibrium length of the lateral cracks l and the index load:

$$l = l_0 \cdot \left[1 - \left(\frac{F_0}{F} \right)^{\frac{1}{4}} \right]^{\frac{1}{2}} \quad [\mathrm{mm}] \tag{6}$$

where F_0 is the apparent threshold value, which can be expressed with the use of the fracture toughness:

$$F_0 = \frac{\zeta_0 \cdot \frac{E}{H} \cdot K_C^4}{A^2 \cdot (\cot g\psi)^{\frac{2}{3}} \cdot HV^3} \quad [\mathrm{N}] \tag{7}$$

where l_0 corresponds to the length of the crack limits defined by the following equation:

$$l_0 = \left[\frac{\zeta_l \cdot (\cot g\psi)^{\frac{5}{6}} \cdot \left(\frac{E}{H} \right)^{\frac{3}{4}}}{A^{\frac{1}{2}} \cdot K_C \cdot HV^{\frac{1}{4}}} \right] \cdot F^{\frac{5}{8}} \quad [\mathrm{mm}] \tag{8}$$

In this equation ζ_0 and ζ_L are dimensionless constants independent of the system material - indenter. These constants determined Marshall et.al. (in 1982), based on the measurements of ceramic materials. The constant ζ_0 corresponds to the value $1{,}2 \cdot 10^3$ and the constant ζ_L to the value $25 \cdot 10^{-3}$. The magnitude A on the previous equations is a dimensionless geometric constant. Using the previous relationships it can be predicted the dependence between the equilibrium length of lateral cracks and indenter ´s load.

7. Crack propagation and chipping formation

During the loading, in the material occurs the crack propagation and may subsequently also occur the chipping formation. The formation of cracks in the index corresponds to the time of contact of the indenter with the surface of the evaluated material and its subsequent entry into the volume of the material.
This process can be simply described as a penetrating process that leads to plastically deformed zones surrounding the Vickers´ indentation prints. There also occurs the formation of the surface tension zones which are adjacent to the contact places between the indenter and the material and occurs the emergence of a field for tensile stresses in the front of the indenter (Antis et al., 1981).

The perforation of the indenter into the sample surface leads to the initiation of subsurface cracks at the indentation´s front. At this place the subsurface tensile stress levels reach their maximum. A bigger load leads to the development of already initiated median type cracks, in both directions: towards the material and towards the surface of the sample - radial/median crack- (see Fig.5).
The median crack remains, in the case of a no sufficient load during the loaded cycle, under the surface of the evaluated material, which means that it is primarily propagated into the volume of the material and only a minority to its surface.
The subsequent intersection of the cracks with the surface of the sample can be detected only after unloading, when the active surface compressive stresses are sufficiently reduced (Sglavo & Green, 1995).

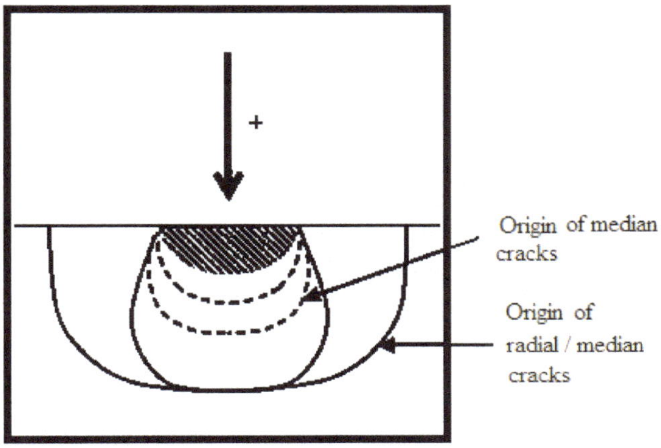

Fig. 5. Schematic representation of the origin of median and radial/median cracks under load

The median crack has a tendency to close during the subsequent unload if the indenter. The full closure will not result because of the action of the fields of internal stresses, which is related to the execution of the elastic-plastic state. Such induced residual stress fields cause the formation of lateral (side) cracks that are initiated at the plastic zone and then propagated in a parallel to the sample surface plane. The development of lateral cracks continues during the indenter´s unloading.
The initiated induced internal stresses affect the next development of subsurface median or eventually radial/median cracks. Dissemination of lateral cracks increases proportionally with increasing the load of the indenter. Their intersection with the surface of the material leads to the chipping formation.

8. Composite materials based on ceramics

As previously mentioned, composite (compound) materials present due to their high resistance to brittle crack propagation, that is because of their high fracture toughness, promising materials for technical applications. Particularly, ceramic nanocomposites meet the demands of modern technologies.
Micro / nanocomposites based on nitride and silicon carbide are a new generation of the ceramic materials developed in the last decade. The range of materials based on Si_3N_4 is recently widely used in the manufacture of cutting tools, for components exposed to friction, combustion engine components, water pumps, industrial pumps, piston engines, gas turbines etc. This type of ceramic material is characterized mainly by the high strength

at normal and elevated temperatures, chemical stability, good hardness, wear resistance, low coefficients of friction and low density.

Ceramics and ceramic composites are nowadays used in various technological fields. The ceramic composites are classified according to their particles size into microcomposites where the secondary phase particle size is measured in microns, and nanocomposites where there is a nanometric size of the secondary phase particles.

In addition to improved mechanical properties of ceramic nanocomposites there were observed new features such as machinability and superplasticity. The aim of the preparation of ceramic composites is to improve not only the "traditional" mechanical properties such as hardness, strength and reliability at room temperature, but mainly to improve the mechanical properties at high temperatures, such as creep resistance to high temperature, strength and corrosion resistance.

As a disadvantage we can see quite a significant financial demand of the preparation of nanocomposites because of the high prices of the nanopowders and the expensive furnace equipment, because the compaction of the nanocomposite requires special methods of sintering with rapid heating and short holding time at high temperature.

The ceramic nanocomposites can be divided into four categories (Fig. 6): *intragranular nanocomposites, intergranular nanocomposites, intra-intergranular nanocomposites* and *nano / nano composites*. Intra and intergranular nanocomposites have a two to five times higher toughness and strength at room temperature than the monolithic materials.

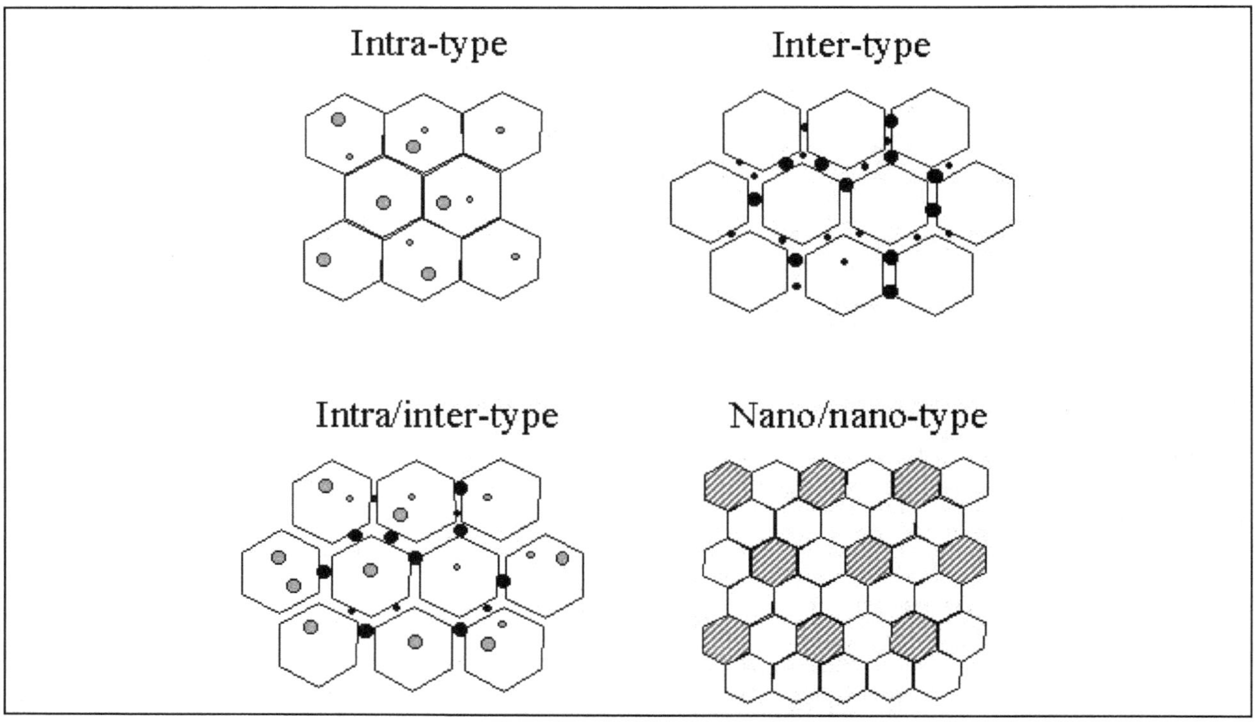

Fig. 6. Types of ceramic nanocomposites (Jonšta et al., 2010)

9. Nanocomposites based on Si_3N_4/SiC

Ceramic materials on the base of Si_3N_4 are recently one of the best candidates for tribological application due to the combination of their good thermal and mechanical properties, chemical resistance and high hardness.

On recent investigations (Bolaños et al., 2010) it was studied the preparation of Si_3N_4/SiC nanocomposites with different rare earth oxide sintering aids (La_2O_3, Sm_2O_3, Yb_2O_3 a Lu_2O_3) and the evaluation of their hardness, indentation fracture toughness and wear resistance at room temperature.

9.1 Analyzed results of the investigation

9.1.1 Density

The density of the samples after the hot pressing process amounted to 99%, so it is considered that the sintering conditions were chosen properly. At temperatures above 1510° C, there was a decrease in the vacuum, ie. CO gas was liberated proving the formation of SiC with the carbothermic reduction. After this reaction and the subsequent increase on the temperature to 1650° C occurred the formation of the liquid phase in the system RE_2O_3 – SiO_2. In the presence of Si_3N_4, the ternary eutectic point is lower, as for binary systems.
The formation of liquid phase allowed the compaction of the ceramic composite. The densities of the sintered samples prepared by hot pressing are shown in table 1.

Sample	Density ρ [g·cm^{-3}]
K3Lu-2	3,378
K3Yb-2	3,348
K3Sm-2	3,322
K3La -2	3,289

Table 1. Densities of the K3Re-2 nanocomposite materials

9.1.2 Microstructure

The microstructure of the hot-pressed composite materials denominated as K3La, K3Sm, K3Yb and K3Lu are shown in Fig. 7 and Fig. 8. According to the observed microstructures, we can conclude that the samples are slightly over etched (Si_3N_4 fine particles below 300 nm are not visible), so visually it appears that glass phase is more than 10% by volume. In all samples the microstructure contains large elongated particles of β-Si_3N_4 (whiskers) which are surrounded by finer particles of β-Si_3N_4 with needle shapes. We call this fine fraction the matrix and we call the larger β-Si_3N_4 whiskers the tougher particles.
It was expected a different evolution of the microstructures because of the used oxide RE_2O_3. A more detailed study of the Fig. 7 and Fig. 8, shows that the microstructure of the sample matrix of K3Yb and K3Lu is fine-grained. The large extended β-Si_3N_4 grains are comparable in thickness with the sample K3La, which has lower viscosity of the glass phase. In the process of sintering of Si_3N_4, some α-Si_3N_4 grains of the primary powder dissolved in the liquid phase and then precipitated, and grew in the form of β-Si_3N_4 needles.
Their growth was mainly controlled by diffusion, and that is why, in systems with a higher viscosity of the liquid phase (system containing Yb and Lu) the transport of material from the melting α-Si3N4 grains to the growing β-Si_3N_4 grains was slower, and that means that the microstructure of the matrix of these materials has then finer grains.
Another possible reason for the fine-grained microstructure of the systems with higher viscosity, is the slower dissolution of the original β-Si_3N_4 grains (about 5 wt.% on the

starting Si_3N_4 powder SN-E10) with small diameter so-called critical diameter (0,1 to 0,2μm). On the systems with the addition of La_2O_3 and Sm_2O_3 the dissolving speed of the Si_3N_4 is slightly higher, and therefore the matrix of these systems is slightly coarse-grained.

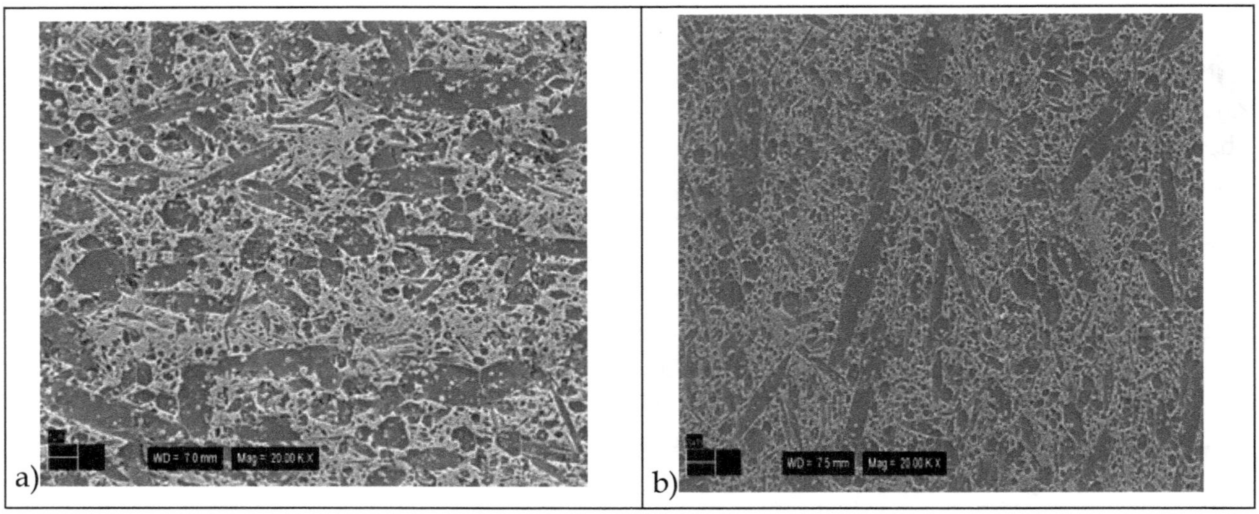

Fig. 7. Microstructure of samples a) K3La, b) K3Sm obtained by SEM at 20000th magn

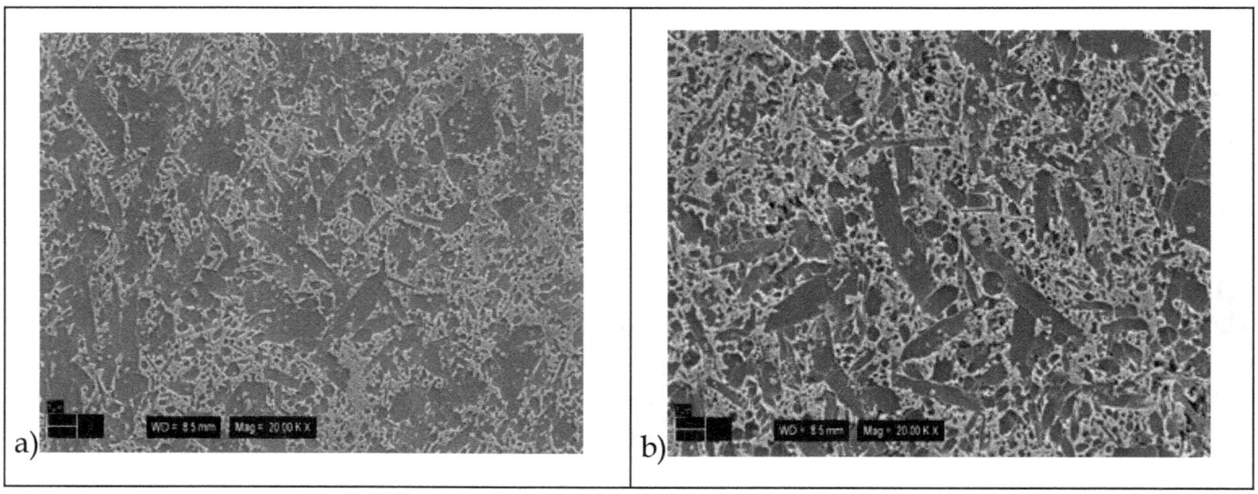

Fig. 8. Microstructure of samples a) K3Yb, b) K3Lu obtained by SEM at 20000th magn

The rare earth metal oxides also affect the thinness of the β-Si_3N_4 particles. The average value of slenderness decreases with the growing ionic radius of the rare earth metals, or with the decrease of the viscosity of the liquid phase.

The growth of elongated β-Si_3N_4 grains is affected not only by the viscosity of the liquid phase, but also by the ability of rare earth elements to adsorb itsefls on the surface of Si_3N_4 crystals. It was observed, with a transmission electron microscopy, that Lu has a higher stability in an oxy-nitride glass (stronger bond with O), while La adsorbs preferentially on the prismatic surface of the Si_3N_4 grains (binding with N), thereby preventing the further build up of dissolved Si in the surface of N, and thus growing on the thickness of the β-Si_3N_4 grains.

From the observation of 15 photomicrographs of each sample of K3RE-2, was showed that with the decrease of the ionic radius of the rare earth elements (La→Sm→Yb→ Lu) the structure is more bimodal, the matrix is finer and therefore has more elongated particles of

β-Si_3N_4. From La to Lu their thickness slightly increases. For a more detailed statistical analysis it is necessary to measure the slenderness of at least 2000 grains of each microstructure of each ceramic material containing K3RE.

The microstructure of the samples contains, together with elongated Si_3N_4 grains, irregularly distributed SiC particles, which are located both at grain boundaries and inside the Si_3N_4 grains. The larger SiC particles are mainly located at the grain boundaries (inter SiC grains), and the finer SiC particles (<50 nm) were observed within the Si_3N_4 grains (intra SiC grains). The SiC nanoparticles can slow up, or may completely stop the abnormal growth of the β-Si_3N_4 grains. Therefore, the overall microstructure of the composites with K3RE was finer compared with the reference samples.

9.1.3 Abrasion resistance

The abrasion resistance tests were performed using the "ball-on-disc" method. The friction curve had a similar pattern for all the tested samples (K3RE-1). At the beginning, the friction coefficient was low, but then increased to a maximum value and then stabilized at a relatively constant value during the test.

The Fig.9 shows the friction coefficients of reference (monolith: Si_3N_4+ 3 mol.% RE_2O_3) and composite samples with K3RE (Si_3N_4+ SiC + 3 mol.% RE_2O_3). The friction coefficient of the monolithic samples was higher compared with the composite samples with the same additive RE_2O_3 (RE = La, Sm, Yb, Lu). The lower friction coefficient of composite materials is attributable to their higher hardness compared to the reference monoliths and the presence of SiC nanoparticles. From the figure 3 it is also clear, that the friction coefficient slightly decreases with decreasing the ionic radius of RE3+. The better sliding properties of the composites with K3Lu and K3Yb compared with K3Sm and K3La may be due to a finer microstructure of the matrix (Fig. 7-8) and to a higher rigidity of the glassy phase at the grain boundaries, because the glass with a smaller radius of the RE3+ ion has a higher hardness, fracture toughness and Young's modulus.

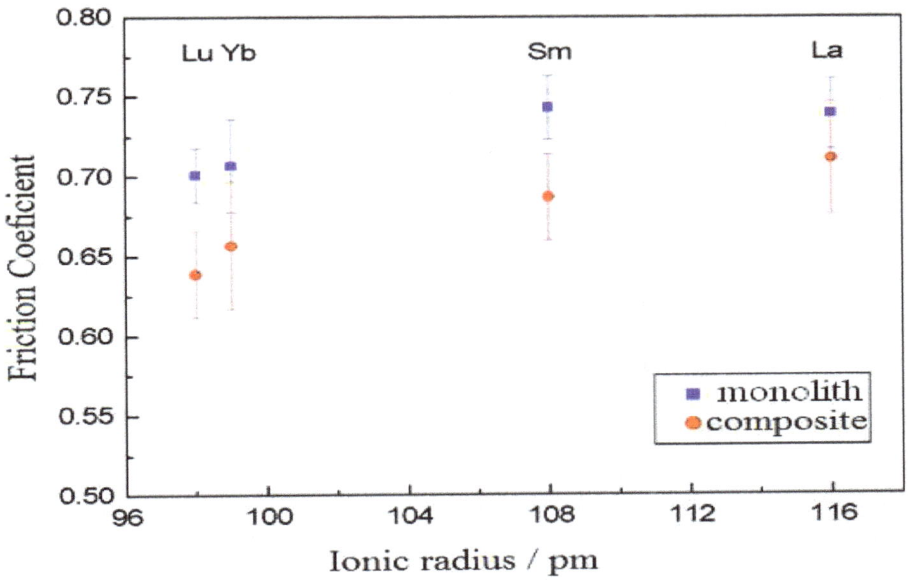

Fig. 9. Effect of ionic radius of RE3+ sintering additives on the friction coefficients of the reference monoliths (Si_3N_4+ 3 mol. % RE_2O_3)
and composite materials K3RE (Si_3N_4+ SiC + 3 mol. % RE_2O_3)

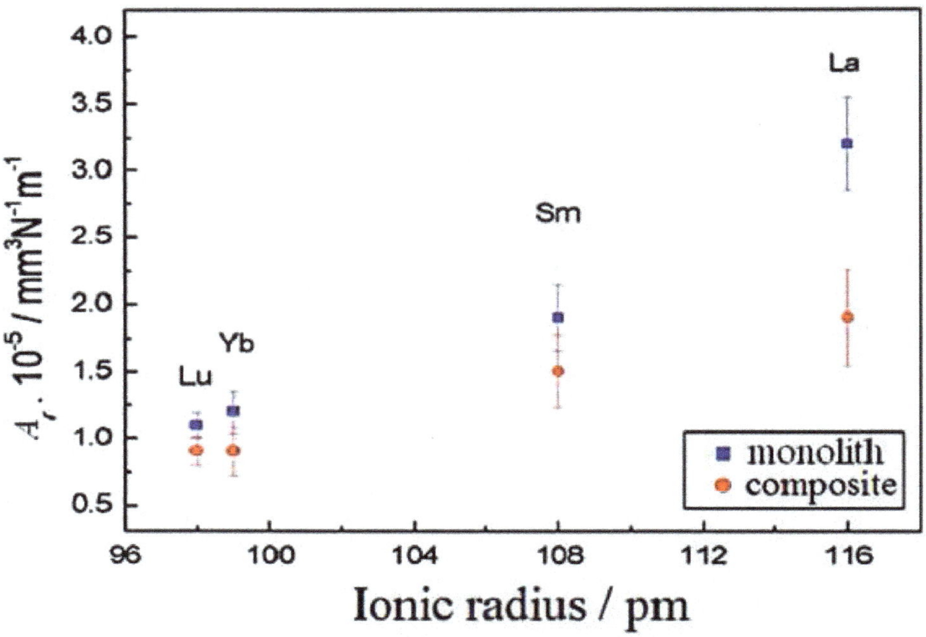

Fig. 10. Effect of ionic radius of RE3+ sintering additives on the abrasion rate (Ar) of the reference monoliths (Si_3N_4+ 3 mol. % RE_2O_3) and composite materials K3RE (Si_3N_4+ SiC + 3 mol.% RE_2O_3)

The Fig. 10 shows the relationship of the abrasion rate (Ar) of the reference monoliths (Si_3N_4+ 3 mol.% RE_2O_3) and the composite materials K3RE (Si_3N_4+ SiC + 3 mol.% RE_2O_3) with the ionic radius of the RE3+ sintering additives. The difference in the abrasion rates is more pronounced than the difference in the friction coefficients; however, the tendency is maintained. The samples with the addition of smaller RE3+ ionic radius (K3Lu and K3Yb) exhibit less abrasion in comparison with the samples containing additives with larger ionic radius (K3Sm and K3La). This is expected to be due to a higher rigidity of the glassy phase of the samples K3Lu and K3Yb because, in general, the abrasion rate of the ceramic materials based on Si_3N_4 with lower strength on the grain boundaries is higher.
Moreover, the Si_3N_4 materials with the addition of Lu_2O_3 and Yb_2O_3 have better high-temperature properties compared with ceramics containing Sm_2O_3 or La_2O_3. Then, during the "ball-on-disc" tests, the tested sample or the standard Si_3N_4 ball could get heated to higher temperatures, because of the lack of lubricant, and this exhibits the high-temperature characteristics of the material.

10. Use of carbon nanotubes as reinforcement material

Recently, CNTs are considered for use as reinforcements in advanced composite materials in view of their high elastic modulus and strength compared to existing fibers. In that respect, being rolled up graphite sheets, their mechanical properties are expected to be equal to or greater than the value for a graphene sheet. Qualitative and quantitative TEM and AFM studies have been performed on individual tubes and have confirmed their superior strength and stiffness. The Young's modulus of a CNT is over 1 TPa and its estimated tensile strength is over 150 GPa, i.e. they are a hundred times stronger than steel, though three to five times lighter.

The properties of carbon nanotubes are a mix of diamond and graphite: strong, thermally conductive like diamond; electrically conductive like graphite; and are light and flexible. Many potential applications have been proposed for carbon nanotubes like: conductive polymers; energy storage and energy conversion devices; sensors; field emission displays; replacing silicon in microcircuits; multilevel chips; probes for SPM (scanning probe microscopy).

The objective of current investigations is to prepare, by sintering in the presence of liquid phase, a dense ceramic material based on SiC with sintering additives, with the addition of CNTs, resulting in a composite with high electrical conductivity, while maintaining its good mechanical properties.

10.1 Sintering in the presence of liquid phase

This type of sintering is one of the used techniques to accelerate the process, while maintaining the mechanical properties of the material. The liquid phase is understood as an additive which is liquid at the sintering temperature. This liquid phase must fulfill two basic conditions: first, it has to drench all of the areas of the solid phase, and second, the sintered material has to have the liquid phase grains partly dissolved.

Sintering in the presence of liquid phase consists in three processes. First, we have the formation of the liquid phase and its flow, which causes changes in the spatial organization of the grains at the solid formation. This process is called relocation and it often leads to a sharp increase on the density of the sintered material. The relocation, which is due to the capillary forces of the liquid phase wetting the solid grains, is dependent on the amount of liquid phase, its viscosity, and the grain size and solubility of the solids in the liquid. The next and most important process for sintering in the presence of liquid phase is the process of increasing the density by dissolving the solid grains in the liquid phase and then the process of the subsequent elimination.

It is thus clear, that the melt should have the lowest possible viscosity at the sintering temperature and the highest possible at the application temperatures. These are conflicting requirements, and are very difficult to joint. The chemical composition of the phases at the grain boundaries is the subject of study of many authors, because grain boundaries have a significant effect on the mechanical and functional properties of the material.

10.2 Formation of the CNTs by CCVD

The CNTs can be added by a method called catalytical chemical vapor deposition (CCVD), where tubular carbon nanofibers are formed on a support by a catalytical deposition of carbon out of the vapor phase. This method is characterized by a catalyst layer, which is deposited without current, and is applied to the support at least in places. Said catalyst layer is thermally activated before depositing carbon out of the vapor phase in a reducing atmosphere. Since it was discovered, that the elements of the iron group, Fe, Co and Ni, are catalytically active, these elements are used as catalytics.

11. Conclusion

In the present paper are discussed issues of ceramic materials as basic elements of composite materials, which present advanced technical materials for a wide practical use. The attention was focused mainly on the causes of brittleness on ceramics and the possibilities of its elimination. There were also discussed issues of composite materials based on ceramics considering their preparation and evaluation of selected properties. The results obtained extend the knowledge about the preparation of the above mentioned advanced materials with the addition of rare earth metals and their selected mechanical properties.

12. Acknowledgment

This paper was created in the project No Cz.1.05/2.1.00/01.0040 "Regional Materials Science and Technology Centre" within the frame of the operation programme "Research and Development for Inovations" financed by the Structural Funds and from the state budget of the Czech Republic. The authors gratefully acknowledge the Ministry of Education, Youth and Sports of the Czech Republic for the financial support of the project MSM 6198910013.

13. References

Anstis, G.R., et al. (1981). A critical evaluation of indentation techniques for measuring fracture toughness: I direct crack measurements, *J.Am.Ceram.Soc.*, vol.64, No 9, pp.533-538

Bolaños, A., E., Mores, M., Hrabalová, M., Jonšta, Z (2010): Vztah mikrostruktury a otěruvzdornosti nanokompozitu Si_3N_4/SiC s přidavkem oxidu vzácných zemin. Zeszyty Naukowe Pol.Opolskiey, Mechanika, z. 97, Nr 337/2010, red. Prazmowskiego, pp. 15-23. ISSN 1429-6055, Opole, Poland, October, 2010. Cook,R.F.,Pharr,G.M.: Direct observation and analysis of indentation cracking in glasses and ceramics. *J.Am. Ceram.Soc.*, 1990, 73, pp.787-817

Evans, A.G., Marshall,D.B. (1981): Fundamentals of friction and wear of materials, ASME, 439-442.

Jonšta, Z., Bolanos, A., E., Hrabalová, M. (2010): Influence Of Metallic Oxides of Rare earth on the Abrasion Resistance and Microstructure of Si_3N_4/SiC Nanocomposites. *Hutnické listy* č.6/2010, roč.LXIII, pp.72-76. ISSN 0018-8069

Jonšta, Z. (1998): Vlastnosti technické keramiky a metody jejich hodnocení. Vyd.Kovosil, Ostrava

Kumar, V., Khamba, J.S. (2006): Experimental investigation of ultrasonic machining of an alumina-based ceramic composite. *J.Am.Ceram.Soc.*, pp.2413-2417

Liang, K.M., Orange, G., Fantozzi, G. (1990): *Jnl. Mat. Sci.*, 25, 207

Marshall, D.B., Lawn, B.R.: Residual stress effecting sharp contact cracking. *Jnl. of Mat.Sci.*, 1979, 14, 8, pp.2001-2012

Marshall, D.B., Lawn,B.R., Evans,A.G. (1982): *J.Am.Ceram.Soc.*, 65, pp.561-566

Matzke, H. (1980): Jnl. Mat. Sci., 15, 739

Matzke, H., Toscono, E.H. (1990): Indent. Techniques, EUR-Rep. 12799 EN

Matzke, H. (1991): *Key Engineering Materials*, 56-57, 365

Pokluda, J., Kroupa, F., Obdržálek, L. (1994): Mechanické vlastnosti a struktura pevných látek. Vyd. PC-DIR spol. S r.o., Brno.

Sglavo, V.M., Green, D.J. (1995): Subcritical growth of indentation median crack in soda-lime-silica glass. *J.Am.Ceram.Soc.*, 78, 3, pp.650-665

8

Mechanical Properties of Kaolin-Base Ceramics During Firing

Igor Štubňa[1], Anton Trník[1,2], František Chmelík[3] and Libor Vozár[1]
[1]*Department of Physics, Constantine the Philosopher University, Nitra,*
[2]*Department of Materials Engineering and Chemistry,*
Czech Technical University, Prague,
[3]*Department of Physics of Materials, Charles University Prague, Prague*
[1]*Slovakia*
[2,3]*Czech Republic*

1. Introduction

Firing of silicate ceramics, which are made of clays with high contents of kaolinite, transforms a green body into a ceramic product [1, 2]. The green body exhibits significant changes of its properties resulting from dehydration at low temperatures, phase changes during dehydroxylation and high-temperature reactions, and densification during sintering [3, 4]. All these changes significantly influence mechanical properties of the fired body as well as its other physical properties.

To save time and energy, it is desirable to conduct the firing in the shortest time possible without damage to the fired ceramic body. Calculating the safe upper limit of the heating or cooling rate of the large ceramic bodies (e.g. high-voltage insulators) is a complex task that requires knowing five material quantities: mechanical strength (MOR), Young's modulus (YM), Poisson's ratio, thermal conductivity, and coefficient of the linear thermal expansion (CLTE). All of these quantities must have to be known as functions of the actual temperature at the firing. The upper limit of the heating rate according to [2] is

$$v_{\max} = \frac{\sigma_f (1-\mu)\lambda}{A\alpha E r^2 c\rho}, \qquad (1)$$

where σ_f is MOR, μ is Poisson's ratio, λ is thermal conductivity, A is the shape factor of the ceramic body, α is CLTE, E is YM, r is the characteristic dimension of the body, c is volumetric heat capacity and ρ is the volume mass density.

The relationships between the temperature and these qualities (MOR, YM, CLTE, μ and λ) can be identified only using experimental approach despite of some theoretical connection between YM and CLTE or between YM and MOR. These theoretical relationships were derived for simple materials and they are only very approximately applicable to complex multiphase material as kaolin-base ceramics. For thin-wall ceramic bodies such as whiteware, a more simple method based on thermodilatometric results can be used for control of the firing [5].

This chapter is devoted to the development of the two most important mechanical parameters, MOR and YM of the green ceramic material during heating and cooling stages of the firing, with the exception of the highest temperature region where the body becomes partially plastic. Poisson's ratio $\mu = (E / 2G) - 1$, where E is YM and G is the elastic shear modulus, can be easily calculated if μ at the room temperature is known. The relationships between YM or shear modulus and the temperature during firing can be assumed as qualitatively identical, i.e. $\mu \approx const$ during the all firing.

2. Experimental

2.1 Samples

Samples from quartz porcelain mixture for the modulated-force thermomechanical analysis (mf-TMA) and measuring MOR were made from 50 % kaolin and clay, 25% quartz and 25% feldspar. The mixture was ground and sieved on a 100 mesh/mm^2 sieve and turned into a plastic material. Then cylindrical samples (diameter 11 mm) were made with the laboratory vacuum extruder. The samples were dried in the open air and finally content ~1 wt.% of the physically bounded water.

2.2 Modulated force thermomechanical analysis (mf-TMA)

The mf-TMA is an effective experimental method to determine temperature dependence of YM of ceramics [6]. Most of the technical solutions of the method are based on continual measuring the resonant frequency of the sample during a defined temperature regime. The resonant frequency *f*, the volume mass density ρ, and the sample dimensions (length l and diameter d) serve for the calculation of YM [7, 8]

$$E = 1.12336 \left(\frac{l^2 f}{d} \right)^2 \rho T \tag{2}$$

Here, T is a correction coefficient, which should be used if $l/d < 20$. Since the ratio for the used samples $l / d = 15 < 20$, and assuming Poisson's ratio $\mu = 0.2$, the coefficient T = 1.06821 can be determined from the table given in [9].
Since the mass and the dimensions of the green sample are not constant during firing, to obtain the correct value of Young's modulus at the actual temperature *t*, the mass and the dimensions at the actual temperature *t* should be substituted into Eq. (2). Then we obtain

$$E(t) = 1.52782 \frac{m_0 \left(1 + \dfrac{\Delta m(t)}{m_0} \right) l_0^3 f^2(t)}{d_0^4 \left(1 + \dfrac{\Delta l(t)}{l_0} \right)} . \tag{3}$$

Relative changes in the mass $\Delta m(t) / m_0$ and dimensions $\Delta l(t) / l_0$ must be determined from the results of thermodilatometric (TDA) and thermogravimetric (TGA) analyses. The values m_0 and l_0 are the initial mass and length of the sample at room temperature. If Eq. (3) is used, simply considering $\Delta m(t) = 0$ and $\Delta l(t) = 0$, a certain error arises. Its relative value for measuring the green sample is shown in Fig. 1.

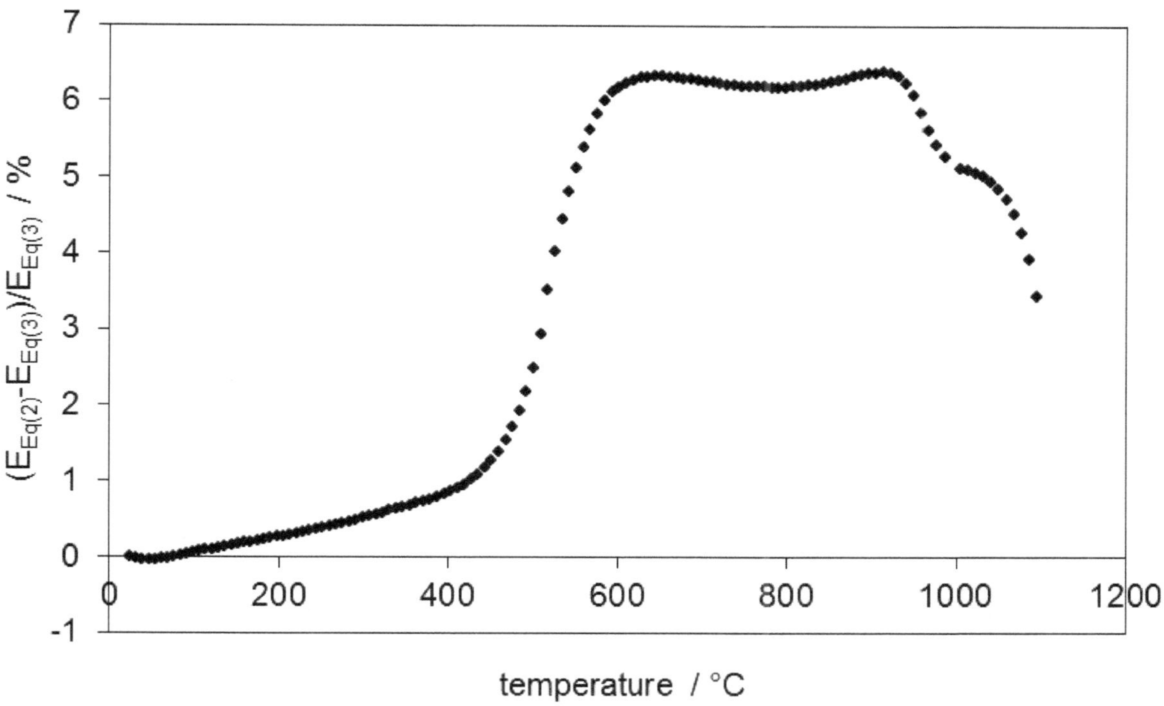

Fig. 1. Relative error $(E_{Eq.(2)} - E_{Eq(3)}) / E_{Eq(3)}$ originated from neglecting the mass and dimensions changes of the sample

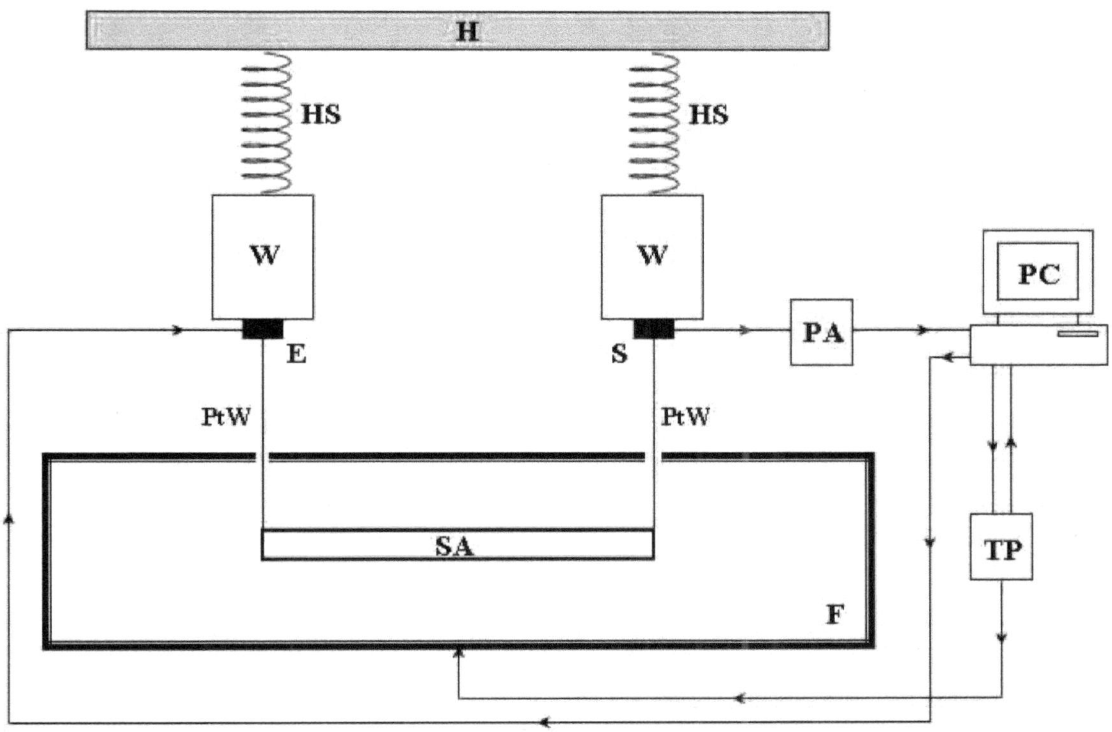

Fig. 2. A scheme of the apparatus for measuring YM.
H – frame, HS – helical spring, W – weight, E – exciter, PtW – platinum wire, S – sensor, PA – preamplifier, SA – sample, F – furnace, PC – computer, TP – temperature programmer

The apparatus for measuring Young's modulus of ceramics at higher temperatures has to satisfy several requirements: a) the sample must be adequately large because some ceramic materials are inhomogeneous and contain crystals of different minerals, glassy phase, and pores; b) green ceramic material is brittle and soft and its mechanical strength is low, and c) the dimensions of the sample contract by up to 8 – 12 % during sintering. We designed an apparatus that satisfies these requirements, see Fig. 2 [10]. It is based on the chart showed in [7, 8] and, in more detail, in [11]. A sample placed horizontally in a furnace is suspended on thin platinum wires (∅0.15 mm) which connect the sample with exciter and sensor located outside the furnace. For the measurement of the sample temperature we use the thermocouple Pt-PtRh10. Its hot ending is in the vicinity of the sample and the cold endings are connected to the temperature programmer.

2.3 Measurement of MOR

Measurements at high temperatures are energy and time-consuming. To make a measurement of MOR more effective, a new apparatus with a heated magazine for 10 samples was designed (Fig. 3). It allows testing of these samples in one heating cycle from 20 to 900 °C using the three-point-bending method.

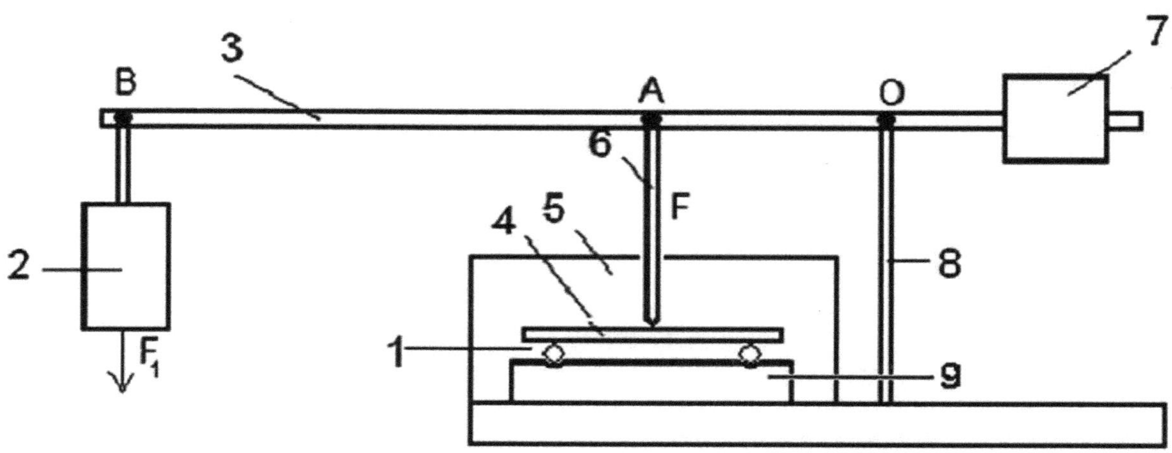

Fig. 3. Apparatus for measuring MOR.
1- supports, 2 - dynamometer, 3 - lever, 4 - sample, 5 - furnace, 6 - alumina rod, 7 - counterweight, 8 – frame, 9 – alumina brick

Temperature of the samples changed linearly at the rate of 5 °C/min. When the temperature reached a chosen value, one sample was broken. The next sample was broken at the next chosen temperature, etc. The loading force increased linearly with a rate of 2 N/s. If the three-point-bending is used, MOR can be calculated from a formula [12]

$$\sigma_f = \frac{8lF_f}{\pi d^3} \tag{4}$$

where d is the diameter of the sample, l is the span between the supports and F_f is the force at the moment of the rupture. However, thermal expansion of the sample diameter $\Delta d(t)$ and the span between supports $\Delta l(t)$ must be taken into account at elevated temperatures. Then Eq. (4) obtains the form

$$\sigma_f(t) = \frac{8[l_0 + \Delta l(t)]}{\pi[d_0 + \Delta d(t)]^3}[aF_{1f}(t) + F_2] \ , \tag{5}$$

Where d_0 and l_0 are the diameter and the span at room temperature, a is a constant of the lever mechanism, $F_{1f}(t)$ is the force registered at the moment of the rupture and F_2 is a small known initial pre-loading force.

3. Heating stage of firing

During heating, several important processes take place in ceramic material. To understand them in more detail it is useful to perform thermal analyses, mainly TGA, DTA and TDA. Their results are shown in Fig. 4 and Fig. 5. Both curves depicted in Fig. 4, TGA and DTA, are of a typical shape for kaolin base ceramics. Since the sample was not fully dried, some physically bounded water was present at the surface defects on walls of kaolinite crystals [13] and in micropores, which caused a subsequent loss of the sample mass at the lowest temperatures. This process is endothermic as confirmed by DTA.

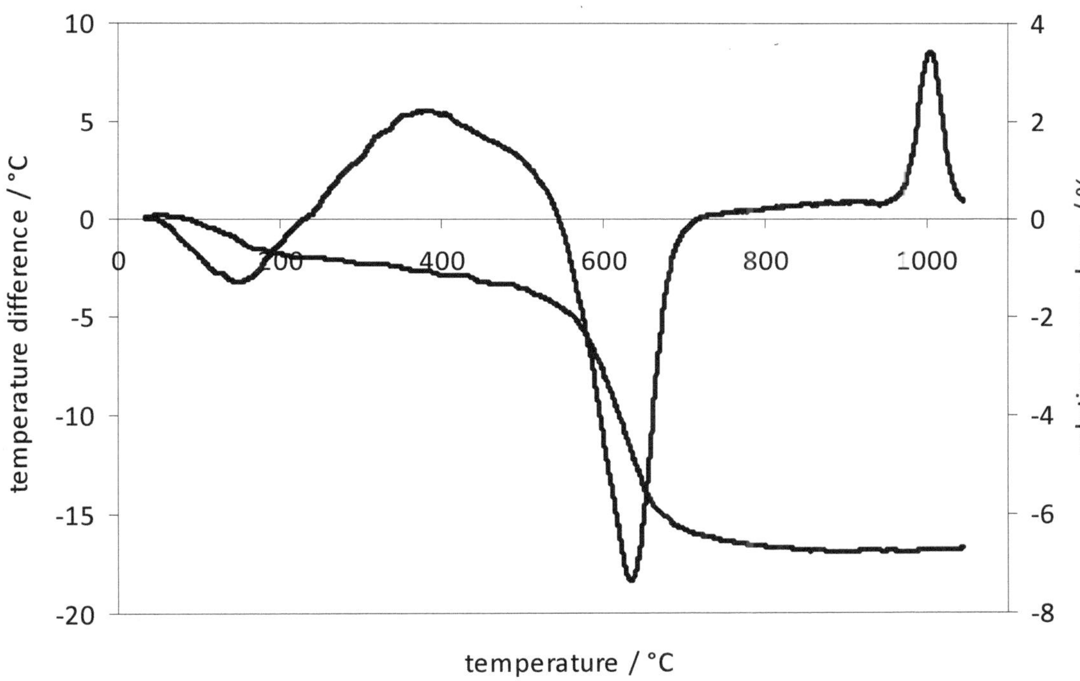

Fig. 4. Relative mass loss (black line) and DTA (gray line)

Fig. 5 and Fig. 6 show the results of TDA and mf-TMA of the green porcelain sample reflecting the same processes as Fig. 4. Comparing Fig. 5 to Fig. 6, the rule "the higher thermal expansion, the lesser YM" is not fulfilled in the temperature range of 20 – 200 °C. Here, two concurrent mechanisms appear. The first is the escaping of the physically bound water from the finest pores and the surface of crystals. This leads to a) a closer contact between crystals and, thereby, to an increasing YM and b) to a small contraction of the sample. Simultaneously, thermal expansion makes YM lower and the dimensions of the sample larger. The net effect of these phenomena demonstrates itself as a steep growth of YM and as a very moderate deceleration of the thermal expansion up to 120 °C.

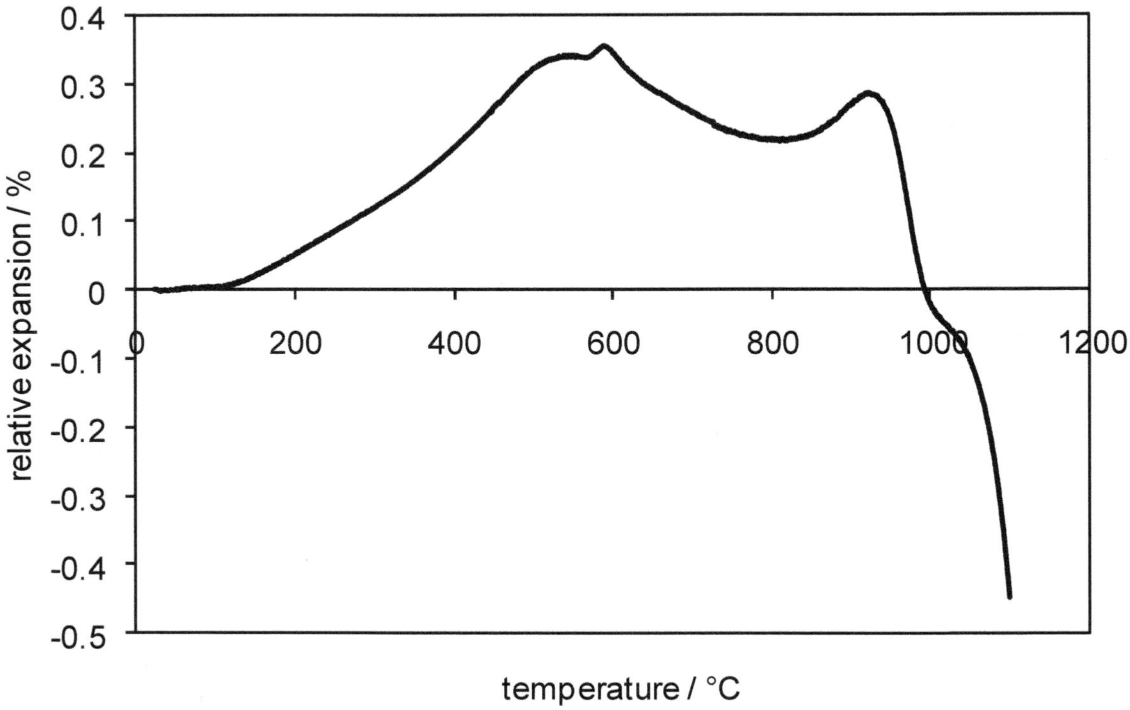

Fig. 5. Thermodilatometric curve of the green sample during heating

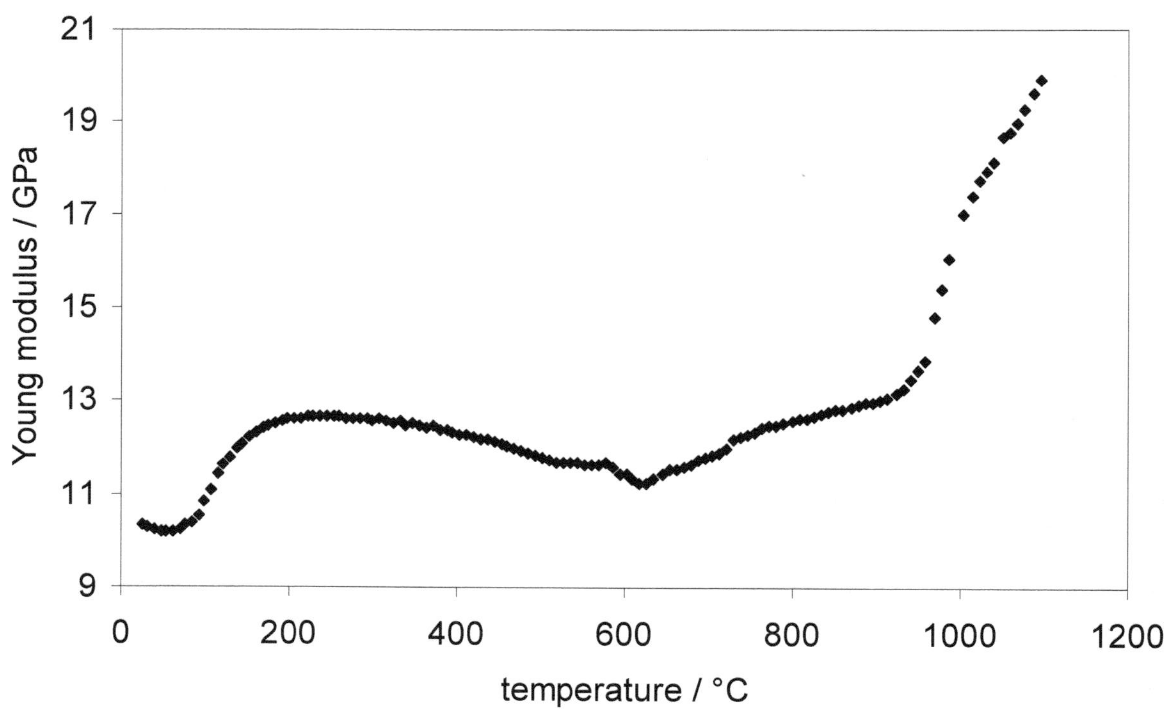

Fig. 6. The results of mf-TMA - Young modulus (calculated according to Eq. (3)) versus temperature

After removing the weakly bound physically water, the sample loses the rest of this water and a very small amount of the organic material, see TGA in Fig. 4. The course of the dilatometric curve represents a simple thermal expansion, as shown in Fig. 5. The values of the thermal expansion increase up to ~500 °C when dehydroxylation occurs, which leads to a contraction of the sample and dropping of its mass. Following the escape of the physically bound water, there is no further change in the structure and composition. Hence, above 200 °C both TDA and mf-TMA results approximately satisfy the rule "the higher thermal expansion, the lesser YM" as long as dehydroxylation emerges at ~450 °C. This process is exhibited in different ways for mf-TMA and TDA. The TDA reflects a) shrinkage of the kaolinite crystals as the result of escaping the crystalline water from their octahedral sheets and b) thermal expansion of the feldspar and quartz.

The mf-TMA is a complex result of four phenomena: a) creation of defects in the kaolinite crystals interiors (which makes YM lower), b) probably the beginning of the solid phase sintering (which makes YM higher), c) loss of the sample mass (which makes YM lower) and d) shrinkage of the sample dimensions (which makes YM higher).

We can conclude that the thermodilatometric behavior of the sample during dehydroxylation is determined primarily by the dehydroxylation process in the kaolinite crystals. The elastic behavior is determined mainly by improving the contacts between the crystals as a consequence of the solid phase sintering. Dehydroxylation plays a minor role in the mf-TMA results (except at a short temperature interval 550 – 620 °C). The temperature interval of dehydroxylation is considered a critical stage of the firing process. It is known that an inadequately high rate of heating or cooling leads to cracking of the ceramic body. This was studied in detail on large cylindrical samples in the dehydroxylation region [14].

The $\alpha \rightarrow \beta$ transition of quartz takes place in the dehydroxylation background, and in spite of a relatively large content of quartz in the sample (25 wt.%), the transition proves only with a small peak at ~570 °C on both curves, TDA and mf-TMA.

Thermal expansion after the completion of dehydroxylation ends at temperature ~950 °C when the collapse of the metakaolinite lattice accompanied by a rapid shrinkage appears. Identically, there is a steep increase in YM in response to a faster solid state sintering and the new structure. A typical exothermic peak on the DTA curve at 1000 °C is characteristic for this process.

Mechanical strength (Fig. 7) approximately follows the course of mf-TMA in Fig. 6. As our preliminary (not yet published) results show, MOR slightly increases in the interval 20 – 200 °C. However, MOR is much more sensitive to defects created by dehydroxylation. Its values run through a clear minimum at ~450 °C. Very similar results were published in [15]. An increase of MOR during the continuing dehydroxylation (above 450 °C) as well as insignificant decrease of YM (Fig. 6) and of the sound velocity (Fig. 8) during dehydroxylation are not easy to explain. Other constituents (quartz and feldspar) do not lose their mechanical properties, and metakaolinite must be considered a mechanically weak material because of the high concentration of defects [16]. The structure of metakaolinite does not change until ~950 °C, so the effect of the MOR increase should be attributed to improving the interfaces between metakaolinite crystals. Solid phase sintering could be a possible relevant mechanism.

It is interesting to compare the results of mf-TMA of the green porcelain sample (Fig. 6) to mf-TMA of kaolin (Fig. 8) [17]. Sound velocity c is related to YM by formula $c^2 = E / \rho$; therefore the temperature dependence of YM and the sound velocity are similar. Comparing

Fig. 6 with Fig. 8 we can conclude that the changes of YM in the three-axial porcelain mixture are predominantly caused by the changes in the kaolin phase and the kaolin crystals interfaces. The same can be stated about MOR. We also made some trial measurements of the acoustic emission during heating. Weak signals were monitored at the temperature range of 250 – 650 °C, which is in accordance with the results of mf-TMA.

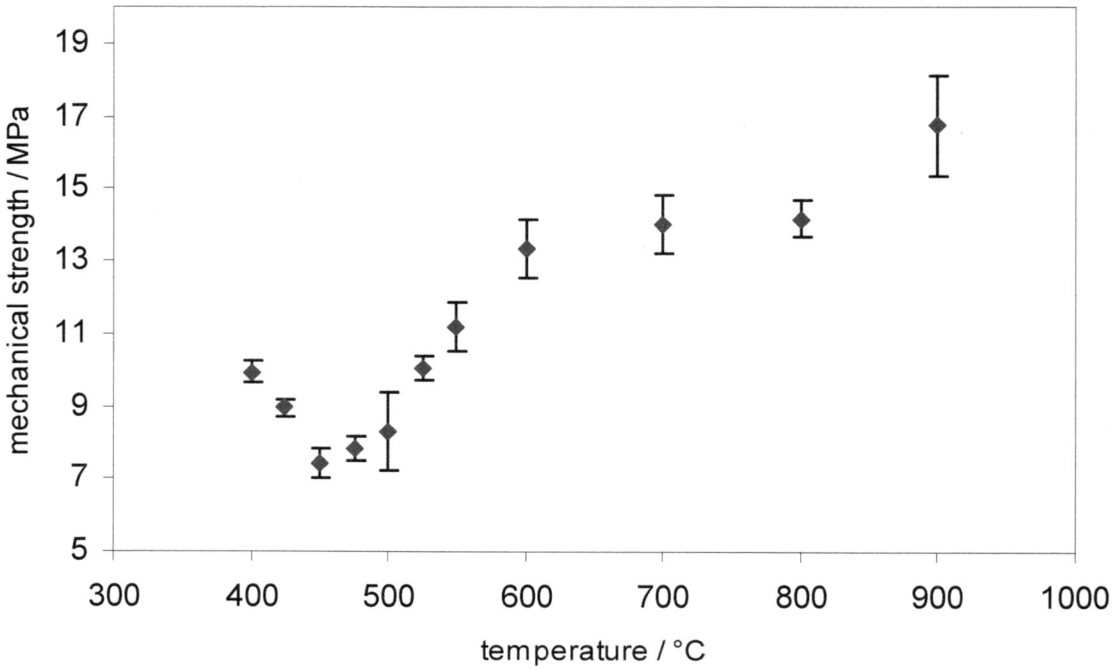

Fig. 7. Mechanical flexural strength measured during heating stage of the firing

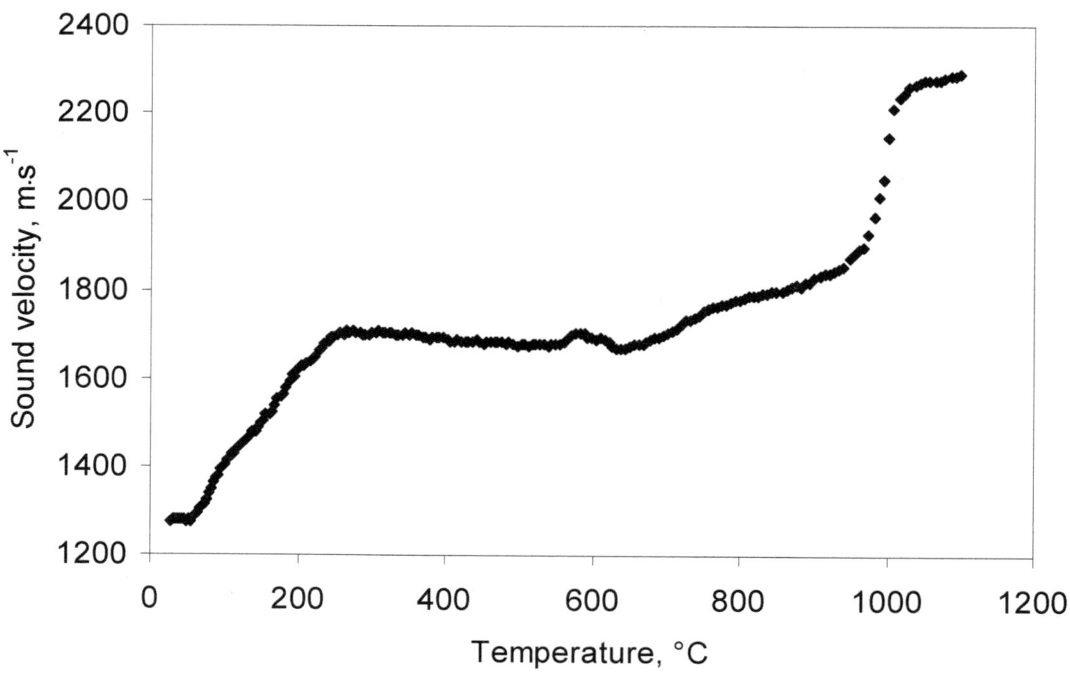

Fig. 8. Sound velocity of kaolin vs. temperature

In theory, the direct implication of Hooke's law is that the relationship between MOR and YM is linear. As can be seen in Fig. 6 and Fig. 7, the courses of YM and MOR are similar, which confirms this theoretical assumption. But there is a considerable scatter of points in this relationship, see Fig. 9, which allows only for a rough estimation of MOR from YM [18].

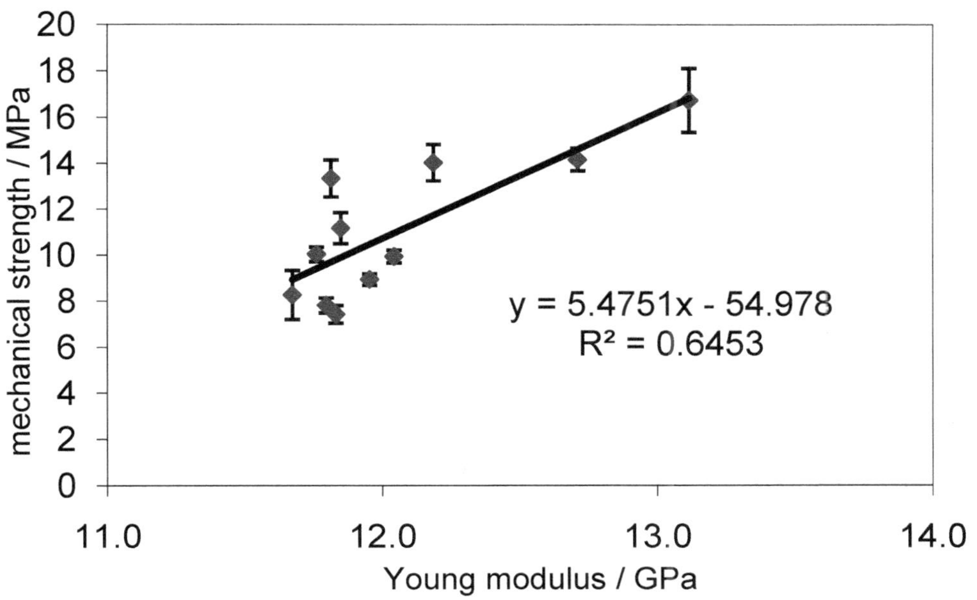

Fig. 9. Relationship between mechanical strength and Young modulus measured at the actual temperature

4. Cooling stage of firing

During heating of the ceramic body from ~1150 °C up to the maximum firing temperature, mullite is created, feldspar melts, the glassy phase occurs and the liquid phase sintering runs. The ceramic body losses elastic properties and becomes plastic. Shrinkage of the sample volume is significant, porosity decreases to a low value [3, 4, 19]. When the body cools down, a viscosity of the glassy phase increases and elastic properties progressively restore. The ceramic material becomes a porcelain.

Numerous studies deal with the influence of the unsolved quartz grains in the porcelain structure. It is recognized that residual quartz grains have a negative influence on the strength of the porcelain. The primary problems associated with quartz in electroporcelain and the ways to decrease the negative effects of quartz were recently described by J. Liebermann [20, 21, 22]. It is commonly believed that the relatively large volume change accompanying the $\beta \rightarrow \alpha$ transformation of the unsolved quartz grains ($\Delta V/V$ = –0.68 % for free quartz grain) is the basic source of microcracking. Investigations performed at room temperature with the help of the electron or light microscopy show circumferential microcracks and (less often) microcracks in quartz grains, see Fig. 10. Based on the model of quartz grain with glassy cladding and the thermodilatometric results (as depicted in Fig. 11), the cooling interval can be divided in two parts. Above 570 °C, the volume of the quartz grain remains approximately constant, and the glassy cladding tends to contract its volume.

The grain is under the compressive stress and no circumferential cracks can arise. But in the narrow temperature interval around the $\beta \rightarrow \alpha$ transition of quartz, the quartz grain volume rapidly contracts and the circumferential microcracks can appear around the grain. The microcracks are the result of the release of the mechanical stress caused by the difference in the thermal expansion between the quartz grains and the glass matrix [21, 23, 24, 25].

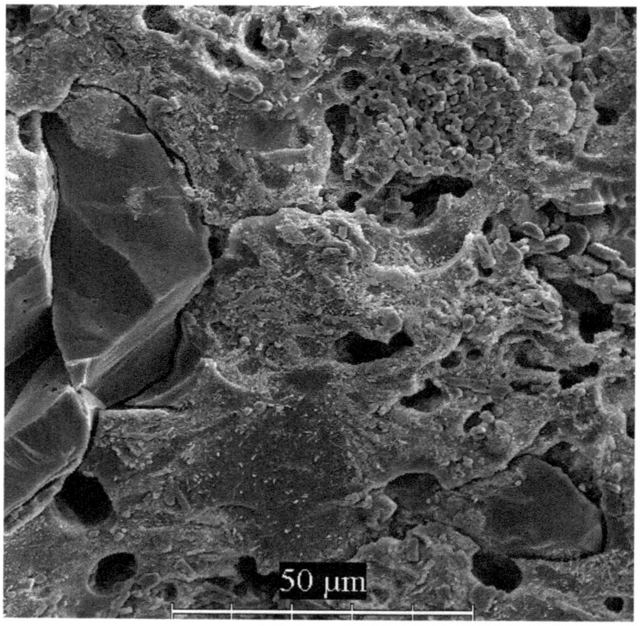

Fig. 10. Fracture surface with quartz grains and circumferential cracks around them. SEM after etching with 2.5 % HF for 15 s. Courtesy of PPC Čab

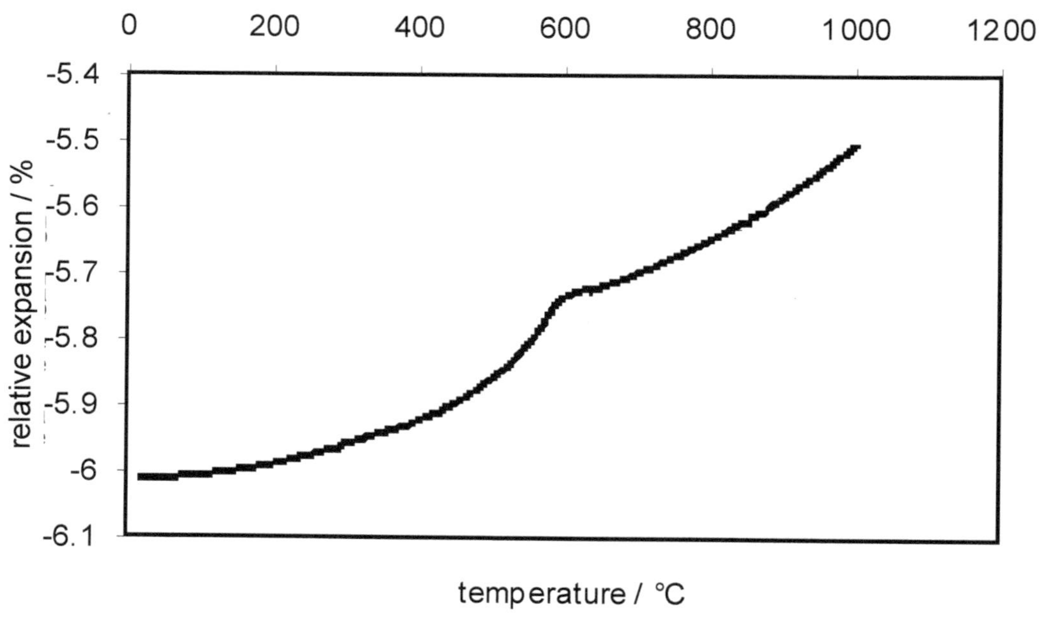

Fig. 11. Thermodilatometric curve of porcelain during cooling stage of firing

Results obtained using the real time non-destructive methods applied during the cooling stage of firing offer another view on the cracking process. An acoustic emission measurement directly registers the formation of the microcracks, and mf-TMA, which measures YM (or sound velocity), is sensitive to the concentration of the microcracks. The acoustic emission was employed in [26] to examine spontaneous cracking in porcelain samples during cooling. Acoustic emission signals were detected in the temperature range of 900 - 800 °C and never at temperatures less than 600 °C. These results were indirectly confirmed in [27] through measuring the sound velocity. Beside that, a decrease in the sound velocity below 570 °C was observed and assigned to the microcracks.

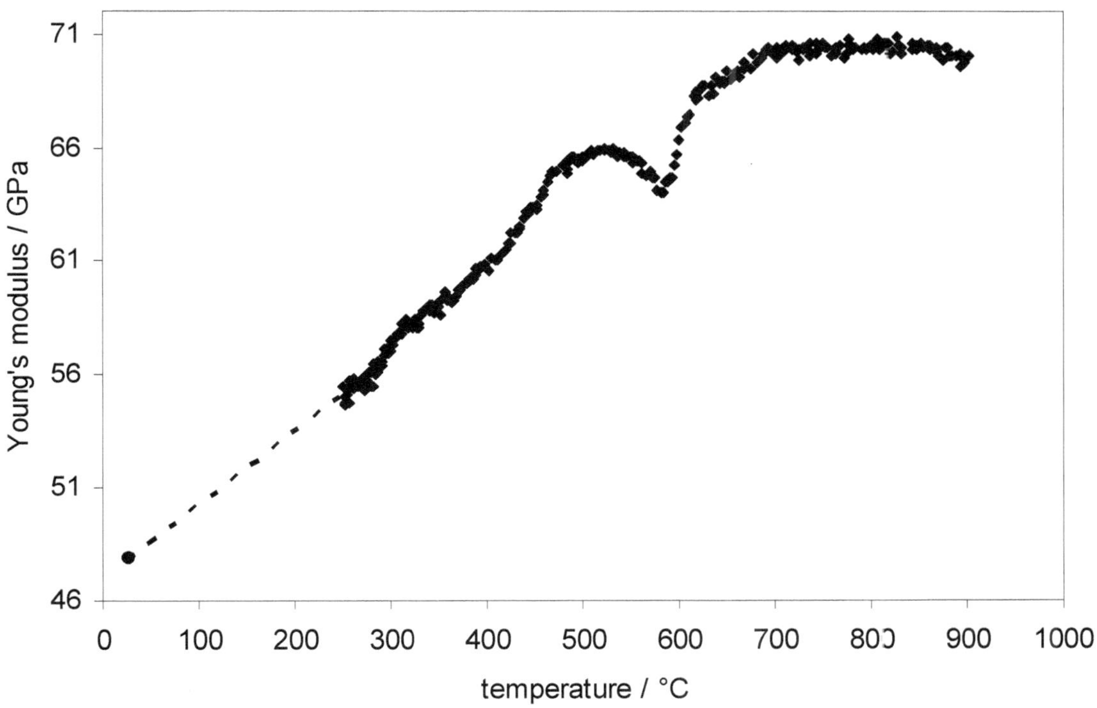

Fig. 12. Relationship between Young's modulus and temperature of green quartz sample during cooling stage of firing from max. temperature 1250 °C

A resonant flexural vibration develops in the sample if viscosity of the glassy phase is high enough for propagation of the mechanical flexural wave. This condition appears during cooling at the temperature of ~1000 °C. The relationship between YM and the temperature during cooling is depicted in Fig. 12. Values of YM increase with a decreasing temperature as long as the viscosity of the glassy phase is low enough to accommodate the mechanical stress generated during cooling. Subsequently, creation of microcracks may begin as a consequence of releasing the stress between phases with different thermal expansion coefficients. This process is demonstrated through descending values of YM below ~750 °C. Then there is a rapid decrease in the values with a sharp minimum at ~580 °C, which belongs to the $\beta \rightarrow \alpha$ quartz transition. During the further temperature decrease, there is a moderate and temporary recovery of Young's modulus values, followed by their restored drop from 500 °C to room temperature. This behavior supports results from the acoustic emission measurement. Intense acoustic signals, connected with the crack creation, were observed from ~800 °C to ~650 °C, then no signals appeared between 600 °C - 500 °C and

weak signals appeared again below 500 °C [28]. Consequently, most of the cracks in the vicinity of the quartz grains are created before the β → α transition and not during this transition as commonly believed.
Crack creation around the unsolved quartz grain can be caused by mechanical stress. According to [19], the total stress σ acting on the quartz particle may be calculated as

$$\sigma = \frac{\Delta\alpha \Delta T}{\frac{1+\mu_g}{2E_g} + \frac{1-2\mu_q}{E_q}} \tag{6}$$

where $\Delta\alpha = \alpha_g - \alpha_q$ is the difference between the thermal expansion coefficients of the glassy matrix and the quartz grain, and ΔT is the temperature range of the cooling process. E_g and E_q are Young's moduli of the glassy matrix and the quartz grain, respectively, and μ_g and μ_q are their Poisson's ratios. The thermodilatometric curves of the quartz and glassy phase and mullite (the last two curves are approximately identical according [26]) are depicted in Fig. 13. As can be deduced from Fig. 13, the quartz grain volume is almost constant and its glassy surrounding contracts its volume during cooling from 1000 °C to 570 °C. The stress applied on the grain according to Eq. (6) is positive (i.e. compressive stress). The glassy phase is under the tensile stress. Below the glass transformation temperature, the glassy matrix becomes brittle, which could result in formation of radial cracks nucleating at the quartz particles [19]. The nucleation and the growth of cracks are demonstrated by the decreasing values of Young's modulus (Fig. 12) below 800 °C. This process is finished by the β → α transition accompanied by a fast decrease in the thermal expansion of the quartz grains.

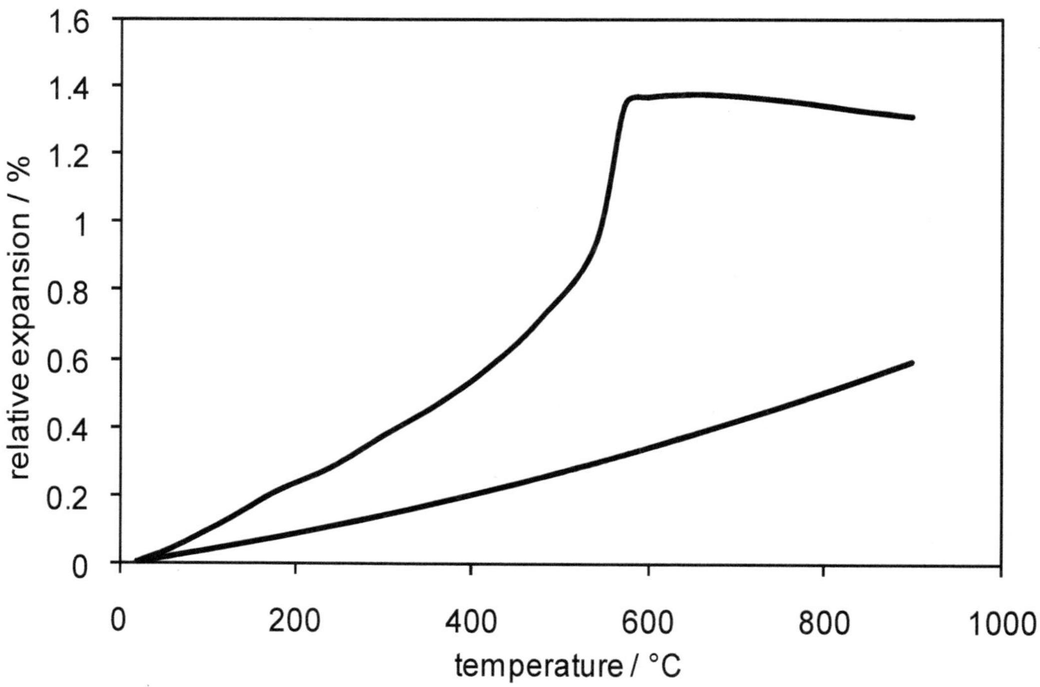

Fig. 13. Thermodilatometric curves of quartz (black line, main values from the curves for c-axis and a-axis [29]) and mullite [30] or glassy phase (gray line)

During the β → α transition of quartz, the contraction of the quartz phase volume proceeds faster than the contraction of the glassy phase, i.e. $\alpha_q > \alpha_g$ and the stress applied on the quartz grains becomes negative according to Eq. (6). During this fast process the tensile stress applied on the glassy matrix was rapidly relaxed, and upon further cooling, the compressive stress built up. This process took place between 573 and 500 °C, demonstrating itself by vanishing of acoustic signals (no new cracks nucleated) and a partial recovery of Young's modulus, likely due to the shrinkage of voids caused by previous cracking.
Below 500 °C, the relation $\alpha_q > \alpha_g$ continues to be valid until the room temperature is reached. After a short interval between 600 °C and 500 °C where no acoustic signals were detected [26, 28] and YM passed over a small maximum, YM continued to decrease its values and weak acoustic signals appeared again [28]. This can be explained by the newly built compressive stresses which induce an interface decohesion that takes the form of cracks around the quartz grains. The acoustic signals disappear at temperatures below 300 °C, which indicates that no further cracking occurs and the ongoing decrease of Young's modulus is caused by the accretion of already existing cracks [28].

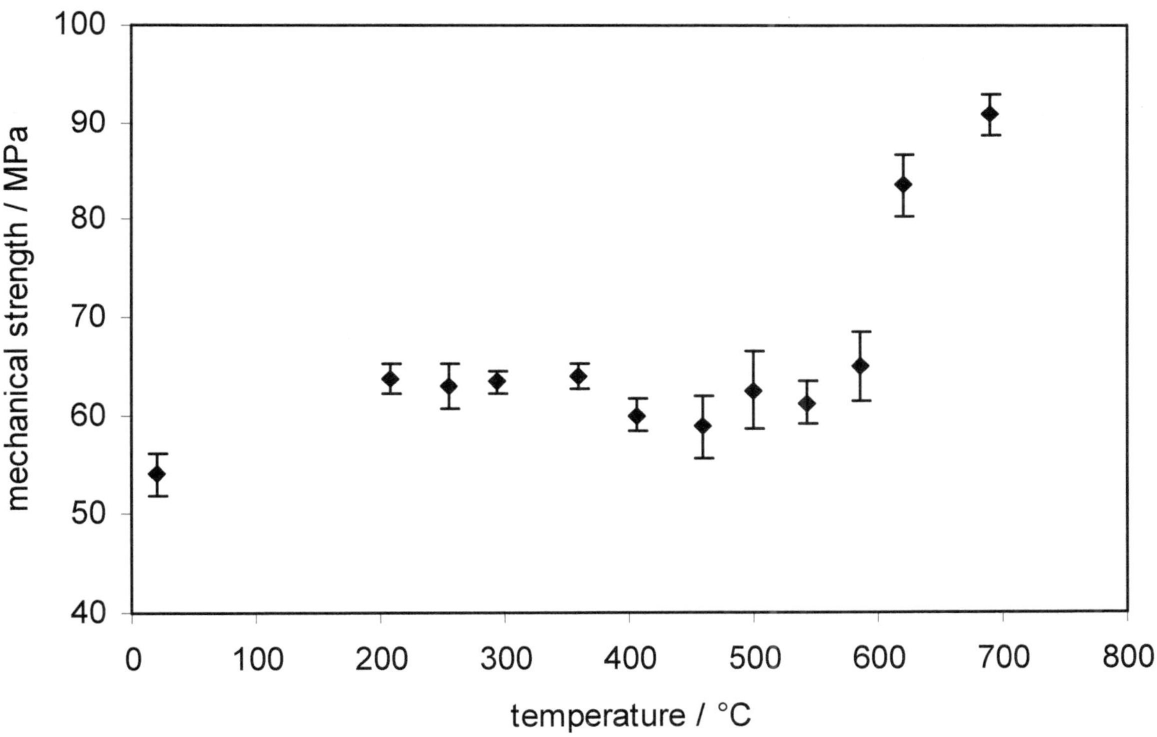

Fig. 14. Mechanical strength measured during cooling stage of the firing

The values of MOR measured in the cooling stage of firing, see Fig. 14, approximately confirm the conclusions shown above. MOR decreases before the β → α transition occurs, which is in agreement with the decreasing YM, and reaches the minimum at ~550 °C. A slightly higher value of MOR at 500 °C can be hardly considered a maximum, although it could correspond to the maximum of YM. Because of a few measured points in this region (450, 500, 550, 600 °C) and relatively large standard deviations of MOR, it is questionable to unambiguously relate the small increase in MOR at ~550 °C to a temporary recovery of YM.

The behavior of MOR and YM below 450 °C differs from the theoretical requirement of the direct proportion between these two quantities. The values of MOR initially increase (between 450 and 350 °C) and then remain constant up to 200 °C. On the other hand, YM decreases its values permanently. Both quantities, MOR and YM, reach their lowest values at room temperature.

5. Conclusions

Elastic properties (Young's modulus, sound velocity and mechanical strength) of the green quartz porcelain body reflect the changes in the structure of the body and the processes in it. They are linked to

a. escape of the physically bound water, dehydroxylation, the $\alpha \rightarrow \beta$ quartz transition, high-temperature reactions and solid-phase sintering during the heating stage of firing,
b. solidifying of the glassy phase, differences between the thermal expansion coefficients of the glassy phase, mullite and quartz grains, and the $\beta \rightarrow \alpha$ transition of quartz during the cooling stage of firing.

Results obtained by mf-TMA during heating showed a direct relationship between the elastic properties of kaolin and green porcelain samples. Young's modulus and the mechanical strength of the green porcelain sample seem to be determined by the properties of the interfaces between kaolinite crystals. Mechanical properties improve between 50 - 250 °C and then between 650 - 1100 °C, which can be explained by solid-phase sintering.

Young's modulus and the mechanical strength in the cooling stage of firing are predominantly determined by the stresses created and relaxed between the glassy phase and the quartz grains. They are a source of microcracks that are mostly located in the vicinity of the quartz grains. As a consequence, Young's modulus and the mechanical strength decrease.

Results obtained using the AE technique reveal several stages of microcracking. This process begins at the temperature of the glass transition. At the temperature of the $\beta \rightarrow \alpha$ transition, cracking is temporarily interrupted and a partial recovery of the structure occurs. Below this temperature circumferential cracks around the particles appear with less intensity, especially when the temperature is decreased below 300 °C.

6. Acknowledgment

This work was supported by the grant VEGA 1/0216/09, Ministry of Education of Slovakia. One of the authors (F.C.) gratefully acknowledges a financial support from the Research Goal MSM0021620834 that is financed by the Ministry of Education, Youth and Sports of the Czech Republic and from the research Grant 108/11/1267 financed by the Czech Science Foundation.

7. References

[1] Pospíšil, Z. - Koller, A.: *Fine Ceramics, Basics of Technology*. SNTL/ALFA, Praha 1981 (in Czech)

[2] Pytlík, P. - Sokolář, R.: *Building ceramics - technology, properties, applications*. Akad. nakladatelství CERM, Brno 2002 (in Czech)

[3] Norton, F.H.: *Fine ceramics – Technology and Application*. McGraw-Hill Book Co., New York 1970

[4] Hanykýř, V. – Kutzendorfer, J.: *Technology of Ceramics*. Silis Praha a Vega Hradec králové, Praha 2000 (in Czech)

[5] Palmour III, H.: Rate-controlled sintering of a whitware porcelain. *Ceramic Engineering and Science Proceedings*, 7, 1986, 1203-1212

[6] Štubňa, I. – Trník, A. – Vozár, L.: Thermomechanical and thermodilatometric analysis of green alumina porcelain. *Ceramics International*, 35, 2009, 1181-1185

[7] *ASTM C 848-88*: Standard test method for dynamic Young's modulus, shear modulus and Poisson's ratio for ceramic whiteware by sonic resonance. (published in 1999, Standard Documents, Philadelphia USA)

[8] *ASTM C 1198-01*: Standard test method for dynamic Young's modulus, shear modulus and Poisson's ratio for advanced ceramics by sonic resonance. (published in 2001, Standard Documents, Philadelphia USA)

[9] Schreiber, E. – Anderson, O. – Soga, N.: *Elastic constants and their measurement*. McGraw-Hill Book Co., New York 1973

[10] Štubňa, I. – Trník, A. – Vozár, L.: Determination of Young's modulus of ceramics from flexural vibration at elevated temperatures. *Acta Acustica + Acustica*, 97, 2011, 1-7

[11] Kashtaljan, J.A.: Elastic properties of materials at high temperatures. *Naukova dumka*, Kiev 1970 (in Russian)

[12] Rath, J. et al.: *Fine ceramics – measurement methods and testing*. SNTL/ALFA Praha, 1988 (in Czech)

[13] Polakovič, J. – Polakovičová, J. – Sokoly, J.: Interaction between kaolinite surface and hematoxyline. In: *Proc. 5th Meeting of European Clay Groups*, Prague, 1983, 279-282

[14] Kozík, T. – Rakovský, O. – Ondrejkovič, E.: Resistivity of ceramic material to thermal shock in dehydroxylation region. *Sklár a keramik*, 28, 1978, 176-180 (in Slovakian)

[15] Kozík, T. - Štubňa, I.: Mechanical strength of ceramic material in dehydroxylation region. *Silikáty*, 25, 1981, 237-241 (in Slovakian)

[16] Freund, F.: Kaolinite-metakaolinite, a model of a solid with extremely high lattice defect concentration. *Ber. Deutsche Keram. Ges.*, 44, 1967, 5-13

[17] Trník, A. – Štubňa, I. – Moravčíková, J.: Sound velocity of kaolin in the temperature range from 20 °C to 1100 °C. *International Journal of Thermophysics*, 30, 2009, 1323-1328

[18] Štubňa, I. – Trník, A. – Šín, P. – Sokolář, R.: Relationship between mechanical strength and Young's modulus in traditional ceramics. *Materiali in Tehnologije*, submitted article

[19] Carty, W.M. – Senapati, U.: Porcelain – raw materials, phase evolution, and mechanical behavior. *J. Amer. Ceram. Soc.*, 81, 1998, 3-20

[20] Liebermann, J.: About the important correlation between microstructure properties and product quality of strength-stressed high-voltage insulators. *Interceram*, 53, 2003, 238-241

[21] Liebermann, J.: Avoiding quartz in alumina porcelain for high-voltage insulators. *Keramische Zeitschrift*, 53, 2001, 683-686

[22] Liebermann, J.: Reliability of materials for high-voltage insulators. *www.ceramicbulletin.org*, (Amer. Ceram. Soc. Bulletin, May 2000)

[23] Porte, F. - Brydson, R. - Rand, B. - Riley, F.L.: Creep viscosity of vitreous china. *J. Amer. Ceram. Soc.*, 87, 2004, 923-928

[24] Carty, W.M. - Pinto, B.M.: Effect of filler size on the strength of porcelain bodies. *Ceram. Engineering and Science Proceedings*, 23, 2002, 95-105

[25] Braganca, S.R. - Bergman, C.P.: Porcelain microstructure and technical properties. *Cerãmica*, 50, 2004, 291-299

[26] Ohya, Y. - Takahashi, Y.: Acoustic emission from a porcelain body during cooling. *J. Amer. Ceram. Soc.*, 82, 1999, 445-448

[27] Štubňa, I. - Trník, A. - Vozár, L.: Thermomechanical analysis of quartz porcelain in temperature cycles. *Ceramics International*, 33, 2007, 1287-1291

[28] Chmelík, F. -Trník, A. - Pešička, J. - Štubňa, I.: Creation of microcracks in porcelain during firing. *J. of the European Ceramic Society* (accepted article)

[29] Prjanishnikov, V.P.: *Sistema kremnezema*. Izd. lit. po stroitelstvu, Leningrad 1971 (in Russian)

[30] Pena, B. - Bartolomé, J. - Requena, J. - Moya, J.S.: Mullite-alumina functionally gradient ceramics, *J. Physique IV*, 1993, 3, 1261-1266

9

Practical Methods for Crack Length Measurement and Fatigue Crack Initiation Detection Using Ion-Sputtered Film and Crack Growth Characteristics in Glass and Ceramics

Gang Deng[1,2] and Tsutomu Nakanishi[1]
[1]*Faculty of Engineering, University of Miyazaki,*
[2]*State Key Lab of Mechanical Transmission, Chongqing University*
[1]*Japan*
[2]*China*

1. Introduction

A crack length measurement method is very important in the investigation of the fatigue crack growth characteristics of a material and in the evaluation of the fatigue strength of a machine element. Many crack length measurement methods have been proposed. Brown et al. (1966) and Nishitani et al. (1985) measured crack length through the changes in the displacement and compliance of the test piece. Ashida et al. (1996) and Nakai et al. (1989) investigated the changes in electrical properties around a crack during its growth. Lee et al. (1985) and JSME (1986) recommended estimating the crack length from the opening or closing displacement of a crack. In addition, crack length can also be measured by using an optical microscope, an optical grid technology (Bucci et al., 1972, Maustz et al., 1976, James et al., 1981), an acoustic emission method (Masuyama et al., 1994) and an ultrasonic method (Shimada et al., 1983). However, the methods proposed up to now require specific measuring apparatuses and complex calculation processes for crack length, and are applicable to long cracks several millimeters in length, not to small cracks shorter than 1mm in length or cracks with a high growth rate. In addition to the above-mentioned methods, a thin metal film, such as a crack gauge (e.g., KV-25B manufactured by Kyowa Co.), has been used for crack length measurement and fatigue crack initiation detection in some machine elements, such as gears, where cracks are up to several millimeters in length (Deng et al., 1991). Since this metal film is formed on a plastic base film with a thickness of about several tens of micrometers, there is a relatively large error between the actual and measured crack lengths. Increasing the measurement accuracy of the film can only be achieved by decreasing the thicknesses of the metal and plastic base films.

An ion-sputtering device, widely used in a scanning electron microscopy system, is a coating device used to form a nanoscale-thickness metal film on a surface. In this chapter, we introduce methods for crack length measurement using an ion-sputtered film, which are considered as ideal and practical methods applicable to micro-cracks or small cracks in glass and ceramics as well as metals. The measurement principle and measuring system are very simple. As an application of the ion-sputtered film to crack length measurement, cracks in soda-lime glass and alumina ceramics were measured and the microcrack growth characteristics of these materials were clarified. In addition, an ion-sputtered film was also applied to the detection of fatigue crack initiation, and a very small crack that initiated from the notch surface of the acrylic test piece during the fatigue test was detected..

2. Crack length measurement using grid pattern ion-sputtered film

2.1 Measurement principle

Conductive films have been used for crack length measurement. Ogawa et al. (1991a, 1991b) and Nakamura et al. (1993) utilized carbon films to measure the crack length on ceramic surfaces, but their methods are not widely used because the presented relationship between crack length and the resistance of the film is applicable only to the film of specific dimensions. A crack gauge is also a sensor for measuring crack length and has been used to measure the crack length of gears (Deng et al., 1991), but the measurement precision is not satisfied owing to the thickness of the gauge.

An ion-sputtering device, as shown in Figure 1, is commonly used to fabricate a conductive with a thickness of several tens of nanometers. The application of an ion-sputtered film to crack length measurement would contribute to increasing the crack length measurement precision.

Fig. 1. Ion-sputtering device

The electrical resistance R of a conductive film is expressed by

$$R = \rho \frac{l}{wt} \quad (1)$$

where l, w and t are the length, width and thickness of the film, respectively. ρ is the specific resistance of the film material. However, expression (1) is applicable only to a film with a simple shape. If there is a crack of length a across the film, attention should be paid on the fact that the electrical resistance cannot be calculated simply by substituting the width w of expression (1) with the residual width w-a; the theoretical expression for the electric resistance of a cracked film is nearly impossible. However, if the length of the film is very large compared with the width of the film, indicating that the shape of the film is similar to that of a grid whose length l is very larger than its width w, the electric resistance will not change before the crack entirely cuts off the film; the electric resistance is expressed by the following expression (Ogawa et al., 1991a).

$$R = \begin{cases} \text{constant} & a \neq w \\ \infty & a = w \end{cases} \tag{2}$$

Thus, at the instant when a crack grows across the film with a grid shape, the electric resistance of the film will increase to infinity and the complex calculation of the electric resistance of the ion-sputtered film with a crack can be avoided; referring to the crack gauge, a grid pattern ion-sputtered film should be available for crack length measurement (Deng et al., 2006). The electric resistance of the grid pattern ion-sputtered film will increase owing to crack growth, crack length can be determined by counting the number of the increasing steps in the electric resistance and measuring the positions of the grids after or before an experiment.

2.2 Bending test piece and crack length measurement

To investigate the precision of the crack length measurement method using a grid pattern ion-sputtered film, three-point bending fatigue tests for acrylic test pieces were performed. Figure 2 shows the loading method, the test piece and the grid pattern ion-sputtered film. In the experiment, first, a notch was introduced into the test piece to easily introduce the fatigue crack from the notch under a cyclic load. The crack growth experiment was performed under a cyclic load for a slow crack growth. The thickness of the grid pattern ion-sputtered film was about 10nm, the grid width was 0.2mm, and the interval between two grids was 0.1mm. The supporting span was 80mm, and the width and thickness of the test piece were 20mm and 10mm, respectively. The cyclic load was a sine wave of 10Hz and the load ratio (P_{min}/P_{max}) was approximately 0.02. A very simple measurement system used for recording the electrical resistance, which consists of a DC power source, a variable resistor and a voltmeter, is shown in Figure 3. The supply voltage was approximately 1-2V, the supply voltage and variable resistor were adjusted for a current below 100mA. The voltage on the film during the fatigue test was recorded by a data recorder. The sampling speed for the voltage was 80Hz.

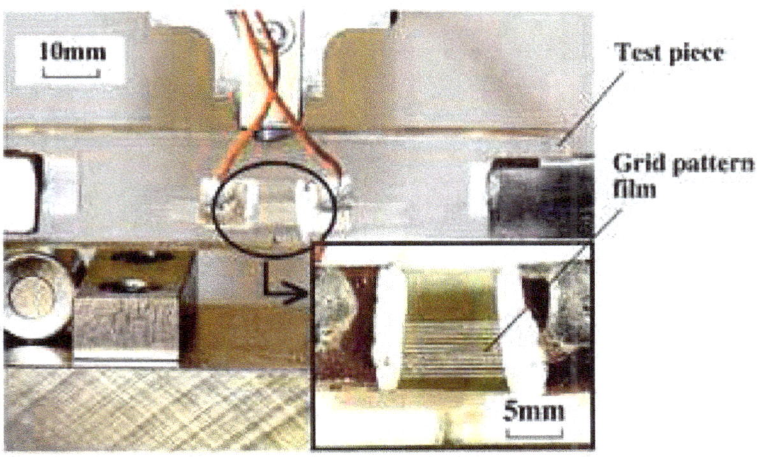

Fig. 2. Test piece and grid pattern ion-sputtered film

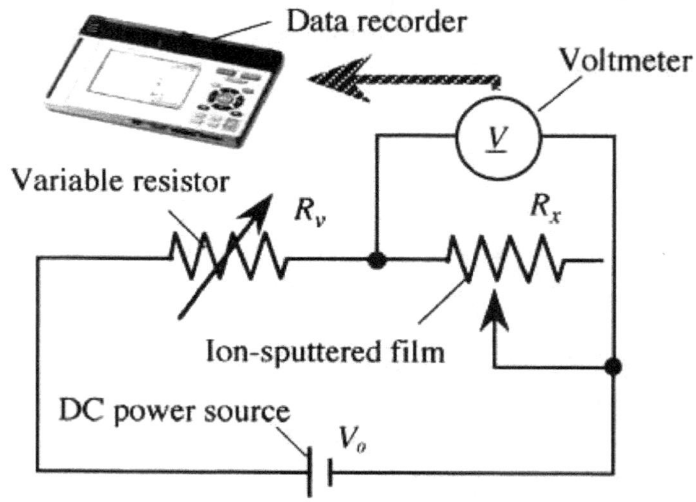

Fig. 3. Measurement system for the electrical resistance of ion-sputtered film

2.3 Precision of crack length measurement with grid pattern ion-sputtered film

The change in the voltage of the grid pattern film during crack propagation is shown in Figure 4. Although fluctuations in voltage signals due to the deformation of the film and the sampling error were observed, a distinct staircase pattern was recorded for the electric resistance. Considering the characteristic of the electric resistance change for a cracked grid as expressed in expression (2), the increasing step in Figure 4 shows the instant of grid snapping. From the positions of the grids that can be measured with an optical microscope before or after an experiment with a very high precision, the crack length at the instant of the increasing step can be accurately estimated. The precision of this method depends on the film thickness and the speed of the measuring system in response to the crack growth. Since the recorded signal upturns vertically and the film is extremely thin, this method can intermittently measure cracks with a very high precision at the instant of grid snapping.

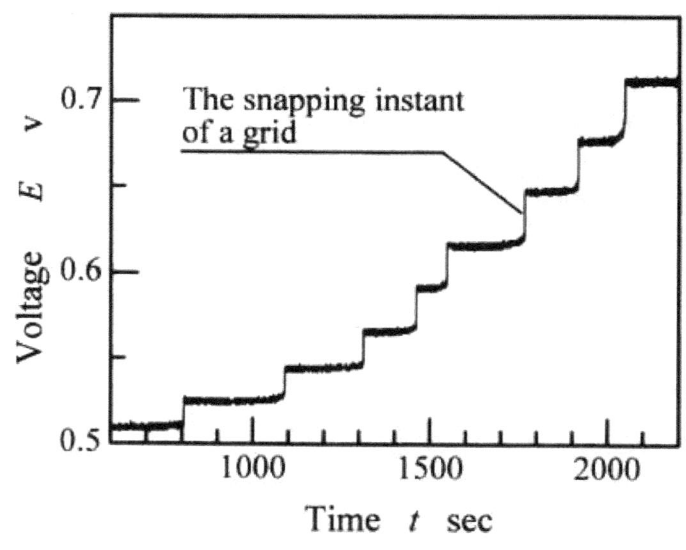

Fig. 4. Change in voltage of grid pattern film during crack propagation

For the application of the crack length measurement method using a grid pattern film to a rapidly growing crack, a fine-pitch grid pattern ion-sputtered film is desirable. Scribing the film using a sharp scriber tip, as shown in Figure 5, can produce grids of 25μm width and 50μm pitch. A film with 40 grids was fabricated and applied to a fatigue test. The change in the voltage of the film during crack propagation is shown in Figure 6, and 40 steps are identified clearly, indicating that 40 points of crack length can be obtained. Tokunaga et al. (2007) have succeeded in fabricating a grid pattern film using an excimer laser micromachining technology where the grid width is about 13μm and the pitch is 20μm for the measurement of cracks in ceramics.

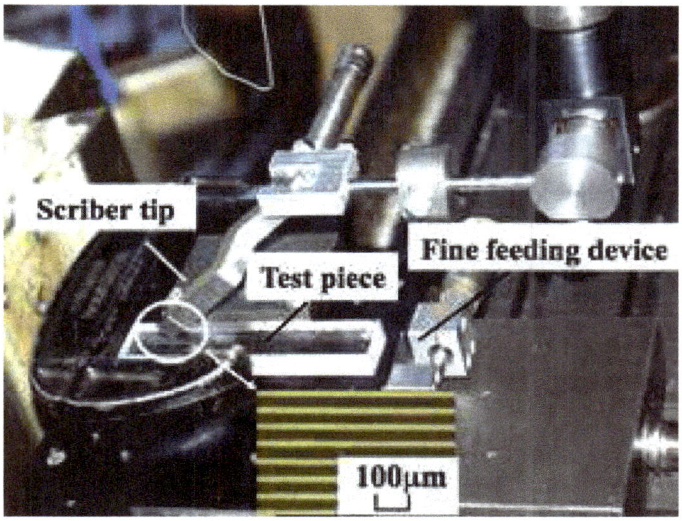

Fig. 5. Fabrication of fine grid film using scriber tip

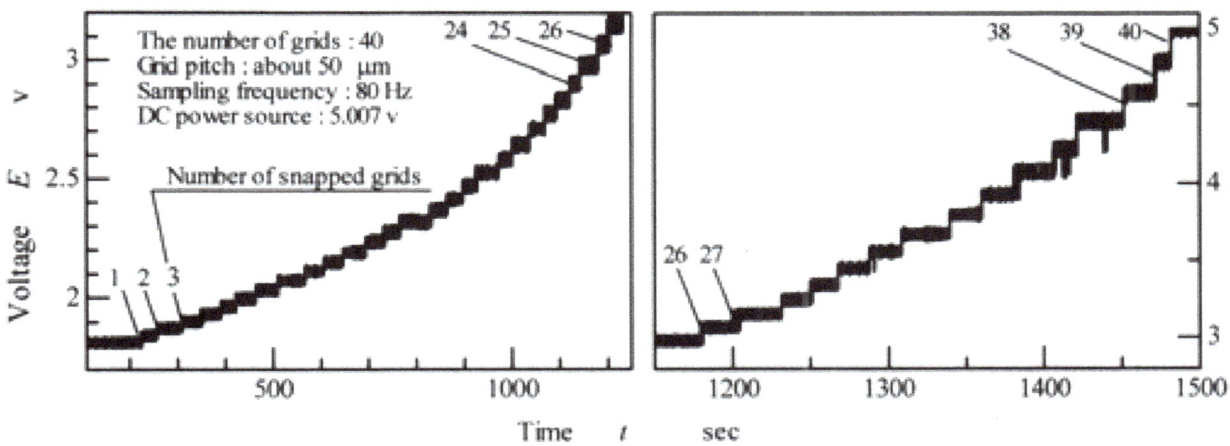

Fig. 6. Change in voltage of fine grid pattern film during crack propagation

3. Crack length measurement using rectangular ion-sputtered film

Although the grid pattern ion-sputtered film can measure crack length with a high precision, the crack length cannot be determined continuously. In addition, an isolation film is necessary for the application of the grid pattern film to a conductive surface; the formation of the grid pattern without destroying the isolation film would be very difficult. Thus, for measuring the crack length continuously and applying the ion-sputtered film to a conductive surface, such as a metal surface, a rectangular ion-sputtered film is desirable.

3.1 Calculation of electric resistance of cracked metal film

In the case of a cracked film under electric potential difference, the feature of electron flow around the crack, which is similar to that of stress concentration, is shown in Figure 7, and the electric resistance calculation becomes very complex. Ogawa et al. (1991a, 1991b) calculated the electric resistance of an across-cracked carbon film by FEM analysis and presented an approximating formula for the electric resistance of a regulated film where the crack length is in the range of 0.2-0.8w (w is the film width along the crack growth direction). Generally, the film width is several millimeters; thus, it is difficult to calculate the electric resistance of a film with a micro-crack shorter than 1mm using Ogawa's formula.

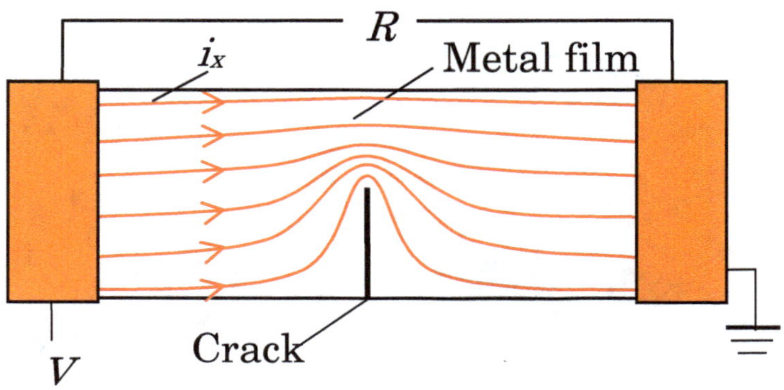

Fig. 7. Electron flows around crack

Electrothermal FEM is a practical method for calculating the electric resistance of the film with a crack (Deng et al., 2007). For example, when the mesh of a rectangular metal film and the boundary condition, shown in Figure 8, are prepared, the current density i_x of each element can be obtained by FEM analysis. The total current in the x direction can be calculated by integrating the current of each element, and the electric resistance can be obtained through the total current and electric potential difference on the basis of the law of ohm. The effect of the element size in FEM analysis on the calculation precision was investigated. Figure 9 shows the relationship between the calculated electric resistance and the number of elements for FEM analysis. Under the conditions that the length l of the film is 20.8mm, the width w of the film is 10.4mm and the crack length a is 0.4w, the electric resistance approaches a constant value when the number of elements is more than 5000. It is strongly recommended that the change in the calculated electric resistance with the increase in the number of elements should be investigated, and that an appropriate number of elements should be chosen. Many software programs, such as ANSYS and Marc, are available for electrothermal analysis.

Figure 10 indicates the electric resistance of a cracked ion-sputtered film on an acrylic piece, where crack length was measured using an optical microscope. The difference between the electric resistances obtained by measurement and FEM analysis is only 2-3%; thus, it can be considered that the electric resistance of a cracked film is calculated correctly by FEM analysis.

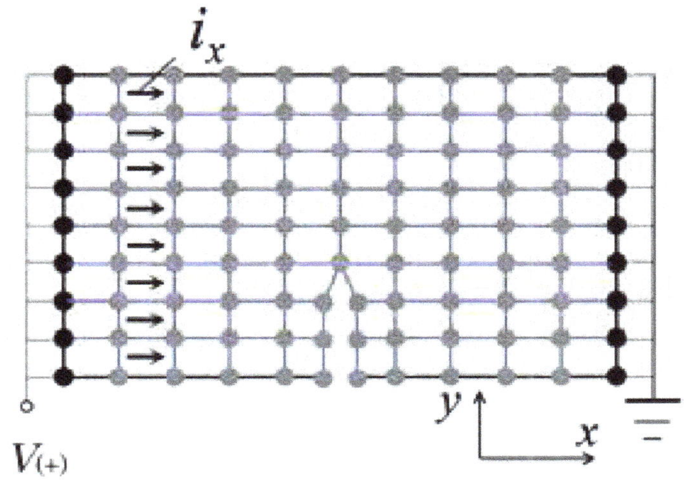

Fig. 8. Mesh of cracked film and boundary condition for electro-thermal FEM analysis

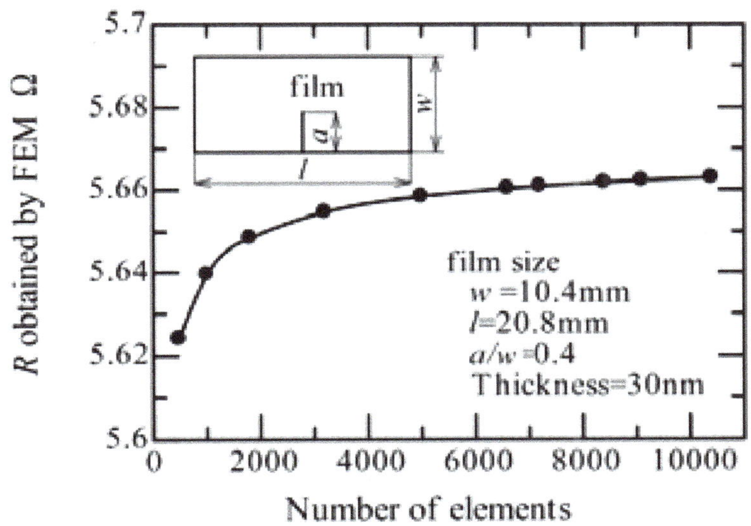

Fig. 9. Electric resistance value obtained under different number of elements

Despite the many approaches that can be used to find the expression for the relationship between the electric resistance and the crack length from 0 to the film width, a satisfactory expression was not obtained. Finally, considering practicability, the following expression is introduced to show the relationship in the crack length range from 0 to a portion of the film width.

$$R = R_0 + ca^b \tag{3}$$

Here, R (Ω) is the electric resistance of the cracked film, a is the crack length (mm), R_0 (Ω) is the electric resistance when a=0, and c and b are constants. Many calculations were carried out to determine the values of b for the various dimensions of the films. The results of the calculations revealed that b depends on the shape of the film. Figure 11 shows the values of b at different ratios of the film length l to the film width w. In the l/w range from 2.0 to 3.5, constant b is approximately equal to 2.268 on average.

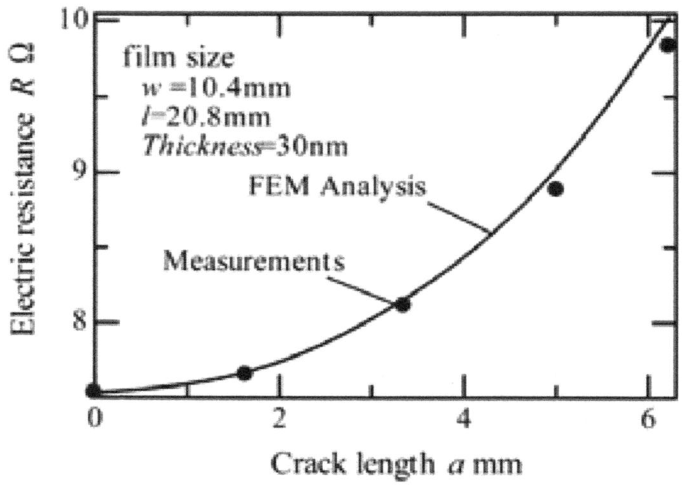

Fig. 10. Electric resistance obtained by measurement and FEM analysis

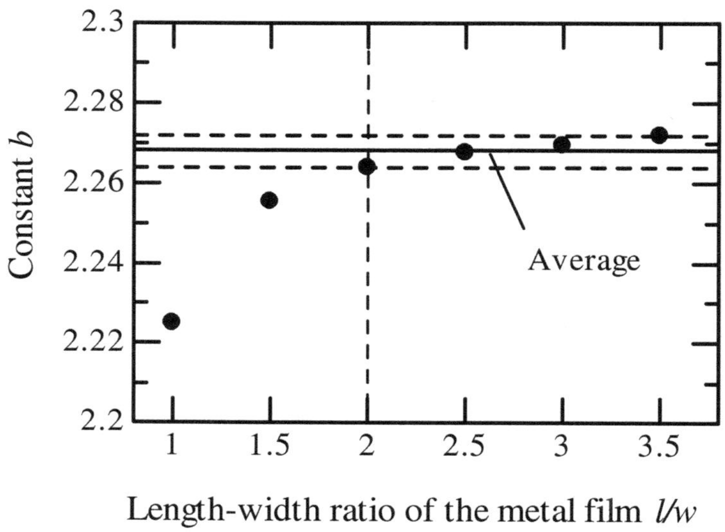

Fig. 11. Values of b under different shape of the film

On the other hands, c can be determined by stopping the experiment before the crack entirely cuts off the film, measuring the crack length a_e using an optical microscope, and substituting the electric resistance R_e and crack length a_e into expression (3). Several ion-sputtered films were used to verify the accuracy of expression (3). On the basis of the good agreement between the crack length estimated with expression (3) and measured using an optical microscope, the following expression is considered practical for estimating the electric resistance of a cracked film.

$$R = R_0 + ca^{2.268} \quad (0 \le a \le 0.6w,\ 2.0 \le \frac{l}{w} \le 3.5) \tag{4}$$

However, the physical meaning of expression (4) has not yet been clarified, and such an expression is only a curve-fitting formula and should be applicable to the condition shown in expression (4). Although this approximating expression can calculate the electric resistance of a cracked film in a specific crack length range, it is better to carry out FEM analysis for a good precision in the electric resistance calculation if a FEM software program is available; the necessary physical and electrical properties for FEM analysis can be obtained by using the initial electric resistance and dimensions of the film.

3.2 Crack length measurement using rectangular ion-sputtered film

Three-point bending tests were performed for acrylic test pieces, whose crack lengths were measured using a microscope and ion-sputtered films. The test piece were 100mm in length, 10mm in depth, and 20mm in height, and the supporting span in the tests was 80mm. A V-notch was introduced into the middle of the supporting span. The load ratio P_{min}/P_{max} was 0.02 and the load frequency was 10Hz. The experimental method and an ion-sputtered film on a test piece are shown in Figure 12. All tests were stopped when the crack grew to about 0.6*w* (*w*: film width) for the determination of the constant *c* in expression (4).

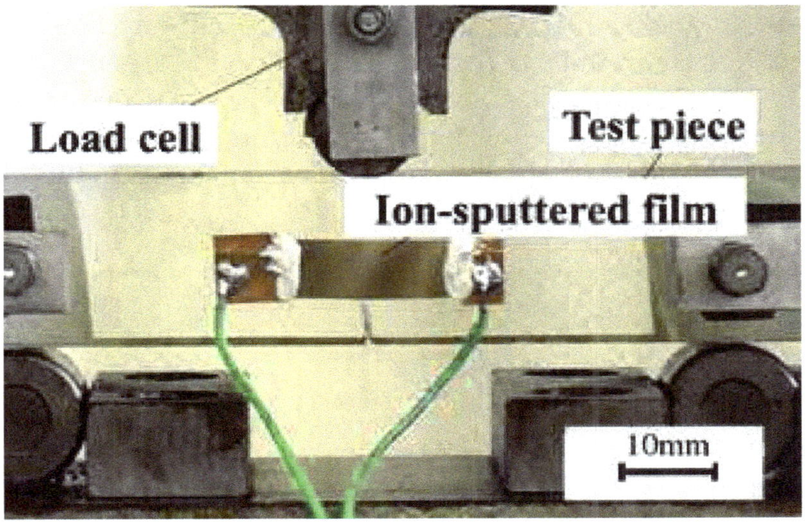

Fig. 12. Rectangular ion-sputtered film on acrylic test piece

The comparison of the crack lengths measured using an optical microscope and an ion-sputtered film is shown in Figure 13. The crack lengths are in very good agreement. Their maximum difference is 0.113mm and their average is only 0.051mm. A relatively large crack length difference occurs when the cracks are very short. This difference is considered to be due to the interruption of the experiment and taking off the test pieces form test rig for crack length measurement using an optical microscope that leads to the changes in the electrical environmental conditions around the test pieces. Although the precision of this rectangle is not as high as that of a grid pattern film, it is greatly higher than that of a commercial crack gauge (Deng et al., 1991). Therefore, a rectangular ion-sputtered film can be used to measure crack length continuously; short cracks or cracks with high growth rate, such as those in glass and ceramics, might be measurable by using a high-frequency sampling system.

4. Evaluation of crack growth characteristics of glass and ceramics based on crack length measurement using ion-sputtered film

It is known that fatigue failures also occur in structure components made of glass and ceramics under a constant load or a dynamic load (Hoshide, 1994). Since a small crack will grow more unstably and crack growth rate may increases more markedly within a small stress intensity factor range in brittle materials than in ductile metals, the understanding of the small crack growth characteristics in brittle materials, such as glass and ceramics, is very important for the prediction of the residual life. Many researches have been performed to clarify the relationship between the crack growth rate and the stress intensity factor for glass and ceramics using fatigue tests (Wakai et al., 1986, Takahashi et al., 1991, Sawaki et al., 1992, Aze et al., 2002, Yoda, 1989, Kimura et al., 1987). However, most of these researches used standard test pieces, such as compact test pieces or double-torsion test pieces, where the crack length is usually over several millimeters; thus, the crack growth characteristics known up to now are considered appropriately restricted to the large cracks that are impractical in the structure components made of glass and ceramics. For investigating the growth characteristics of a small crack with a length shorter than 1mm in glass and ceramics, a high-precision and high-speed crack length measurement method is necessary. Other methods, such as microscopy observation, the compliance method, the opening displacement method and the estimation methods based on the changes in electric properties around the crack are not applicable. However, the presented crack length measurement method using an ion-sputtered film makes it possible to measure the crack growing in glass and ceramics, since the ion-sputtered film is extremely thin that its electric resistance may change owing to a small increase in crack length, and a high-speed sample system is used for recording the electric resistance of the film.

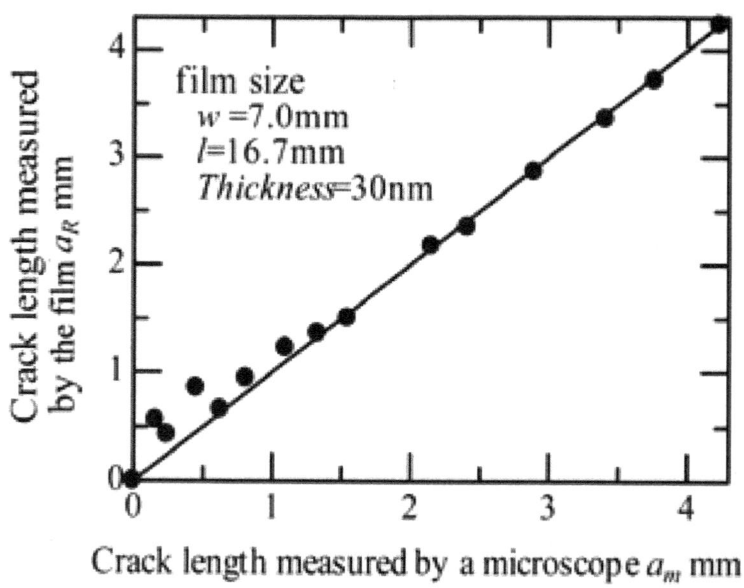

Fig. 13. Comparison of crack length obtained by microscope and ion-sputtered film

In this section, a concrete example of the application of a rectangular ion-sputtered film to clarifying small crack growth characteristics, namely, the relationship between the crack growth rate and the stress intensity factor, in soda-lime glass and alumina ceramics is discussed in detail.

4.1 Four-point bending test piece and fatigue test method

Four-point bending fatigue tests were performed under a constant load. The test pieces were made of soda-lime glass and alumina ceramics. The main chemical components and typical properties of the materials are shown in Table 1. The dimensions of the test pieces are shown in Figure 14. The test piece of soda-lime glass is 100mm long, 12mm wide (*B*) and 9.6mm thick (*W*), and the test piece of alumina ceramics is 100mm long, 12mm wide and 10mm thick. A precrack was introduced at the center of the lower surface of the test piece using a diamond indenter, and the ion-sputtered film was prepared traversing the crack growth directions. The film w and l, precrack length c_0, and distance s were measured using an optical microscope before the fatigue tests. The measurement error in the electric resistance of the cracked film was about 2-3% for the film of l/w=2.0 based on the results mentioned in section 3.1.

Material	Soda-lime glass		Alumina ceramics	
Main components wt%	SiO_2	72.0	Al_2O_3	99.5
	CaO	10.0		
	Na_2O	14.0		
Grain size μm	Amorphia		10	
Young's modulus GPa	71.6		363	
Bending strength MPa	49.0		441	

Table 1. Main chemical components and properties of the test materials

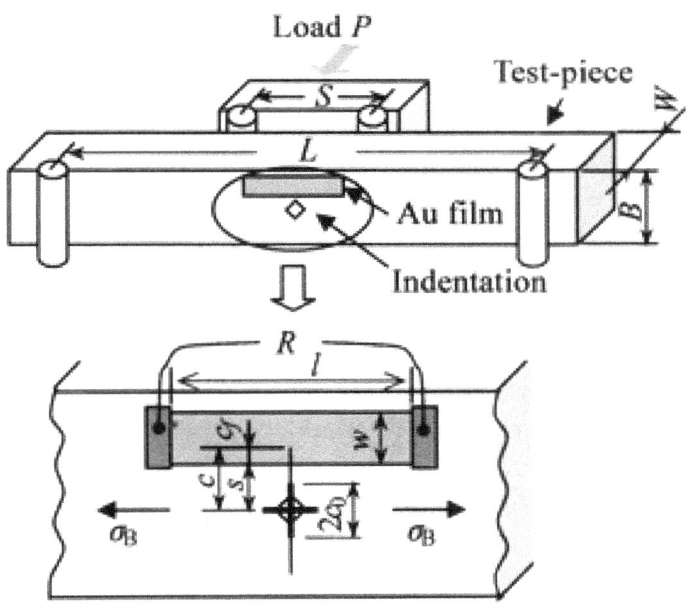

Fig. 14. Dimensions of four-point bending test piece

Static fatigue crack growth can be observed in glass and ceramic materials owing to stress corrosion. A constant load was applied to the test piece in this experiment. The loading and supporting methods are shown in Figure 14. The upper loading span (*S*) and lower supporting span (*L*) are 30mm and 90mm for the soda-lime glass, and 10mm and 25mm for the alumina ceramics, respectively. The simple system shown in Figure 3 was used to record the voltage on the ion-sputtered film, and the electric resistance of the film was calculated with the voltage and the set values of the other components in the measurement system. Since the crack grew at a high speed in glass and ceramics under a constant load, the recording frequency was 125kHz; thus, the crack length could be measured nearly continuously.

4.2 Calculation of stress intensity factor for surface crack and crack growth rate

A surface crack from a diamond indenter pit takes the shape of a semi-ellipse as shown in Figure 15. The stress intensity factor varies with the position at the edge of a crack. The maximum stress intensity factor at point *A*, which is used in the following discussion, is expressed as below (Kunio et al., 1984),

$$K_I = \frac{3.3P(L-S)\sqrt{\pi c(a_0 / c_0)}}{2BW^2\sqrt{1+1.46(a_0 / c_0)^{1.65}}} \tag{5}$$

here, P, S, L, W, B, a_0, and c_0 are the load, upper loading span, lower supporting span, test piece thickness, test piece width, initial crack length in the depth direction, and initial surface crack length, respectively. In expression (5), the crack length a, as shown in Figure 15, is substituted by $c(a_0/c_0)$ on the basis of the assumption that the aspect ratio λ, defined as the ratio of the crack length (depth) a to the surface crack length c namely, $\lambda=a/c$, is a constant during crack propagation. The aspect ratio λ can be obtained using the initial crack length a_0 and c_0 measured in the break section after the fatigue test using an optical microscope, and c is the surface crack length measured using the ion-sputtered film.

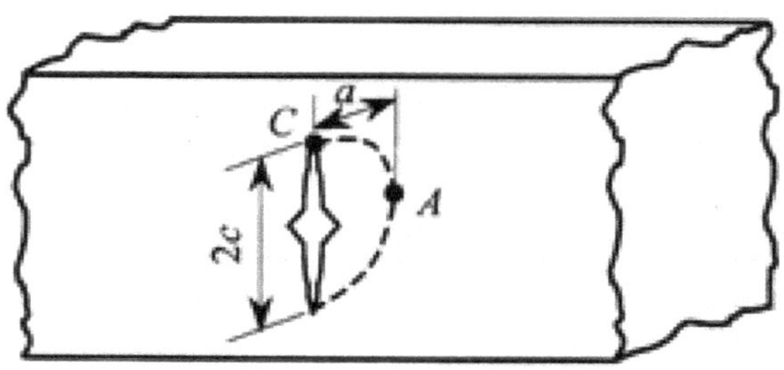

Fig. 15. Semi-ellipse surface crack

The surface crack length was derived from the electric resistance of the film based on the relationship between the crack length and the electric resistance of the ion-sputtered film obtained through the above-mentioned FEM analysis. To remove the effects of the sampling errors included in the electric resistance of the ion-sputtered film, as shown in Figure 16, on the surface crack length, the simple moving average method was used to obtain a smooth relationship between the surface crack length and the loading time, as shown in Figure 17. On the basis of the assumption that the aspect ratio λ is constant during crack propagation, the crack length a in the depth direction of the test piece was measured as the product of the surface crack length c multiplied by the aspect ratio λ, namely $a=c\cdot\lambda$, and then, the crack growth rate da/dt was determined by differentiating the relationship between the crack length a and the loading time t.

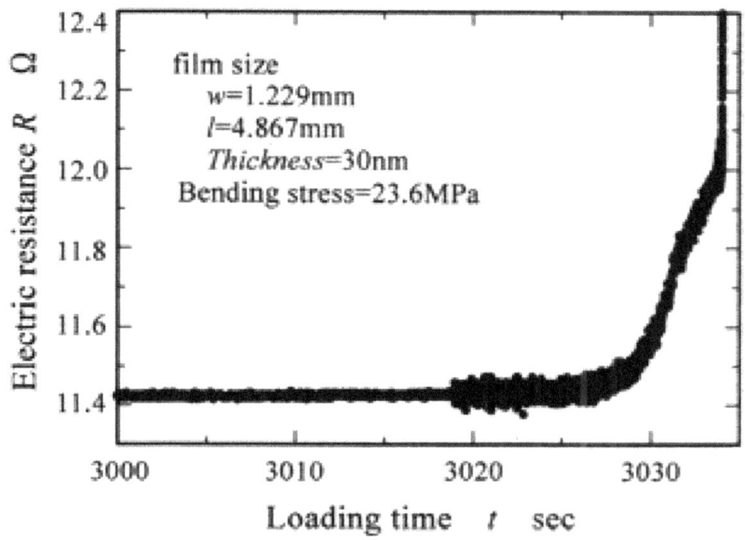

Fig. 16. Electric resistance of the ion-sputtered film during fatigue test for soda-lime Glass

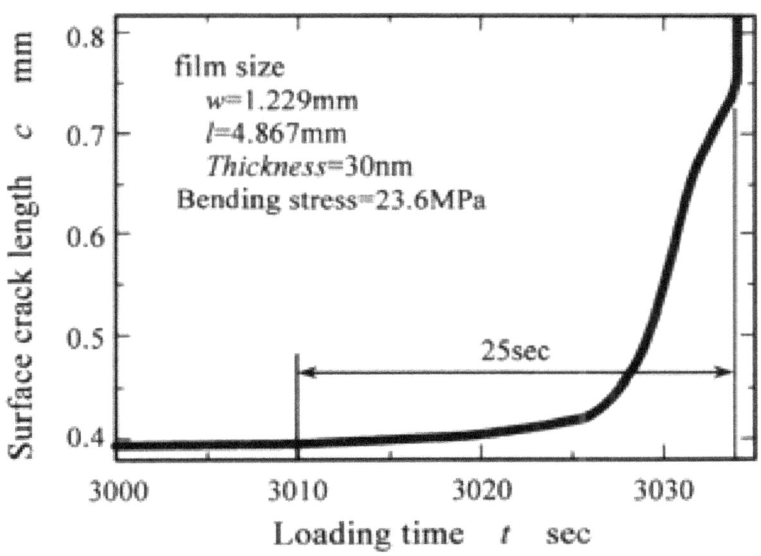

Fig. 17. Crack growth during fatigue test for soda-lime glass

4.3 Crack growth characteristics of soda-lime glass

The growth of a surface crack in the soda-lime glass test piece is shown in Figure 17. Fracture occurred when the crack length reached about 0.8mm; thus, the critical or unstable crack growth is expected to start from a small crack under a constant load. Moreover, crack growth time was about 25sec from the total fatigue time of about 3,040sec; the crack growth stage occupied less than 1% of the entire fatigue process.

Figure 18 shows the relationship between the crack growth rate and the stress intensity factor for the soda-lime glass under the bending stress of 22.8 and 23.6MPa. It has been very difficult up to now to measure the growth rate of a small crack, particularly the low and high (unstable) growth rates in brittle materials. In this experiment, since the ion-sputtered film can measure a small crack with high precision, and the sampling frequency of the measurement systems is so high that the surface crack can be measured almost continuously, a complete relationship including the three regions of crack growth was obtained successfully. In the stress intensity range from 0.45 to 0.85MPa•m$^{1/2}$, the crack growth rate changed from very low to very high in the soda-lime glass; thus, the unstable crack growth occurs easily in the soda-lime glass under a constant load.

Test piece		●	○	△
Bend stress MPa		23.6	22.8	22.8
Film size	w mm	1.229	0.946	0.495
	l mm	4.867	3.137	2.053
Film thickness nm		30	30	30

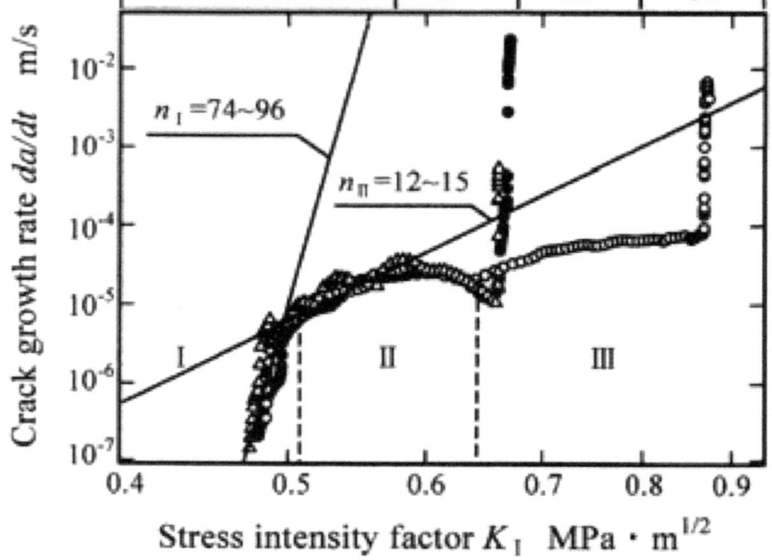

Fig. 18. Crack growth characteristics in soda-lime glass

On the basis of the Paris law indicating that the crack growth rate is proportional to the *n*th power of the stress intensity factor, that is $\frac{da}{dt} = cK_I^n$, the constant *n* was determined to range from 74 to 96 in region I, and from 12 to 15 in region II for the soda-lime glass. Yoda (1989) and Wiederhorn (1967) obtained *n* values ranging from 10 to 16 in the region of the crack growth rate from 1x10^{-6} to 1x10^{-4}, which agree with those obtained in region II of this experiment. The results of this experiment consequently clarify the crack growth characteristics of the soda-lime glass not only in region II but also in regions I and III; the crack length measurement method using a rectangular ion-sputtered film might be practical for investigating the crack growth characteristics of brittle materials, such as soda-lime glass.

4.4 Crack growth characteristics of alumina ceramics

The surface crack lengths measured during the test for alumina ceramics are shown in Figure 19. Brittle fracture occurred when the crack length reached about 0.55mm; thus, the critical or unstable crack growth is expected to start from a small crack under a constant load. Moreover, crack growth time was about 130sec from the total fatigue time of about 15,760sec; the crack growth stage occupied less than 1% of the entire fatigue process.

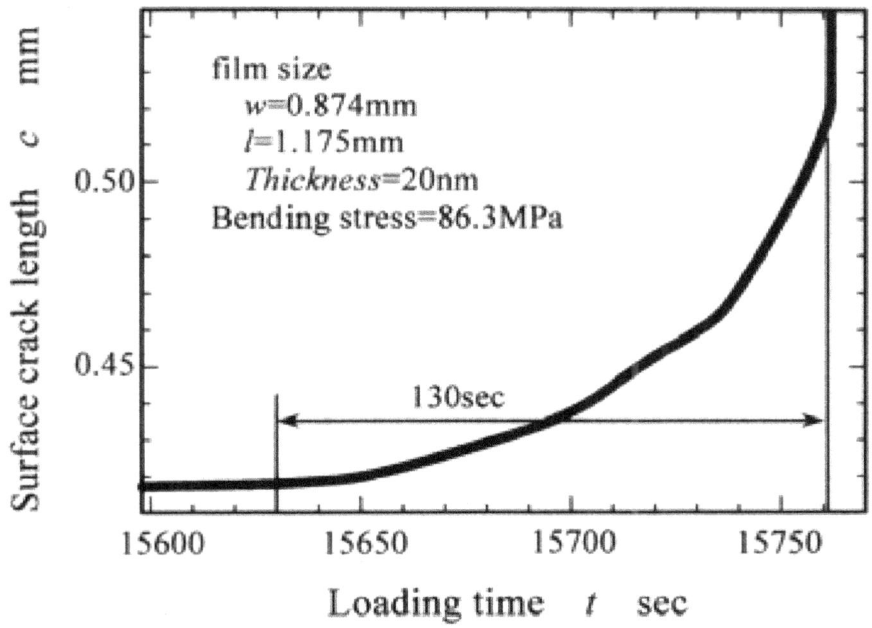

Fig. 19. Crack growth during fatigue test for alumina ceramics

Figure 20 shows the relationship between the crack growth rate and the stress intensity factor for the alumina ceramics under the bending stress of 86.3MPa. The surface crack length was also measured almost continuously, resulting in the successful determination of a complete relationship in the three regions of crack growth. The stress intensity factors range where the crack growth rate changes from low to high (unstable) in the alumina ceramics is very small (about 0.2 to 0.4 $MPa\cdot m^{1/2}$), which is similar to that of soda-lime glass; thus, the unstable crack growth occurs easily under a constant load. On the basis of the fact that the stress intensity factor for the crack growth in alumina ceramics is much higher than that in the soda-lime glass, alumina ceramics are considered to have a much higher fatigue strength than soda-lime glass.

Considering that the n values range from 530 to 580 in region I, and from 17 to 25 in region II for the alumina ceramics, the crack growth rate increases markedly owing to the small increase in stress intensity factor, particularly in region I. The n values obtained by Wakai et al. (1986) using double-torsion test pieces range from 84 to 130 in the crack growth rate range of $2x10^{-7}$ to $1x10^{-3}$, which is different from those obtained in this experiment. Considering that the crack lengths in this experiment are below 1mm, much shorter than those in the experiments of Wakai et al., the difference in n might be due to the fact that the crack growth characteristics depend on crack length.

The circle in region II in Figure 20 also indicates a decrease in crack growth rate. The shapes of the surface cracks in the soda-lime glass and alumina ceramics are observed using a microscope and shown in Figure 21. The crack grew linearly in the soda-lime glass but meandered in the alumina ceramics. Since the ion-sputtered film can only measure the crack length in the traversing direction of the film, the meandering of the crack would be seen as a decrease in crack growth rate.

Test piece		●	△
Bend stress MPa		86.3	86.3
Film size	w mm	0.743	0.874
	l mm	1.632	1.175
Film thickness nm		20	20

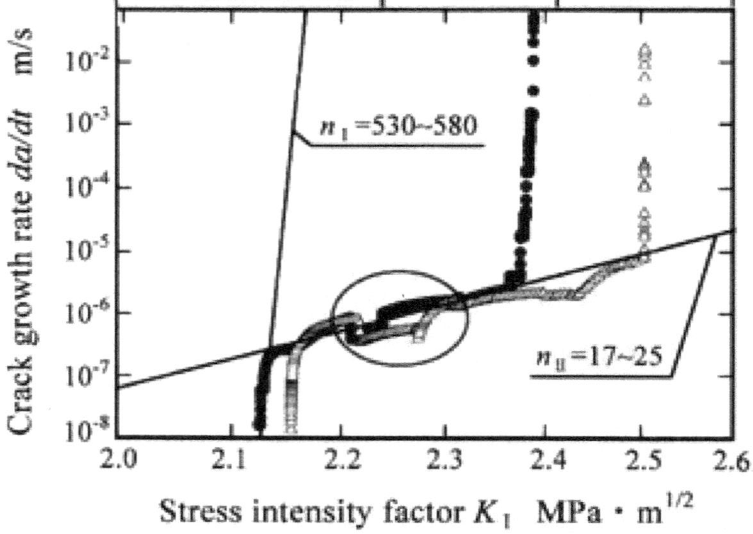

Fig. 20. Crack growth characteristics in alumina ceramics

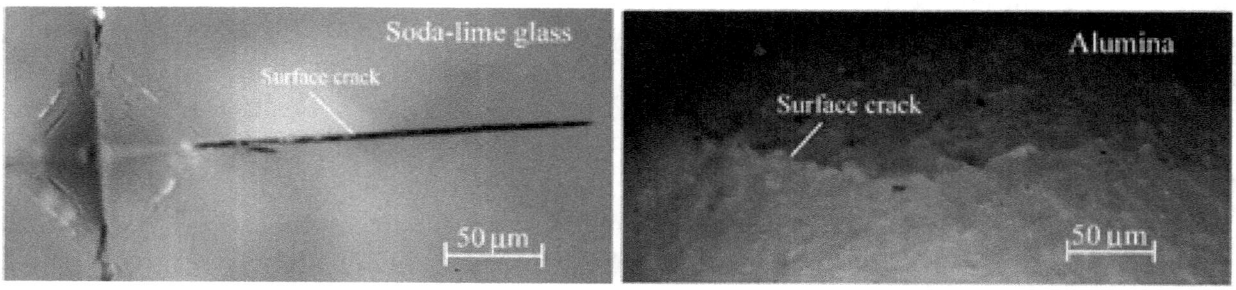

Fig. 21. Appearances of surface cracks in soda-lime glass and alumina ceramics

5. Application of crack length measurement using ion-sputtered film to metal surface

Ion-sputtered films can also be applied to crack length measurement for a metal surface if an insulating film could be fabricated between an ion-sputtered film and a metal surface. The insulating film should be as thin as possible for a high measurement precision. Many methods, such as the vacuum evaporation of silicon and glass, spray pyrolysis using ZnO, painting with quick-drying glue (Krazy Glue) and varnishing using an insulating spray paint, have been used to fabricate a thin insulating film on a steel surface. The method of varnishing using an insulating spray paint was selected finally for the high insulating resistance, thinness and fabrication convenience. The spray paint used in this experiment is the coating agent AY-302 (Sunhayato Co., Tokyo, Japan), which is commonly used in the insulation of printed circuit boards. The main component of the coating agent is polyvinyl resin. Wiping the steel surface with absorbent cotton wetted with the coating agent produces a good-quality and thin isolating film of about 1-2μm thickness. The insulating electric resistance between the ion-sputtered film and the steel surface was measured; its value was as high as several hundreds of ohm for an ion-sputtered film of about 5mm×10mm.

A sectional view of the insulating film, the ion-sputtered film and the wring method used for the crack length measurement on a steel test piece are shown in Figure 22. The insulating film should be fabricated after the setting of the terminal and wiring, and the ion-sputtered film should be fabricated immediately before the experiment.

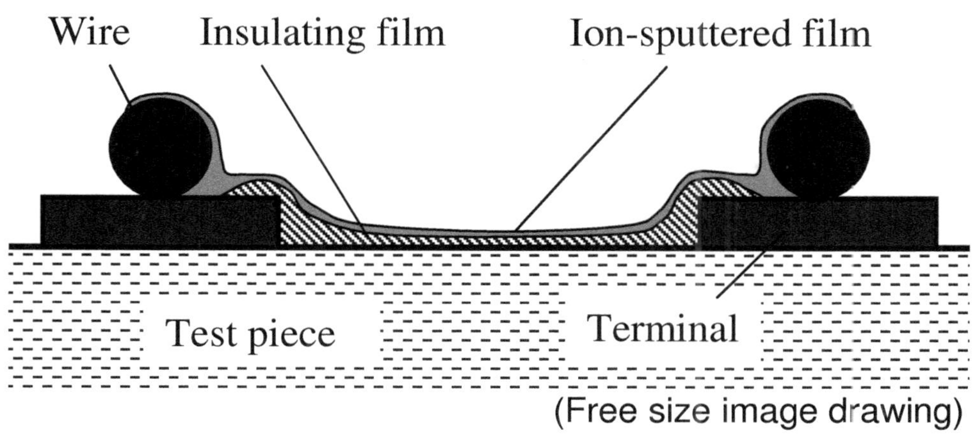

Fig. 22. Method applying ion-sputtered film to metal surface

6. Fatigue crack initiation detection method using ion-sputtered film

Many studies have been performed on the detection method for fatigue crack initiation that is very important in the evaluation of fatigue strength. Makabe et al. (1992, 1994a, 1994b) used a strain waveform for the detection of a fatigue crack with a length of 0.5 to 1mm on a round test piece with a small hemispherical pit. Katayama et al. (1996) and Lee et al. (2002) detected crack initiation using an AC potential method on a sharply notched test piece. Tohmyoh et al. (2001) detected the fatigue crack initiation of a test piece with a surface slit during rotating bending tests using surface SH (shear-horizontal) waves. Papazian et al. (2007) and Zilberstein et al. (2001, 2003, 2005) performed many studies on the application of a meandering winding magnetometer array sensor, ultrasonic techniques and an electrochemical fatigue sensor for the detection of crack initiation for many different types of fatigue test pieces. In addition to the above-mentioned methods, since the ion-sputtered film is very thin, a very short crack can change the electric resistance of the ion-sputtered film; thus, the ion-sputtered film would be practical for detecting the instant of fatigue crack initiation during a fatigue test.

6.1 Principle of fatigue crack initiation detection method

The ion-sputtered film is as thin as several tens of nanometers; thus, a small surface crack is expected to crack the film and change the electric resistance of the film, as shown in Figure 23. The starting point of the increase in electric resistance corresponds to the instant of crack initiation.

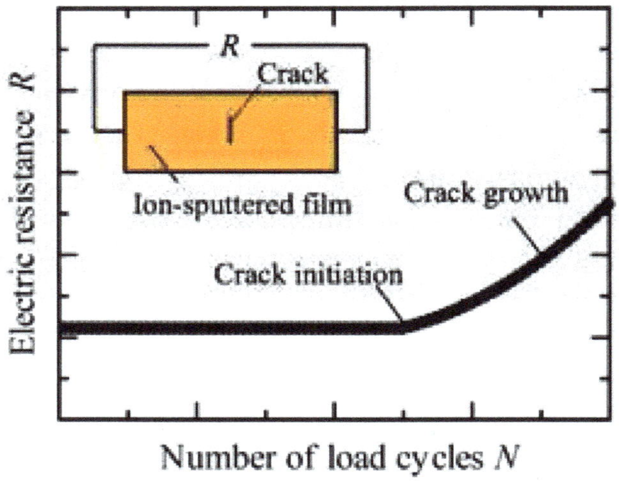

Fig. 23. Change in electric resistance of ion-sputtered film due to crack initiation and crack growth

6.2 Test piece for fatigue crack initiation detection

Three-point bending fatigue tests were performed to investigate the effectiveness of the ion-sputtered film in crack initiation detection. Considering that the crack initiates from stress concentration regions in machine elements, such as shafts and gears, and the convenience of fabricating an ion-sputtered film, the notch in the test piece should be round instead of being V- or U-shaped. The notches of the test pieces were in the shape shown in Figure 24, and crowned so that the crack could initiate at their centers. The test pieces were 100mm long, 10mm thick and 20mm high, and the supporting span in the tests was 80mm. The radius of the round notch was 4mm, the depth of the notch top *T* was about 5mm and the crowning radius *R* was about 5mm.

The test pieces were made of acrylic. The ion-sputtered film was formed on the notch surface. The length of the film in the direction of the test piece length was about 10mm, and the width and the thickness of the film were about 5mm and 10-50nm, respectively. The initial electric resistance of the film was lower than 20ohm. As shown in Figure 3, the system was used to record the voltage on the ion-sputtered film.

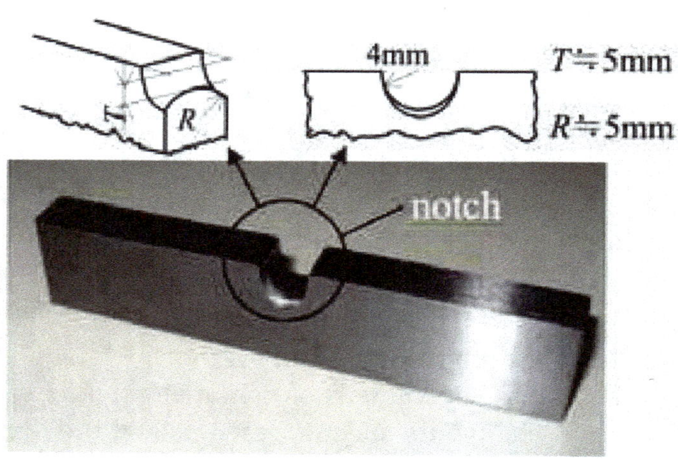

Fig. 24. Test piece for crack initiation detection test

6.3 Durability of ion-sputtered film

Whether the change in electric resistance or physical damage to the film occurs during the fatigue test before crack initiation was investigated to confirm the suitability of the film. The film was subjected to an endurance test under a light load without fatigue breaks. Figure 25 shows the electric resistance of the film during the test. It is obvious that no physical damage occurred in the film during the test, although a slow decrease in the electric resistance of about 0.5ohm was observed under load cycles of 1×10^7 (about 90hr). Gradual changes in electric resistance also occurred during some of the subsequent experiments; the reasons for these changes have not yet been clarified. However, considering that the film was made of pure gold (Au) without deteriorations of material properties due to chemical reactions under these test conditions, and that the changes were much slower than that due to crack initiation, the gradual changes might be caused by some unknown environmental factors or the electric shifts of the apparatuses in the measurement system. On the basis of the above-mentioned results, the gradual changes are considered not so severe that the judgment of crack initiation is prevented.

6.4 Change in electric resistance due to crack initiation

The electric resistance of the ion-sputtered film on an acrylic test piece during the fatigue test is shown in Figure 26. The maximum load was 400N, the loading frequency was 10Hz and the film thickness was about 30nm. The experiment was stopped manually after recognizing a rapid increase in voltage signal on the monitor of the data recorder. Figure 26 clearly shows an increase in the electric resistance of the film, indicating crack initiation. Despite the fact that gradual changes in the electric resistance of the film sometimes occur, as shown in Figure 25, the increase in electric resistance due to crack initiation is clear compared with the gradual changes shown in Figure 25, and the crack initiation point can be confirmed from Figure 26 by observing the upward tendency of the electric resistance. The point where the electric resistance leaves the extension line of the mean electric resistance is considered to be the start point of crack initiation. The notch after the experiment is shown in Figure 27, and a very small round crack, the crack in the primary stage of fatigue, can be observed clearly.

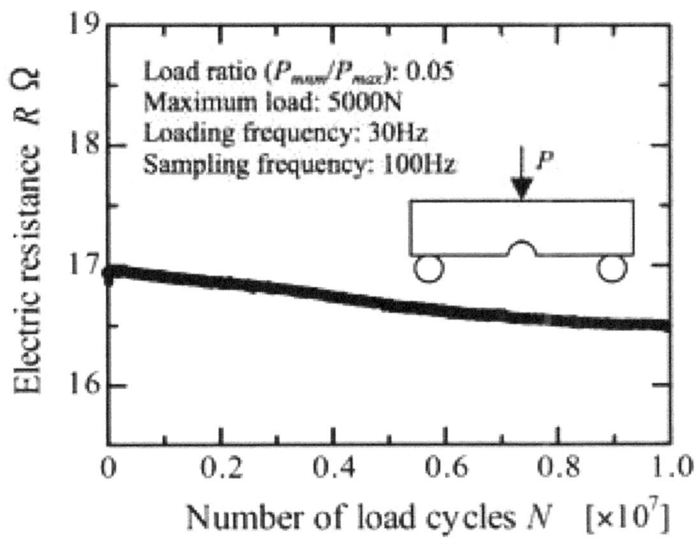

Fig. 25. Electric resistance of ion-sputtered film during the test without crack initiation

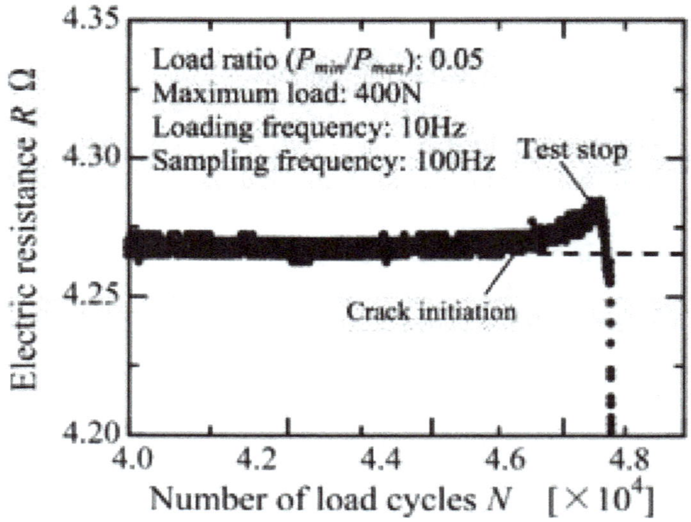

Fig. 26. Change in electric resistance of ion-sputtered film when fatigue crack initiation

Fig. 27. View of micro-crack initiated on notch surface

6.5 Shape of detected crack in primary stage of fatigue

A small sample around the notch was quarried out from the test piece to determine the crack front shape. The sample was cut bit by bit from the side using a milling machine. The depth of the sample d and the crack length a on the side surface were measured using a micrometer and an optical microscope, respectively. The shape of the crack front is shown in Figure 28. It can be seen that the shape of the crack front is approximately half an ellipse, the length of the crack on the notch surface is about 0.35mm and the maximum depth is about 0.2mm. The half ellipse shape indicates that the crack is in the primary stage of fatigue. It is slightly difficult to say that the crack shown in Figure 28 is the just-initiated crack since the dimensions of the crack are rather large. The large dimensions are due to the time lag between the judgment of crack initiation from the change in electric resistance and the termination of the test, not due to the sensitivity of the ion-sputtered film because the measurement error of the film is as small as 0.068mm (Deng et al., 2007). From the change in the electric resistance of the ion-sputtered film and the shape of the detected crack shown in

Figures 26, 27 and 28, it can be concluded that the ion-sputtered film can be used to detect the initiation of a fatigue crack. Considering the convenience to prepare the detecting system, this method is simple and practical for monitoring fatigue crack initiation.

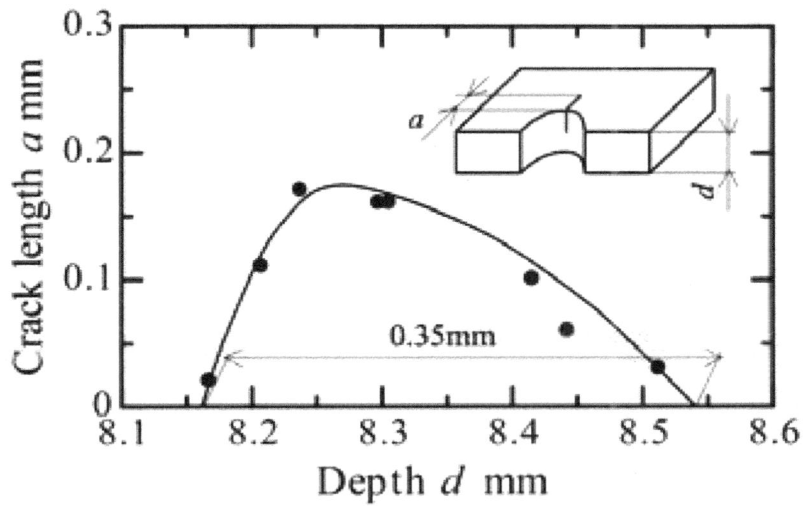

Fig. 28. Shape of crack initiated in acrylic test piece

7. Conclusions

The methods for crack length measurement and fatigue crack initiation detection using an ion-sputtered film based on the changes in the electric resistance of the film due to crack growth or crack initiation are presented. On the basis of that the experimental results of crack length measurement and crack initiation detection, and considering the convenience to prepare the measurement and monitoring systems, the presented methods are very practical for investigating crack growth characteristics and crack initiation life.

One of the crack length measurement methods is the use of a grid pattern ion-sputtered film, which is similar to a crack gauge. Since the electric resistance of the film changes only at the point when a grid is entirely snapped by a crack, the calculation of the electric resistance is not necessary, the crack length can be obtained by counting the number of increasing steps in the recorded electric resistance and measuring the positions of the grids. Because the thickness of the ion-sputtered film is only several tens of nanometers, the change in the electric resistance of the film due to grid snapping is more immediate than that of the commercial crack gauge. Grid pattern ion-sputtered films were applied to the crack length measurement for a three-point bending test piece. Very obvious staircase changes in the electric resistance of the film were recorded.

Another crack length measurement method is the use of a rectangular ion-sputtered film, which can measure crack length continuously. The relationship between the electric resistance of the film and crack length is obtained by electrothermal FEM analysis, and an approximate expression is proposed to calculate the electric resistance of a cracked film using the results of electrothermal FEM analysis. Since the film is extremely thin, it is very sensitive to the increase in the length of a small crack and practical for investigating a small crack with high growth rate. As the application of the crack length measurement method using a rectangular ion-sputtered film, cracks in soda-lime glass and alumina ceramics were measured and the crack growth characteristics in these materials were clarified continuously from the crack growth region I to the crack growth region III.

In addition to crack length measurement, an ion-sputtered film can also be applied to the detection of fatigue crack initiation, which is a very important technique for evaluating the fatigue strength of a material. Since the film is extremely thin and can be fabricated directly on a surface, a very small crack can change its electric resistance. As an attempt to confirm the fatigue crack initiation life, an ion-sputtered film was formed on the round notch surface of a three-point bending acrylic test piece; a very small crack that initiated from the notch surface during the fatigue test was detected. From the change in the electric resistance of the ion-sputtered film during the test and the shape of the detected crack, it can be concluded that the ion-sputtered film can be used to detect the initiation of a fatigue crack.

8. References

Ashida, U., Ishihara, T., Fujihara, H. & Tobita, H. (1996). Automatic Measurement of Fatigue Crack Length by a D.C. Electrical Potential Difference Method, *Journal of The Kansai Society of Naval Architects (in Japanese)*, 225, 181-185

Aze, A.D., Chevalier, J., Fantozzi, G., Schehl, M. & Torrecillas, R. (2002). Crack Growth Resistance of Alumina, Zirconia and Zirconia Toughened Alumina Ceramics for Joint Prostheses, *Biomaterials*, Vol.23, No.3, 937-945

Brown, T.F. & Srawley, J.E. (1966). In Plane Strain Crack Toughness Testing of High Strength Metallic Materials, *ASTM STP410*

Bucci, R.J., Paris, P.C., Hertzberg, R.W., Scbmidt, R.A. & Anderson, A.F. (1972). Fatigue Threshold Crack Propagation in Air and Dry Argon for a Ti-6Al-4V Alloy, *ASTM STP513*, 125-140

Deng, G., Inoue, K., Takatsu, N. & Kato, M. (1991). Evaluation of the Strength of Carburized Spur Gear Teeth Based on Fracture Mechanics, *Trans. Jpn. Soc. Mech. Eng. (in Japanese)*, C-57-535, 909-913

Deng, D., Tokunaga, H., Daniel-Redda, T., Ikeda, K., Nakanishi, T. & Kaizu, K. (2006). Crack Length Measurement with an Ion Sputtered Film, *Trans. Jpn. Soc. Mech. Eng. (in Japanese)*, 72-715, 997-1002

Deng, D., Nasu, K., Kuroia, S. & Nakanishi, T. (2007). Crack Length Measurement with an Ion Sputtered metal Film, *Trans. Jpn. Soc. Mech. Eng. (in Japanese)*, 73-727, 849-854

Hoshide T. (1994). Fatigue in Ceramics I, *Journal of The Society of Materials Science*, Vol.43, No.490, 902-908

James, L.A. (1981). Specimen Size Considerations in Fatigue Crack Growth Rate Testing, *ASTM STP738*, 45-57

JSME (1986). JSME Standard, *Jpn. Soc. Mech. Eng.*, S-001, 8-9

Katayama, Y., Sakane, M. & Ohnami, M. (1996). Surface Crack Detection by A. C. Potential Drop Method; Experiment and FEM Considerations, *Trans. Jpn. Soc. Mech. Eng. (in Japanese)*, Vol.62-602, 2216-2223

Kimura, Y., Yamamoto, S., Sekiya, M. & Kunio, T. (1987). Crack Growth Characteristics of Vickers Indented Glass Plate Specimen in Water, *Journal of The Society of Materials Science*, Vol.36, No.401, 160-165

Kunio, T., Nakazawa, K., Hayashi, I. & Okamura, H. (1984). *(Methods for Fracture Mechanics), (in Japanese)* , 251-252

Lee, J.H. & Kobayashi, H. (1985). Detection and Closure Measurement of Short Fatigue Crack Initiated at Notch Root, *Trans. Jpn. Soc. Mech. Eng. (in Japanese)*, A-51-461, 142-147

Lee, Y. & Sakane, M. (2002). Multiple Surface Crack Detection Using A. C. Potential Drop Method, *Trans. Jpn. Soc. Mech. Eng. (in Japanese)*, Vol.68-672, 1220-1227

Makabe, C., Nishida, S., Kaneshiro, H. & Tamaki, S., (1992). Method of Detecting Fatigue Crack Initiation through Analysis of Strain Waveform, *Trans. Jpn. Soc. Mech. Eng. (in Japanese)*, Vol.58-551, 1191-1195

Makabe, C., Kaneshiro, H., Nishida, S. & Urashima, C. (1994a). Detection of 1mm Fatigue Crack Initiation Using Strain Waveform, *Journal of engineering Materials and technology, Trans. of ASME*, Vol.116, 483-487

Makabe, C., Nishida, S., Urashim, C. & Kaneshiro, H. (1994b). Detection of Fatigue Crack Initiation under a Random Load, *Trans. Jpn. Soc. Mech. Eng. (in Japanese)*, Vol.60-576, 1753-1760

Masuyama, T., Inoue, K. & Kato, M. (1994). Acoustic Emission during Fatigue Crack Growth in Carburized Gear Tooth, *Trans. Jpn. Soc. Mech. Eng. (in Japanese)*, C-60-575, 2456-2461

Mautz, J. & Weiss, V. (1976). Mean Stress and Environmental Effects on Near Threshold Fatigue Crack Growth, *ASTM STP601*, 154-168

Nakai, Y., Akagi, H., Kitamura, Y. & Ohji, K. (1989). Measurement of Short Surface Crack Length by an AC Potential Method, *Trans. Jpn. Soc. Mech. Eng. (in Japanese)*, A-55-511, 543-548

Nakamura, M., Saitoh, K., Ikeyama, M., Kozuka, T. & Shigematsu, I. (1993). A Study on the Utilization of an Electro-Conductive Surface Layer on a Si_3N_4 Ceramic Formed by Ion Implantation as a Crack Gage and a Strain Gage, *Journal of the Ceramic Society of Japan (in Japanese)*, 101-1, 139-142

Nishitani, H. & Chen, D. (1985). A Consideration on the Unloading Elastic Compliance Method, *Trans. Jpn. Soc. Mech. Eng. (in Japanese)*, A-51-465, 1436-1441

Ogawa, T. & Suresh, S. (1991a). Surface Film Technique for Crack Length Measurement in Nonconductive Brittle Materials: Calibration and Evaluation, *Engineering Fracture Mechanics*, Vol.39-4, 629-640

Ogawa, T. (1991b) Fracture Resistance Curve and Fatigue Crack Growth in Polycrystalline Magnesia, *Trans. Jpn. Soc. Mech. Eng. (in Japanese)*, A-57-535, 492-499

Papazian, J. M., Nardiello, J., Silberstein, R., Welsh, G., Grundy, D., Craven, C., Evans, L., Goldfine, N., Michaels, J.E., Michaels, T.E., Li, Y. & Laird, C. (2007). Sensors for Monitoring Early Stage Fatigue Cracking, *International Journal of Fatigue*, Vol.29, 1668-1680

Sawaki, Y., Nagase, Y., Yoshida, H., Inoue, A. & Fujiwara, T. (1992). Crack Propagation Behaviour in Alumina Ceramics under Static and Cyclic Loading, *Trans. Jpn. Soc. Mech. Eng., Series A*, Vol.58, No.552, 1333-1338

Shimada, H. & Date, K. (1983). Crack Depth Measurement with High Accuracy by Ultrasonics, *The Iron and Steel Institute of Japan (in Japanese)*, 69-2, 196-202

Takahashi, M. & Mutoh, Y. (1991). Static and Cyclic Fatigue Crack Growth in Several Ceramic Materials, *Trans. Jpn. Soc. Mech. Eng., Series A*, Vol.57, No.543, 2615-2621

Tohmyoh, H., Ochi, Y. & Matsumura, T. (2001). Study on Detection and Quantitative Evaluation of Fatigue Cracks Using Surface SH Waves, *Trans. Jpn. Soc. Mech. Eng. (in Japanese)*, Vol.67-661, 1508-1513

Tokunaga H., Ikeda K., Kaizu K. & Kinishita H. (2007). Micro-Crack Length Measurement Method in Ceramics by Using Grid Pattern Metal Film, *CD Proceedings of International Conference an Advanced Experimental Mechanics 2007*, 1-5

Wakai, F., Sakuramoto, H., Sakaguchi, S. & Matsuno, Y. (1986). Evaluation of Crack Propagation in Ceramics by Double-Torsion, *Journal of The Society of Materials Science*, Vol.35, No.395, 898-903

Wiederhorn, S. M. (1967). Influences of Water Vapor on Crack Propagation in Soda-lime Glass, *Journal of American Ceramics Society*, 50(8), 407-414

Yoda, M. (1989). Subcritical Crack Growth Characteristics of Soda-Lime Glass Evaluated from CT Specimen and Bend Specimen with Small Indented Crack, *Journal of the Ceramic Society of Japan*, Vol.97, No.9, 960-964

Zilberstein, V., Schlicker, D., Walrath, K., Weiss, V. & Goldfine N. (2001). MWM Eddy Current Sensors for Monitoring of Crack Initiation and Growth During Fatigue Tests and In Service, *International Journal of Fatigue*, Vol.23, S477-S485

Zilberstein, V., Walrath, K., Grundy, D., Schlicker, D., Goldfine, N., Abramovici, E. & Yentzer, T. (2003). MWM Eddy-current Arrays for Crack Initiation and Growth Monitoring, *International Journal of Fatigue*, Vol.25, 1147-1155

Zilberstein, V., Grundy, D., Weiss, V., Goldfine N., Abramovici, E., Newman, J. & Yentzer, T. (2005). Early Detection and Monitoring of Fatigue in High Strength Steels With MWM-Arrays, *International Journal of Fatigue*, Vol.27, 1644-1652

10

Luminescence Properties of AlN Ceramics and Its Potential Application for Solid State Dosimetry

Laima Trinkler and Baiba Berzina
Institute of Solid State Physics, University of Latvia
Latvia

1. Introduction

Aluminum nitride (AlN) is a wide band material (Eg = 6.2 eV) with a wurtzite structure. It has already found practical application in microelectronics as substrate, insulator and packaging material due to combination of the uppermost qualities, such as high thermal conductivity, good dielectric properties and thermal expansion coefficient comparable with that of silicon. There are different modifications of the material used for application and investigation purposes: single crystal, powder, nanostructures, thin films and so on. AlN ceramics is one of them, it is easy to produce and to handle, thereat in the form of ceramics AlN maintains the main properties of the material.

The objective of the study is investigation of luminescence properties of AlN ceramics including elucidation of luminescence centers and mechanisms as well as estimation of potential application of the material in the field of solid state dosimetry. This chapter is based on our papers devoted to luminescent and dosimetric properties of AlN ceramics published in 1998-2009 and quite recent measurements performed at low temperatures.

2. Experimental details

2.1 Samples of AlN ceramics

Samples of high-density AlN ceramics were obtained in Institute of Inorganic Chemistry, Riga Technical University, Latvia (Palcevskis et al., 1997; 1999). AlN powder with specific surface area in the range 20-40 m^2/g and Y_2O_3 powder (30 m^2/g), both produced by plasma synthesis, were used for production of ceramics. Directly after synthesis AlN powders were kept in organic solvent. AlN and Y_2O_3 powders (2-9 wt.%) were mixed in organic media. Then powder mixtures were dried and pressed in cylinders (12 mm in diameter) at 200 MPa. After burning out organic components in vacuum at 500 °C cylinders were embedded into alumina crucibles and placed in a tungsten resistance furnace. The sintering was carried out in nitrogen atmosphere (0.12 MPa) at temperature 1600-1800 °C during 300-900 min. The produced polycrystalline ceramic cylinders were sliced into 1 mm thick tablets, suitable for optical measurements.

Sintered samples were characterized by the producers as follows (Palcevskis et al., 1999). Size of polycrystalline grains is around 4 μm. Depending on sintering conditions (temperature and time) density varies from 3.2 to 3.4 g/cm^3 and thermal conductivity - from 80 to 260 W/mK. AlN grains constitute more than 90% of the volume of the sintered samples; the rest volume belongs to secondary phases: $Y_4Al_2O_9$, $AlYO_3$ and $Y_3Al_5O_{12}$.

Though nominally pure, AlN always contains oxygen impurity as a natural dopant, which substitutes nitrogen in regular lattice sites and forms oxygen-related defects of various types. The total oxygen content in the studied AlN ceramic samples determined by the neutral activation analysis is from 0.1 to 0.7 wt.%. Alongside with oxygen impurity aluminum vacancies are generating for maintenance of charge balance. Besides there are detected other uncontrolled impurities of small concentration (< 1000 ppb) such as Fe, Cr, Ni and Zr. The further spectral measurements have proved the presence of Mn impurity, too. Oxigen and other impurities contribute to luminescence properties of the material.

In some of experiments we used AlN ceramics samples subjected to oxygen ion implantation. This procedure was done in Taiwan National University with doses 10^{14} and 10^{15} ion/cm^2.

2.2 Methods of photoluminescence, thermoluminescence and optically stimulated luminescence

Luminescence properties of AlN ceramics were studied using methods of photoluminescence (PL), thermoluminescence (TL) and optically stimulated luminescence (OSL). Photoluminescence is luminescence of material revealed under irradiation with ultraviolet (UV) or visible light. We have used such characteristics of PL as emission spectra obtained under irradiation with a fixed wavelength from UV region and excitation spectra - dependence of luminescence intensity at a fixed emission wavelength upon the varied wavelength of excitation light.

Irradiation of wide band gap materials with UV light and ionizing radiation results not only in immediate response in the form of luminescence but also in ionization - transfer of previously bound charge carriers (electrons and holes) into the conduction band. After relaxation charge carriers can be trapped on trapping centers. Trapping levels remain occupied until supply of additional stimulation energy in the form of heat or light (visible or infrared) releases charge carriers, which can participate in further retrapping or recombination processes with light emission. Such light emission is called TL or OSL depending on type of stimulation energy supplied.

Optically stimulated luminescence is characterized with OSL emission spectra - dependence of emission intensity versus emission wavelength at fixed irradiation wavelength (energy) and dose, OSL excitation spectra - dependence of emission intensity of irradiation wavelength or energy at fixed emission and stimulation wavelength and OSL stimulation spectra - dependence of emission intensity on wavelength of stimulation light at fixed emission and irradiation wavelength. TL is characterized with TL glow curves - dependence of TL emission intensity on heating temperature, TL emission spectra - dependence of emission intensity versus emission wavelength at fixed irradiation wavelength (energy) and dose, TL excitation spectra - dependence of emission intensity of irradiation wavelength or energy at fixed emission wavelength.

2.3 Equipment

Experiments on PL and OSL as well as some TL experiments in the 0-300 °C temperature range were carried out in Institute of Solid State Physics, University of Latvia, Latvia, using an experimental setup for spectroscopic measurements. The setup was equipped with a deuterium lamp LDD-400 as a source of UV light and a grating monochromator MDR-2 (LOMO) in the excitation channel. The luminescence signal was analyzed either with a prism monochromator SPM-2 (Carl Zeiss Jena) and detected with photo multiplier tubes H7468-03, H7468-20 (Hamamatsu) and FEU-100 (earlier experiments), or with a grating monochromator SR-303i-B (Shamrock) equipped with a CCD camera DV420-BU2. For low temperature measurements samples were inserted into a closed cycle helium refrigerator (Janis) providing stable temperatures in the 8-325 K range. TL measurements were done using a small home-made oven with linear heating up to 300 °C, used as a sample holder in the experimental setup for spectral measurements.

Beta, gamma, X-ray and UV light induced TL and partly OSL measurements were fulfilled in Riso National Laboratory, Denmark, using the available equipment: Riso model TL/OSL readers with linear planchet heating and Alnor Dosacus TLD reader, operating with hot nitrogen heating. For most of irradiations a $^{90}Y/^{90}Sr$ beta source build in the Riso TL reader and a ^{60}Co standard gamma calibration facility were used. UV irradiation was performed with a metal halide lamp Sol-2 (K.Honley Gmbh), which simulates a solar spectrum at ground level from 300 nm upwards.

Some of TL measurements were done in University of Nice-Sophia Antipolis, LPES-CRESA, France, using a home-made TL reader with linear heating and an optical multichannel analyzer (Princeton Instruments). UV irradiation was carried out with a deuterium lamp (50W) with attached interference filters.

3. Photoluminescence properties of AlN ceramics

Luminescence spectra of AlN were studied using different excitation types; however, excitation with UV light, producing photoluminescence, appeared to be the most informative. PL emission of AlN ceramics was observed in the 300-1100 nm region, which is spectral area of luminescence of defect centers. It contains several emission bands, and for each of them excitation spectrum was measured. Excitation spectra were measured beginning from the fundamental absorption edge (in other words band-to-band transition area) at 200 nm and up to 600 nm. The luminescence region of exciton emission (below 250 nm) is not available with the equipment used and will not be discussed here.

The general view of PL emission spectrum could be illustrated by our recent results on PL measurements of AlN ceramics fulfilled in the 8-300 K temperature range. The following emission bands can be distinguished in the PL spectra at 300 K in Fig.1, a.: a broad UV-blue band in the 300-500 nm region, a narrower band at 600 nm and minor bands at 700 nm and 1000 nm. At low temperature the composition of PL spectrum remains the same, except for emergence of an additional minor band at 500 nm, while relative intensity of emission bands under different excitation wavelengths changes, see Fig.1, b. Contribution of individual bands varies from sample to sample and depends on composition and sintering conditions of ceramics. Let us discuss properties and origin of individual bands.

3.1 Oxygen-related emission

The main attention should be devoted to the broad complex UV-blue band. It is composed of two subbands – those at 400 and 480 nm each of them having its own excitation spectrum. As it is seen from Fig.2. taken from (Berzina et al., 2009), 400 nm emission band has excitation band at 240-250 nm, while 480 nm emission band has excitation band centered at 280-290 nm; both of them are excited also in the fundamental absorption edge at 200-220 nm.

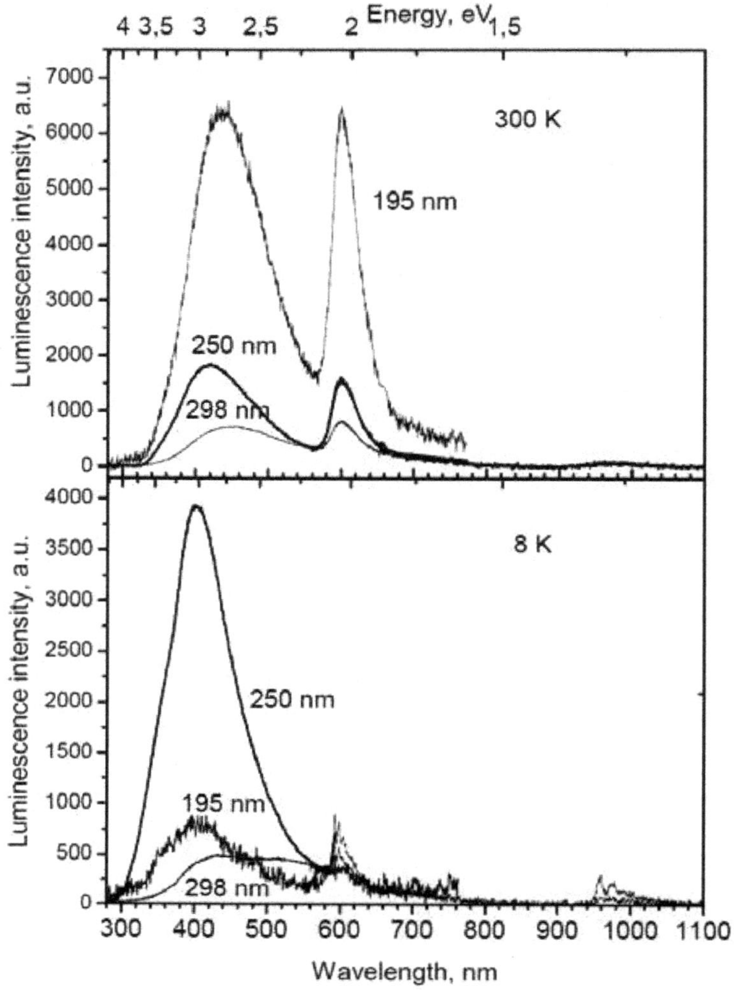

Fig. 1. PL spectra of AlN ceramics at 300 and 8 K, excitation wavelengths shown on the graph

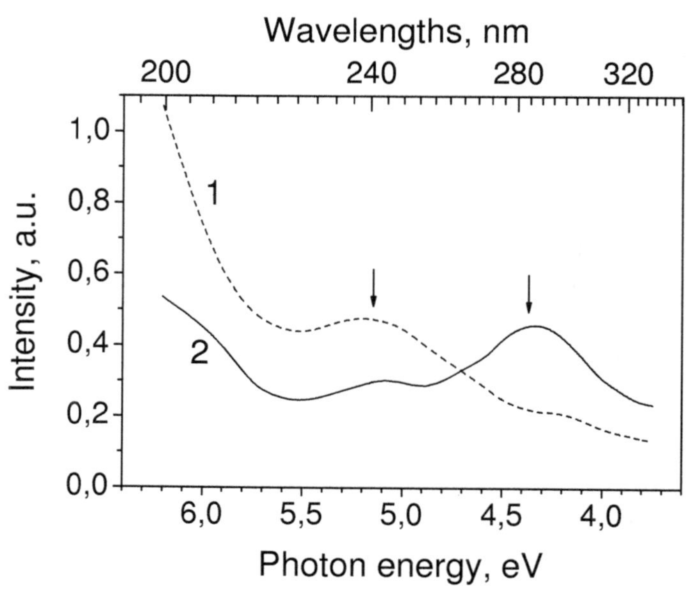

Fig. 2. Excitation spectra of 400 nm (1) and 480 nm (2) emission bands of AlN ceramics at 300 K

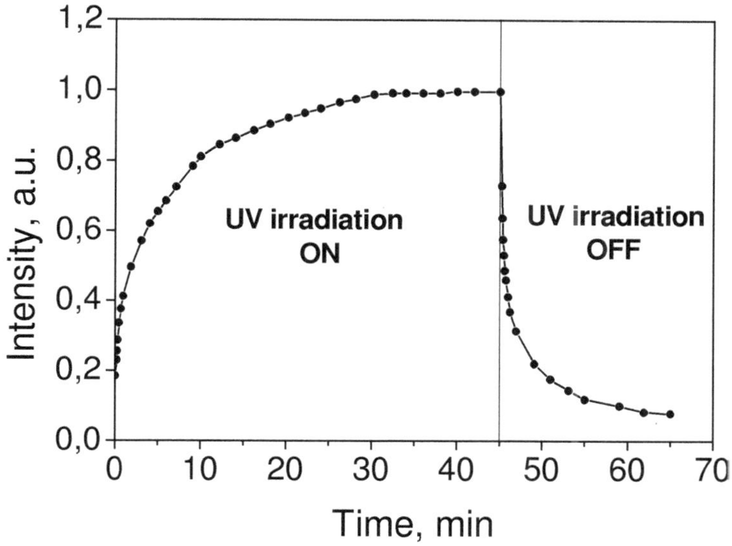

Fig. 3. Time dependence of intensity of the UV-blue emission at start and stop of UV irradiation at 300 K

It was observed in (Berzina et al., 2001) that intensity of the UV-blue emission band depends upon duration of the continuous UV irradiation. As seen from Fig.3, at the beginning of the UV irradiation with 245 nm it is low, and then gradually increases until reaching saturation level after approximately 30 min. After ceasing of UV irradiation the gradually decreasing afterglow emission is observed for several minutes. In the same time position of the maximum of emission band shifts to the longer wavelengths.

Many publications are devoted to elucidating the origin of luminescence centers responsible for the UV-blue band. In (Slack, 1973; Youngman & Harris, 1990 and Harris at al., 1990) it was shown that this band is connected with emission of oxygen–related defects. Later our works (Trinkler et al, 1998, 2001a, 2003, 2005, 2007a, 2007b) have shown that this band is revealed also in TL and OSL emission spectra. Presence of the emission band in PL, TL and OSL as well as afterglow emission and time characteristics of its intensity speak in favor in its recombination character. The nature of the 400 nm subband was determined by the EPR and ENDOR studies (Schweizer et al, 2000) as recombination luminescence of two defect centers: a donor - (O_N-v_{Al}) - a complex defect center formed by an oxygen ion substituting for nitrogen ion O_N in a regular site and a neighboring aluminum vacancy v_{Al} and an acceptor – another closely situated oxygen ion O_N.

The luminescence mechanism responsible for the 400 nm emission band in AlN ceramics, was formulated in papers (Berzina et al., 2002, 2009). The corresponding energy band diagram is shown in Fig. 5. According to it the (v_{Al}-O_N) center forms its energy level within the band gap above the valence band (level 5). Irradiation of AlN with 245 nm light from the own absorption band ($h\nu_{def}$) results in its excitation (level 6). The excited state is situated close to the bottom of the conduction band resulting in ionization of the (v_{Al}-O_N) center and an electron transfer into the conduction band. An electron from the conduction band can be captured by one of numerous electron traps (levels from 7 up to 11). One of them is an O_N defect (level 7) – a close neighbour of the ionized (v_{Al}-O_N) center . Recombination of the (v_{Al}-O_N)$^+$ and (O_N)$^-$ centers with emission of a light quantum ($h\nu_{lum}$) results in appearance of the 400 nm emission band.

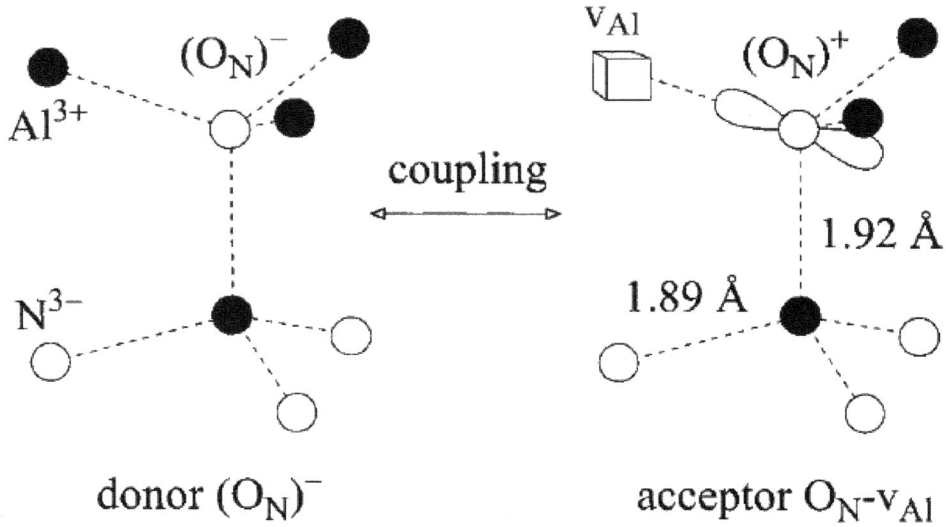

Fig. 4. Model of the coupled donor and acceptor pairs in AlN ceramics (Schweizer et al., 2000)

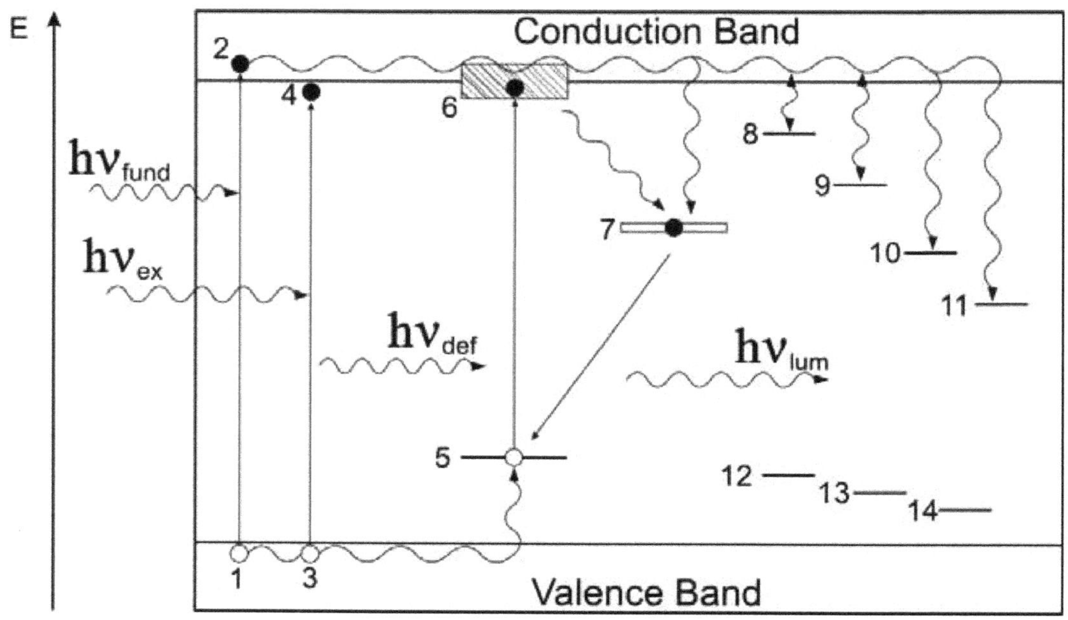

Fig. 5. Energy band diagram of AlN (Berzina et al., 2009). Explanation is given in the text

An electron from the conduction band can be repeatedly recaptured and released by trapping centers, before finally being trapped by O_N defect, thus causing delay of emission rise under continuous excitation with 245 nm, shown in Fig.3.

The similar recombination process can occur under irradiation with light from the fundamental absorption edge region, corresponding to electron-hole pair ($h\nu_{fund}$) or exciton ($h\nu_{ex}$) generation. In this case energy is transferred from host lattice to oxygen-related defects resulting in formation of $(v_{Al}$-$O_N)^+$ and $(O_N)^-$ centers and 400 nm emission.

The 480 nm subband is also connected with oxygen impurity, which most probably is a part of a complex defect. Tentatively 480 nm emission is ascribed to recombination luminescence with participation of (v_{Al}-$2O_N$) center (Nappe et al., 2009 and references therein). Such centers are comprised of two close oxygen ions substituting for nitride ion and an aluminum vacancy. In the energy band diagram the corresponding level is not shown, though it is clear that it should be located a little higher than that of (v_{Al}-O_N) center and participate in similar processes. Our studies of nanostructured AlN (Berzina et al., 2009 and references therein) have shown that contribution of the 480 nm into the total UV-blue emission increases when size of AlN grains diminishes, and hence, specific surface area expands. It allows supposing that these oxygen-related defects responsible for the 480 nm emission band are located mainly on surface. Besides, the relative increase of the 480 nm band was observed in the samples of AlN ceramics after implantation of oxygen ions (Trinkler et al. 2007a, 2007b). Ion implantation causes damage of the surface layers of AlN and increases oxygen concentration, thus making generation of (v_{Al}-$2O_N$) center more probable compared to untreated samples.

3.2 Manganese-related emission

The 600 nm band was observed in PL, afterglow and TL emission spectra of all studied samples of AlN ceramics. Our high resolution measurements, illustrated by Fig. 6 show the presence of fine structure, which can be distinguished at room temperature (RT) and becomes more clearly seen at low temperature. This fine structure is characterized with equidistant peaks with separation energy around 20 meV. This emission band usually is interpreted as luminescence of tetravalent manganese ion Mn^{4+} substituting for Al ion, and equidistant peaks are ascribed to phonon replicas. For the first time this band was observed and interpreted in (Karel et al., 1966) and later studied in other works (Benabdesselam et al., 1995). However, some authors (Miyajima at al., 2006) basing on the results of X-ray absorption fine structure measurements, state that this band is caused by Mn ion having charge between +2 and +3. Other interpretations of the band ascribed to intrinsic defects of AlN are also found in literature, for example, defects formed from nitrogen vacancies (v_N) and excess Al (Sarua et al., 2003).

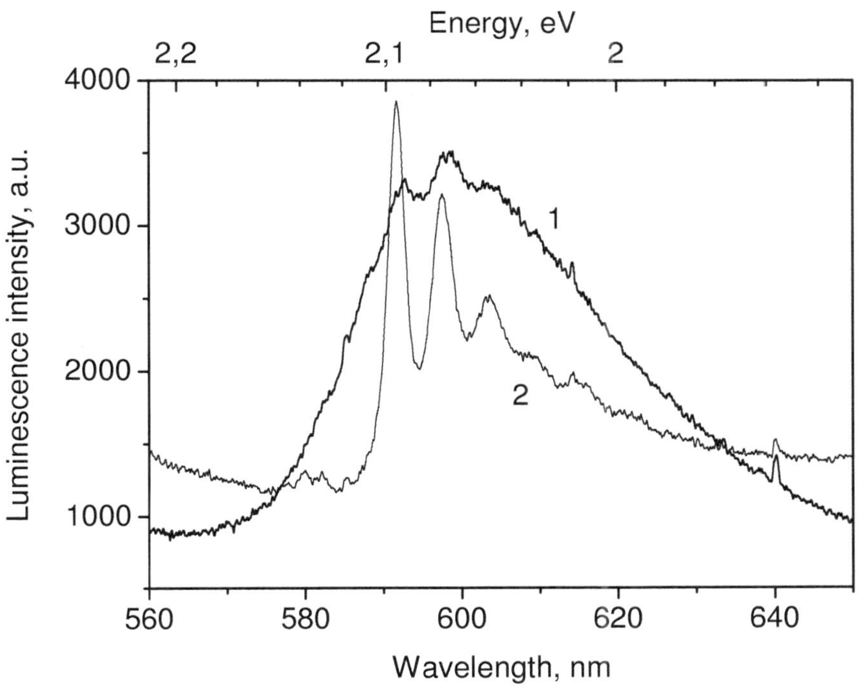

Fig. 6. 600 nm emission band of AlN ceramics measured under excitation wavelength 243 nm at temperature 300 K (1) and 8 K (2)

Excitation spectrum of this emission band is shown in Fig. 7. together with that of 400 nm band. Besides the fundamental absorption edge at 200 nm two excitation bands at 255 and 400 nm are observed. The latter excitation band covers the spectral region of the UV-blue emission band; this results in reabsorption and conversion of UV light into red light. Presence of the 600 nm band in afterglow and TL emission spectra confirms that the corresponding luminescence centre also participates in recombination process.

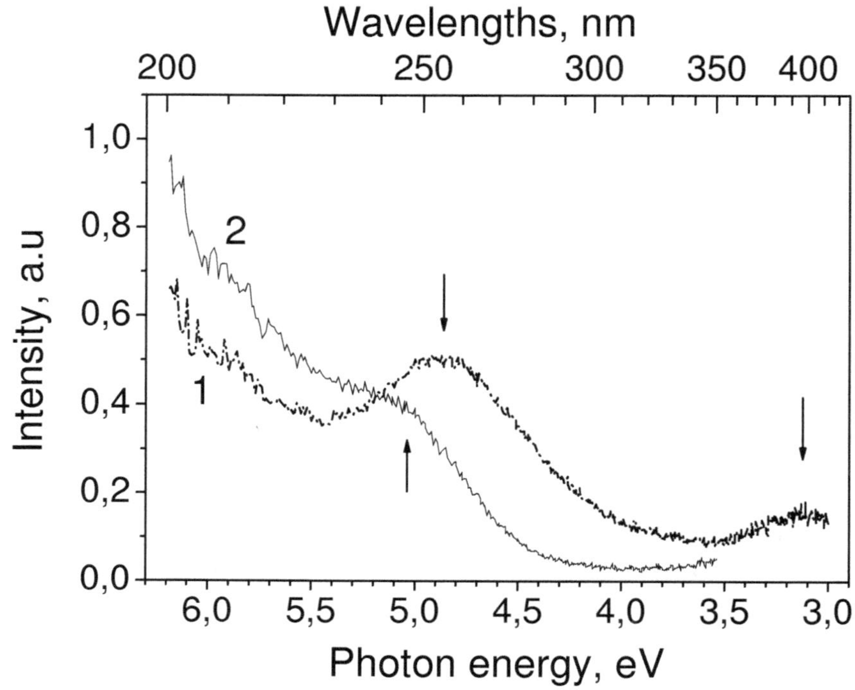

Fig. 7. Excitation spectra of 600 nm band (1) and 400 nm band (2) of AlN ceramics at 300 K

3.3 Other minor PL bands

Other PL luminescence bands observed in AlN ceramics are connected with other uncontrolled impurities. They are the following: a minor band at 700 nm, excited at 280 and 400 nm excitation bands and tentatively ascribed to chromium impurity (Nappe et al., 2011 and references therein) and an infrared band around 1000 nm, connected with iron impurity. At low temperatures the 1000 nm emission band reveals fine structure seen at Fig.1, which is explained as 3d-3d transition ${}^4T_1(G) \rightarrow {}^6A_1(S)$ of Fe^{3+} ion substituting for Al in a regular site and its phonon replicas (Baur et al., 1994).

Summarising the data on PL of AlN ceramics, a conclusion can be drawn that the observed spectral properties of this material are determined by presence of uncontrolled impurities - oxygen, manganese, iron and probably chromium, which form luminescence centres taking part in inter-center and recombination luminescence.

4. AlN ceramics as potential material for solid state dosimetry

There is a number of materials widely used for TL dosimetry as thermoluminescence detectors (TLD) of ionizing radiation, such as LiF:Mg,Ti; LiF:Mg,Cu,P and Al_2O_3:C. Al_2O_3:C is used also as an effective OSL dosimeter of UV light. Nevertheless a search for new prospective dosimetric materials continues. One of such new dosimetric materials could become AlN ceramics due to its properties revealed under exposure to ionizing radiation and UV light.

As it has been already mentioned irradiation of AlN ceramics causes not only the immediate response to irradiation in luminescence form, but also provides the delayed response in the form of optically and thermally stimulated luminescence. Both TL and OSL processes may be applied in dosimetry for recovering the received radiation dose. Processes of stimulated luminescence in AlN have been studied already before our works (Tale et al., 1982; Benabdesselam et al., 1995). Our studies of the processes of storage and release of radiation induced luminescence signal in AlN ceramics began in 90th and continues up to now. We have found that AlN ceramics is sensitive both to ionizing radiation and UV light, producing TL and OSL response in each case. Properties revealed by AlN ceramics after preliminary irradiation with ionizing radiation (X, β, γ rays) and UV light are similar, but they have also some differences. They will be discussed individually according to irradiation type.

4.1 Stimulated luminescence processes induced by ionizing radiation

4.1.1 Thermoluminescence induced by ionizing radiation

There are several important TL characteristics of material, which determine possibility of its practical application in the area of solid state dosimetry: intensity of TL signal (or in other words sensitivity of material to irradiation), characteristics of TL glow curve, TL emission spectrum, need for additional treatment for use and reuse of a sample, linearity of the dose response, stability of the stored signal and others.

TL properties of AlN ceramics induced by ionizing radiation were studied in comparison with other widely known TLD using the same experimental conditions and irradiation doses (Trinkler et al., 1998). Particularly, it was found that AlN ceramics exhibits amazingly high TL sensitivity. Thus, after irradiation with beta rays (^{90}Sr/^{90}Y beta source) for a typical routine read out procedure with the Riso TL reader using a heating rate 10 °C/s the ratio of the TL lightsum (sum of recorded electronic counts) per unit detector mass and radiation dose constitutes 54:27:7:1 for AlN ceramic sample, LiF:Mg,Cu,P, Al_2O_3:C and LiF:Mg,Ti, respectively, see Fig.8.

The TL glow curve of AlN ceramics is characterized with one intensive broad peak extending up to 500 °C, see Fig.8, curve 1. The thermal position of the peak maximum varies from 200 to 300 °C depending on heating rate. No difference in glow curve was observed whether beta particles, γ-rays or X-rays were used for irradiation. The TL signal is completely erased during a single reading using a heating temperature up to 500 °C. Thus no additional annealing or other procedure is necessary for re-use of the sample. A repeatability test showed that reduction of sensitivity after 10 successive irradiation-TL reading cycles was less than 1 %.

Dose response curve of AlN ceramics was measured in dose range $5 \cdot 10^{-3}$ - $5 \cdot 10^{-2}$ mGy (irradiation with ^{60}Co source) in comparison with other mentioned TLD. The curves shown in Fig.9 were obtained by Alnor TL reader using a nitrogen temperature of 300 °C for all samples except for LiF:Mg,Cu,P, which was heated up to 260 °C. AlN ceramics sample revealed linear dose response in the range of five orders of magnitude, like other acknowledged TLD materials, which must be confessed as a very good result.

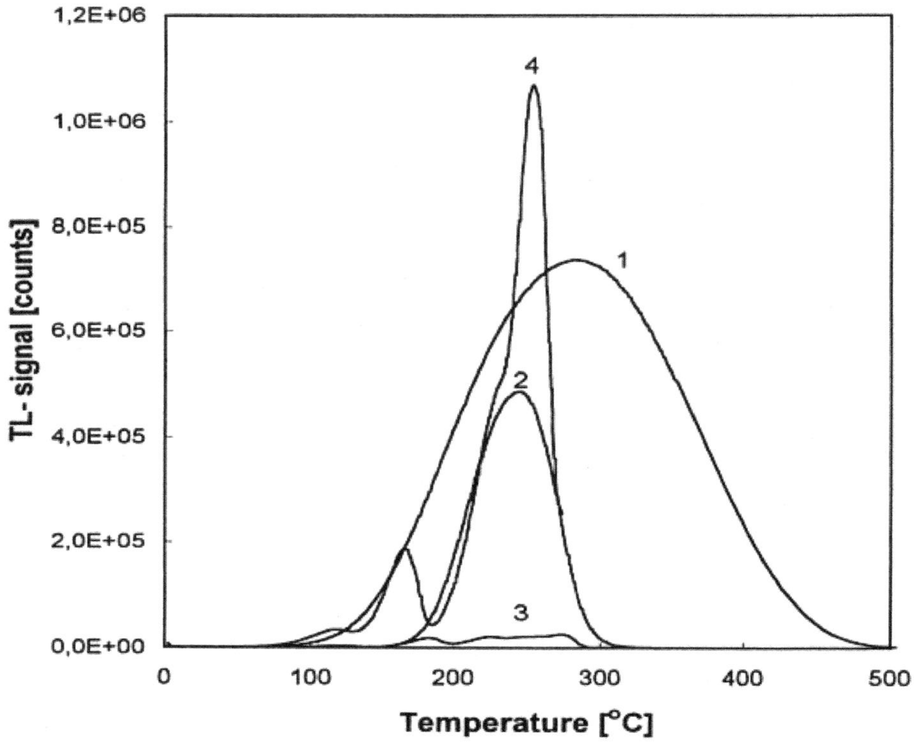

Fig. 8. Glow curves of different types of TL dosimeters irradiated with 28 mGy beta radiation: 1 – AlN ceramics; 2 - Al_2O_3:C; 3 - LiF:Mg,Ti, 4 - LiF:Mg,Cu,P (Trinkler et al, 1998)

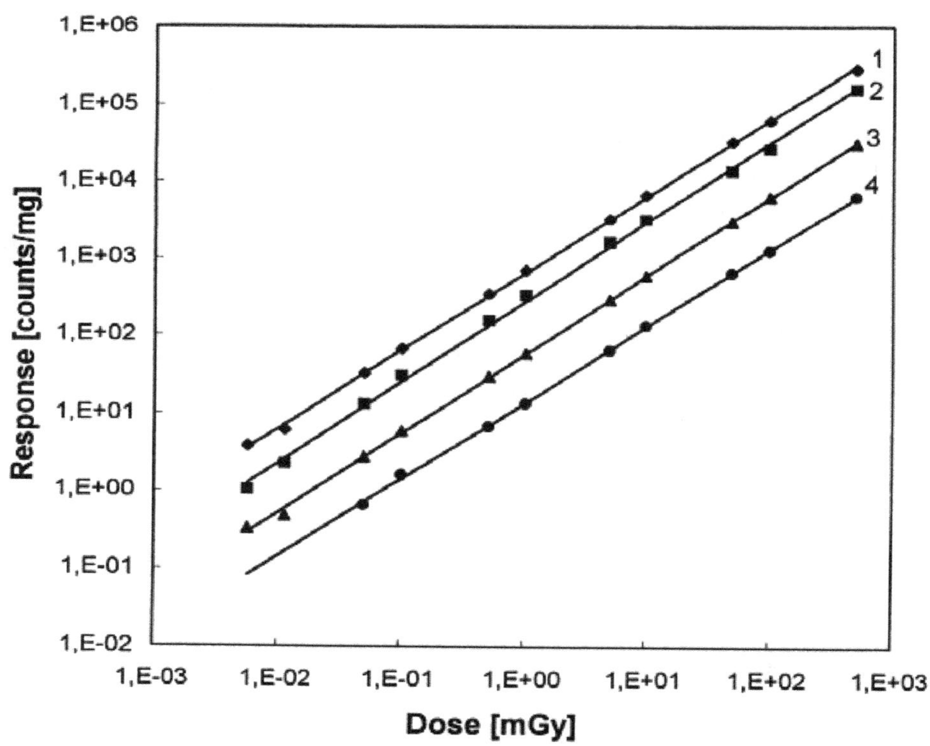

Fig. 9. Gamma ray (^{60}Co) dose response of AlN ceramics (1), LiF:Mg,Cu,P (2), Al_2O_3:C (3) and LiF:Mg,Ti (4) obtained with Alnor hot-nitrogen heated TL reader

TL emission detection region was limited by spectral region of 300-500 nm, in order to discriminate thermal glow emission of the heater, which is observed at high temperatures in the red region. In the same time it is an optimal spectral region for detection with usual photoelectron tubes and other light detectors. In this spectral range AlN emission was observed with a maximum position around 400 nm, which shifted to longer wavelengths with rise of heating temperature. The observed emission corresponds to UV-blue luminescence band of AlN, which is ascribed to oxygen-related defects. Other luminescence bands of AlN ceramics can not be observed in these experimental conditions.

Influence of heating rate on TL signal for AlN ceramics was measured for heating rates from 0.5 to 20 °C/s in comparison with TLD Al_2O_3:C, whose response is critically sensitive to heating rate, hampering routine reading measurements of this TLD and resulting in uncertainty in the dose measurements if significant fluctuations of heating rate occur. Contrary to Al_2O_3:C, TL response of AlN sample shows only a small influence on heating rate, as it is illustrated in Fig.10.

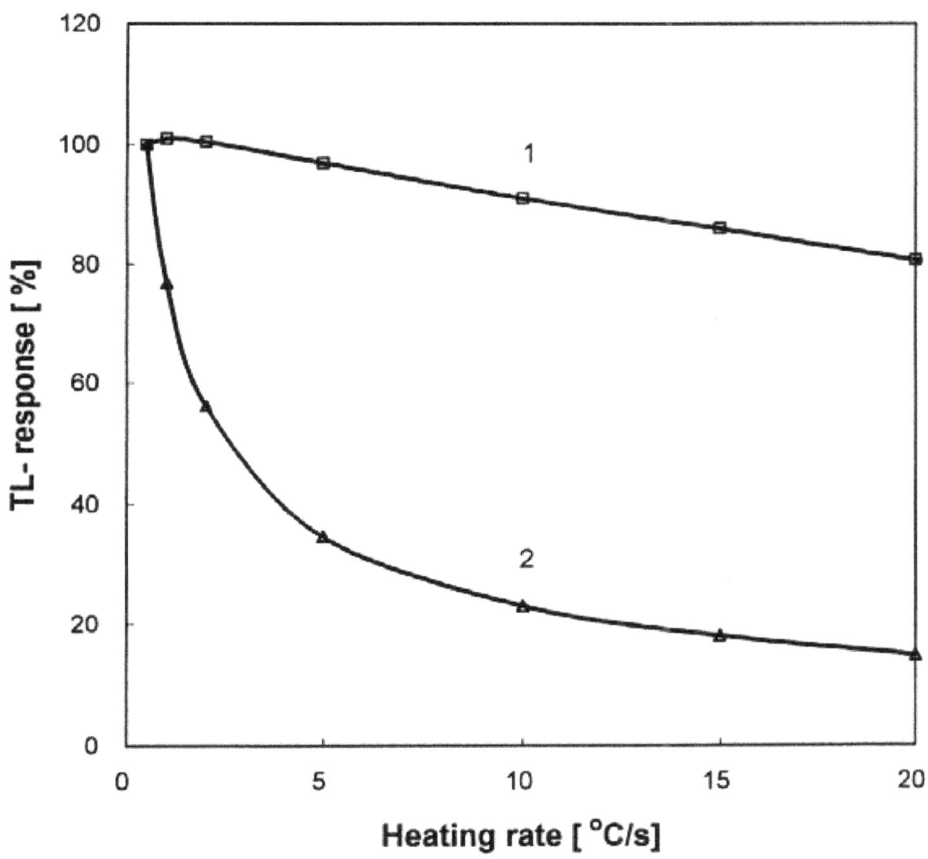

Fig. 10. Influence of heating rate on the TL response of AlN ceramics (1) and Al_2O_3:C (2) irradiated with 28 mGy beta radiation

Considerable fading of the TL signal was observed for AlN ceramics sample on storage at RT. Change of TL glow curve on storage at RT during 114 hours is illustrated in Fig.10. The TL signal is decreasing mainly due to disappearance of the low temperature part of the glow curve, but the high temperature part is also decreasing considerably. Different pre-heat procedures were tried varying preheat temperature and duration: as a result the fading rate was decreased but not eliminated. However, at the same time the TL yield was essentially reduced by this procedure.

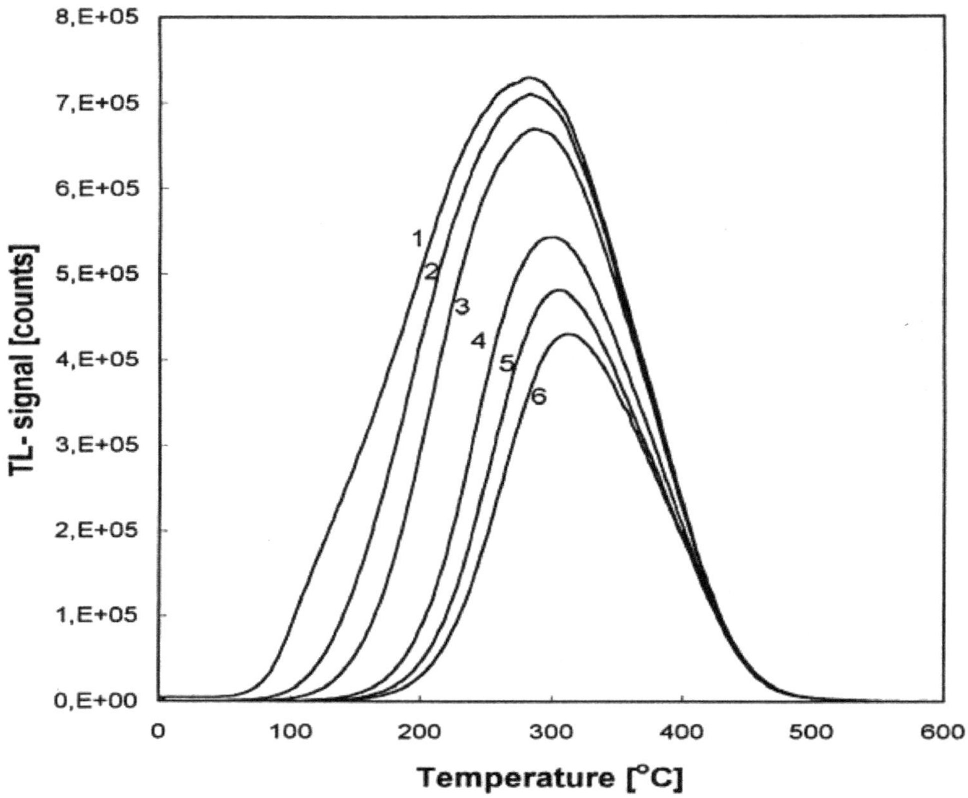

Fig. 11. TL glow curves of AlN samples recorded for different storage periods at RT after irradiation with 42 mGy beta radiation: just after irradiation (1); after a storage period of 0.17 (2); 1 (3); 24 (4); 72 (5) and 114 (6) hours

Neither the pre-irradiation with high dose (1 kGy of ^{60}Co gamma irradiation) with subsequent annealing nor a special procedure of very slow annealing (1 °C/min) resulted in improvement of fading characteristics of AlN ceramics. The pronounced fading of the TL signal in AlN ceramics is a serious obstacle for its application as TLD material.

Sample No	Sintering time (min)	Sintering temperature (°C)	TL response after irradiation (counts.mg^{-1})	TL response after 17 h (counts.mg^{-1})	Fading (%)	TL peak position (°C)
612	210	1645	1334	920	32	380
592	210	1705	13053	7826	40	310
572	210	1745	27673	14639	47	260
552	210	1780	36165	18160	50	250
541	120	1780	32252	15240	53	250
511	40	1780	12601	3238	74	210

Table 1. Influence of sintering conditions of AlN ceramics on TL properties. The TL response is measured immediately after and 17 h after irradiation with 1.7 R X-rays

An attempt was made to study the effect of ceramics sintering conditions on the fading rate (Trinkler et al., 1999). Samples under investigation were produced from the same initial material varying sintering temperature and time. The results of the study are summarized in Table 1. It was found that rise of the sintering temperature leads to the strong increase of the TL yield and also to a shift of the TL peak position to lower temperature with increasing fading rate. An increase of sintering time at constant sintering temperature also causes a rise of the TL yield, but in this case the thermal peak shifts to higher temperature, and hence, the fading rate decreases. The reasons why the TL peak shifts may be explained by generation of oxygen-related defects of different types, whose balance depends on sintering conditions. The optimal sintering conditions are found using the lowest sintering temperature and the longest sintering time, but even this procedure does not eliminate completely fading of the TL signal.

4.1.2 Optically stimulated luminescence induced by ionizing radiation

Optically stimulated luminescence is detected as a pulse of luminescence light decaying in time under continuous exposure of stimulation light. An OSL signal bleaches because stimulation light empties trapping levels. Typical OSL decay curve of AlN contains an intensive fast component and a weak slow component. Each of them has its own specific fading rate, fast component fades faster. Intensity of OSL is characterized with a lightsum – integral of electronic counts detected during a pulse up to its complete depletion. The following characteristics of OSL are important for characterization of material and estimation of its potential application as an OSL dosimeter: OSL emission spectrum, OSL stimulation spectrum; linearity of dose response, fading of the stored signal and others.

OSL induced by ionizing radiation in AlN ceramics (Trinkler et al., 1999) was studied using Riso TL/OSL reader, which is not designed for spectral measurements, but at the same time allows comparison of TL and OSL properties of material under the same conditions. Irradiation was performed with beta source, while stimulation was done by infrared and blue light sources, all sources integrated in the reader. Luminescence was recorded in the 300-500 nm range, corresponding to oxygen-related luminescence of AlN.

It was found that OSL sensitivity to ionizing radiation is approximately 40 times lower than TL sensitivity after the same beta dose. Not all the light yield produced by the irradiation can be released by optical stimulation: after the complete optical bleaching a residual TL signal was observed in the 300-500 °C temperature region. Study of the preheat influence showed that OSL is much more sensitive to preheat treatments than TL: after a preheat up to 150 °C during 300 s only 10% of OSL yield is left compared to 65% of TL yield in the same conditions.

The OSL decay curve induced by ionizing radiation has a dominating fast component, and its fading rate is high. It was found that OSL fading characteristics are worse than those of TL: after 150 hours of storage at room temperature only 10% of the initial OSL response is left compared with 45% of the TL response.

All the above mentioned results show that only a part of the light yield produced by ionizing radiation can be released by optical stimulation. The light-sensitive part of the stored signal can be ascribed to the trapping of the charge carriers in the shallowest trap levels located in the band gap of AlN, while the rest part of the stored signal is more stable due to trapping of charge carriers on deeper levels, which can be emptied only by heating up to sufficiently high temperatures.

It may be concluded that AlN can not be recommended for use as OSL detector of ionizing radiation because a typical OSL response is much weaker and more unstable than the TL response for the same dose.

4.2 Stimulated processes induced by UV light

Detection of UV radiation doses is an important area of dosimetry. UV radiation usually is divided into three spectral regions namely: UV-C (shorter than 280 nm), UV-B (280-315 nm) and UV-A (315-400 nm). Radiation of UV-C region is used in industry for characterization of materials and products and in disinfection procedures due to its bactericidal effect. UV-B region corresponds to the part of the spectral distribution of the Sun's rays reaching ground level and which is potentially harmful to human beings and other organisms. Uncontrolled exposure of naked human bodies to Sun rays from UV-A region also can cause some problems. Monitoring of UV light intensity and measuring of UV doses is important in all of these regions.

A number of our works is devoted to UV light induced stimulated processes in AlN ceramics (Trinkler et al., 2000, 2001b, 2002). It was found that this material is extremely sensitive to UV radiation producing high response both in TL and OSL. Some experiments were done in comparison with dosimetric material Al_2O_3:C, commonly used for UV dose measurements (Akselrod et al., 1990). We have measured full spectral characteristics of UV light induced TL and OSL processes: excitation spectra, emission spectra and stimulation spectra (for OSL), as well as TL glow curves, including 8-300 K temperature region.

4.2.1 Properties of UV light induced TL and OSL compared to those induced by ionizing radiation

A study was done when TL and OSL measurements were performed with AlN ceramics and Al_2O_3:C dosimeter in the same conditions after exposition to ionizing radiation and UV light, simultaneously comparing sensitivity of materials to radiation exposure and effects of different type of radiation on both materials (Trinkler et al., 2000). For this study the Sol-2 lamp in combination with interference filters was used as a UV light source, while TL and OSL response was measured with Riso TL/OSL reader.

The most attractive feature of AlN ceramics is the very high sensitivity to UV radiation, which is demonstrated by Fig.12. The TL response from AlN ceramics irradiated by beta particles is about 20 times higher than that of Al_2O_3:C (curves 1 and 3) and about 6000 times higher in the case of UV irradiation with wavelength 350 nm (curves 2 and 4). On this figure for the used TL yield scale the curve 4 is undistinguishable from the temperature axis. TL signal from the UV dosed AlN ceramics is characterized with a very broad and structureless glow curve (curve 2) with a maximum around 320 °C, which is significantly higher than that for beta irradiated AlN, corresponding to 230 °C (curve 1). A lower fading rate of the integrated TL signal was observed for AlN samples exposed to UV light (40 % fading over 24 h storage) than for samples exposed to beta radiation (66 % fading). This is explained by the shift of the TL curve to the higher temperature.

The TL response of AlN ceramics to UV radiant exposure was obtained by varying both the irradiation time and the irradiance, changing the distance between the sample and the excitation lamp and using sieve filters of different density. Figure 13 shows a linear relationship between the TL yield and UV radiant exposure over at least five decades. The deviation from the linear slope observed at the highest UV doses is explained by fading effects, due to the relatively long irradiation times required.

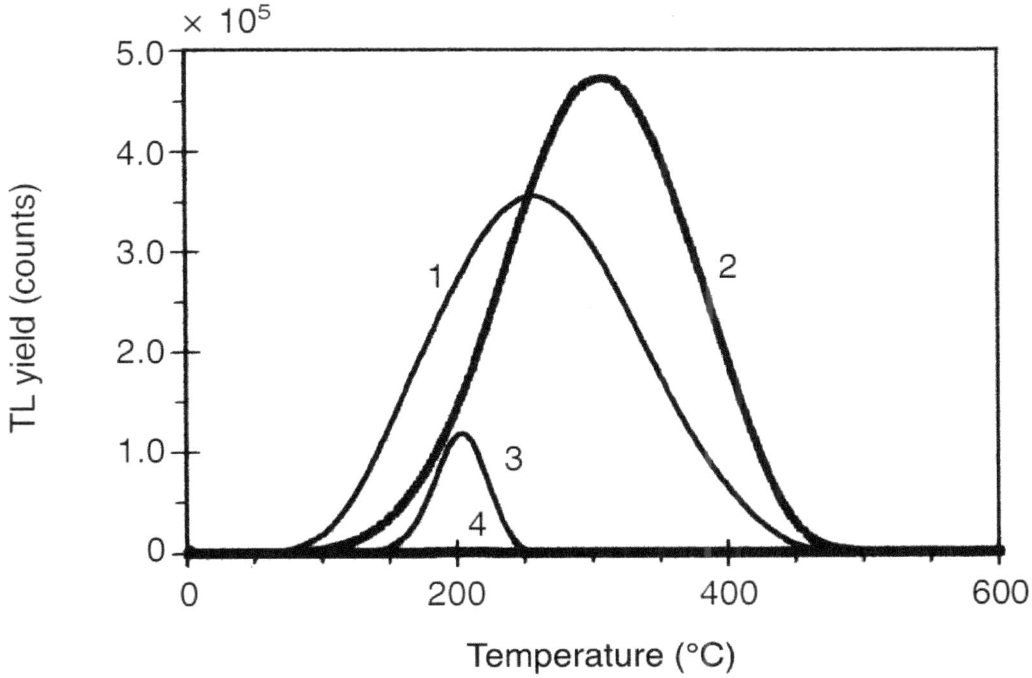

Fig. 12. TL curves obtained from AlN ceramics and Al_2O_3:C after irradiation with 100 mGy beta dose (1 and 3) and 1 min exposure to UV light produced by the lamp Sol 2 and using an interference filter with peak transmission at 350 nm

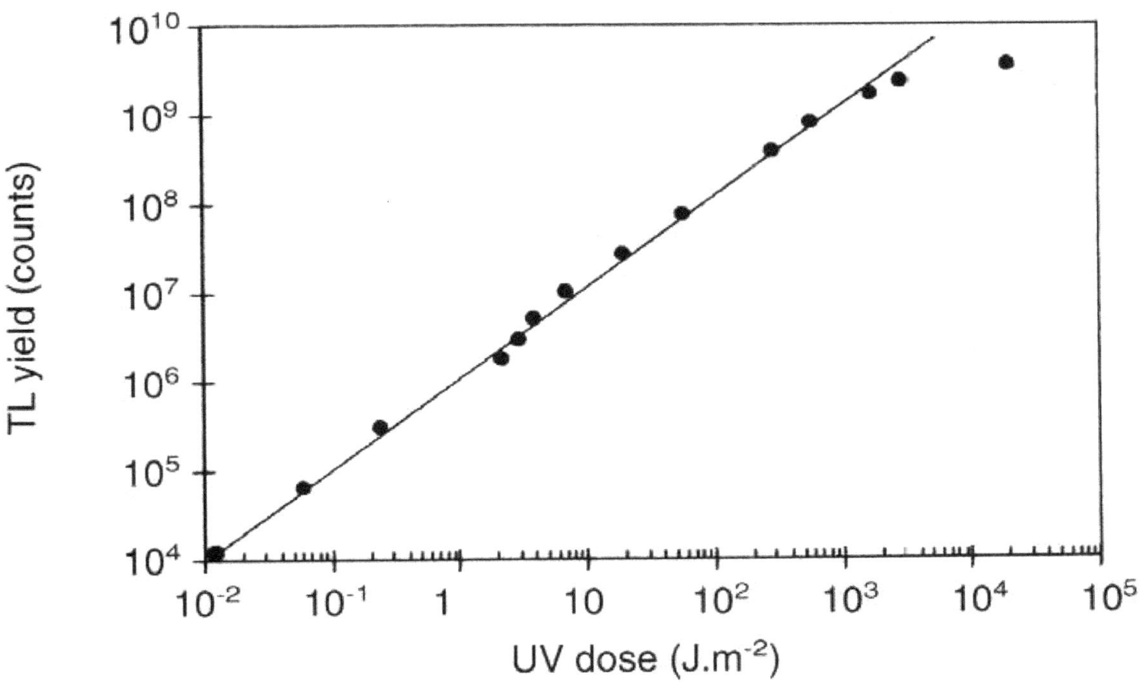

Fig. 13. UV light dose dependence of AlN

The OSL decay curves obtained from AlN and Al_2O_3:C after both beta and UV radiation are shown in Fig.14. In contrast to Al_2O_3:C showing the same shape of decay curves for both type of irradiation (curves 3 and 4), AlN ceramics has two different decay curves: an intensive fast component that dominates in beta-induced OSL (curve 1) and a slow component, which dominates in UV-induced OSL (curve 2). Although the initial OSL intensity of the AlN sample is significantly higher than that of Al_2O_3:C when exposed to beta radiation, this very fast AlN component yields less signal. Alternatively, when irradiating AlN with UV light of 350 nm, its OSL signal is about 60 times higher than that of Al_2O_3:C. Ratio of the UV light induced TL and OSL signals from AlN and Al_2O_3:C depends on the wavelength of UV irradiation, because each of the materials has its own spectral sensitivity, revealed by TL and OSL excitation spectrum. Thus at irradiation 200-230 nm corresponding to the highest sensitivity of Al_2O_3:C is the OSL response of this material is higher than that from AlN, at other wavelength it is lower.

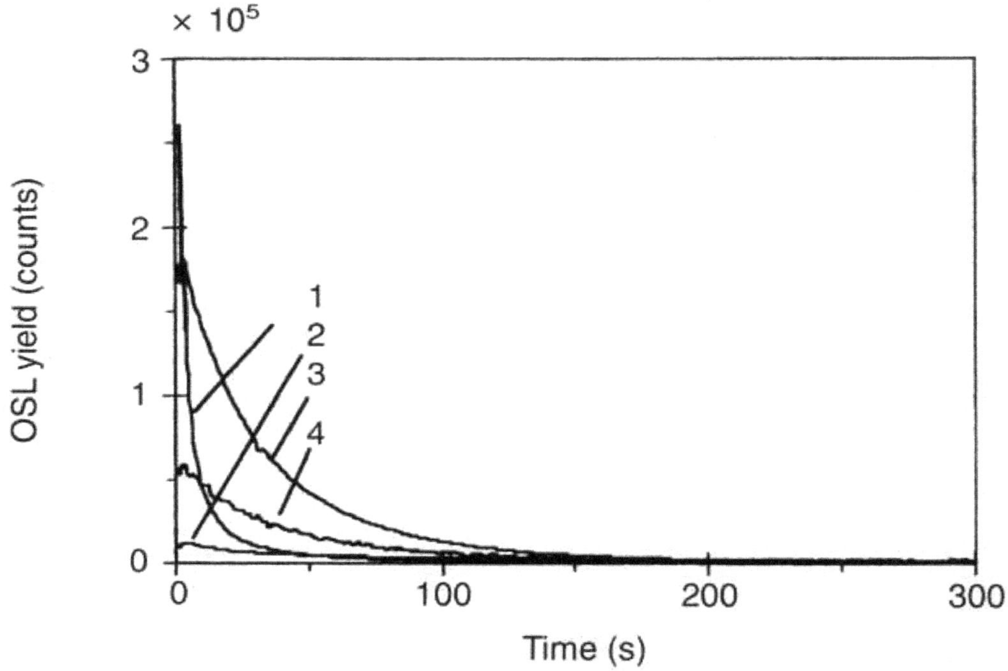

Fig. 14. OSL decay curves obtained from AlN ceramics and Al_2O_3:C after irradiation with 100 mGy beta dose (1 and 3) and 1 min exposure to UV light produced by the lamp Sol 2 and using an interference filter with peak transmission at 350 nm. Stimulation was performed with 470 nm light. Curve 4 is multiplied by a factor of 100

The initial part of the OSL pulse, which consists mainly of the fast component, fades faster during the storage of the dosed sample than the tail of the pulse, consisting of the slow component. That is why the fading effect is higher for the beta induced OSL, which has a dominant fast component of the decay curve: only 10% of the initial response is left after 24 hours of storage. The effect of fading is less pronounced for UV induced OSL having a dominant slow component: 45% of the initial response is left after 24 h.

The properties observed in the UV induced AlN samples such as the peaking of the TL glow curve at a higher temperature, absence of an intensive fast component of the OSL decay curve, lower fading rate of both the TL and OSL signals, assume that the UV radiation induced stimulated luminescence involves a larger part of deep traps compared to those created by beta radiation. The deeper traps need higher energy for depopulation.

The OSL response dose dependence taken as OSL yield versus UV irradiation time up to 60 min deviates from a linear slope. This is explained by fading of the OSL signal during irradiation and insufficient OSL readout time for the higher doses.

4.2.2 Spectral characteristics of UV induced TL and OSL

Spectral sensitivity of AlN ceramics was determined from excitation spectrum of TL and OSL, which appeared to be similar and covering almost the whole UV region from 200 to 350 nm, with maximum effect at 240-270 nm, corresponding to absorption of oxygen-related centers. This characteristic for OSL process is shown on Fig. 15, curve 1 (Trinkler et al, 2002).

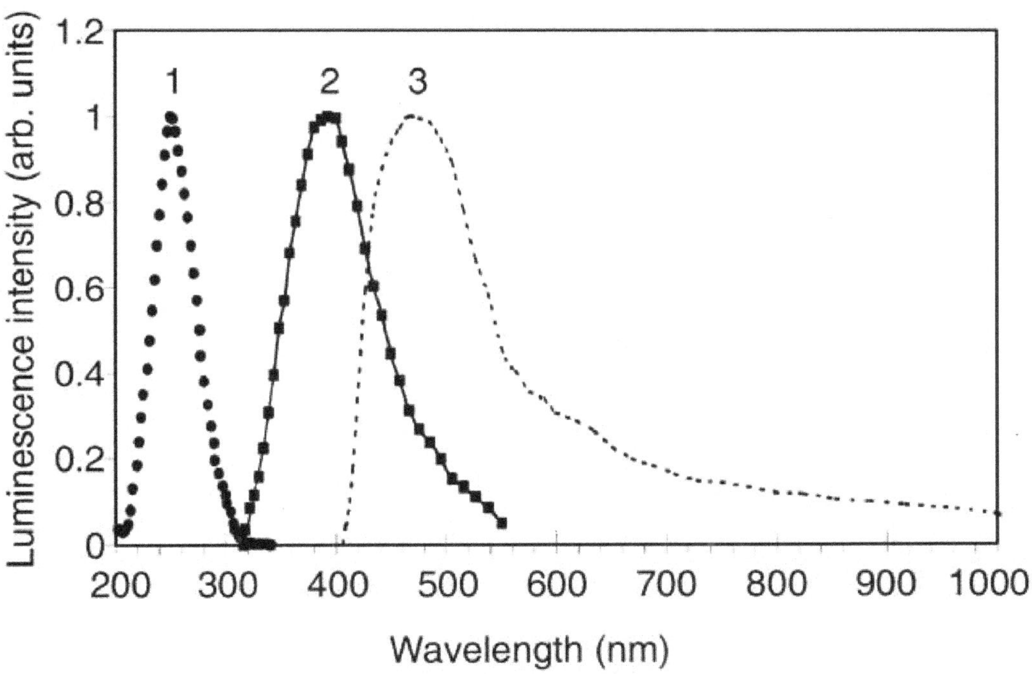

Fig. 15. Spectral characteristics of UV light induced OSL of AlN ceramics: OSL excitation spectrum (1), OSL emission spectrum (2), OSL stimulation spectrum (3)

It is remarkable that spectral sensitivity of AlN is similar with spectral sensitivity of human skin in the 280-340 nm range (see Fig.16), which corresponds to UV-B region of Solar emission potentially dangerous for human beings. This property makes AlN ceramics useful for application as material for personal UV dosimeter.

OSL emission spectrum is presented with the same UV-blue band as PL band ascribed to oxygen-related centers, as it is seen from Fig. 15, curve 2. In order to separate emission and stimulation light the detection region is limited and nothing could be said about presence of long wavelength bands in OSL emission. TL emission spectrum shown on Fig. 17. (Trinkler et al., 2003) also has the same UV-blue band, whose intensity and position of maximum shifts depending on excitation wavelength. Besides, it contains also 600 nm band. TL and OSL emission spectra fall into the 300-600 nm range, which is suitable for detection with common light detectors.

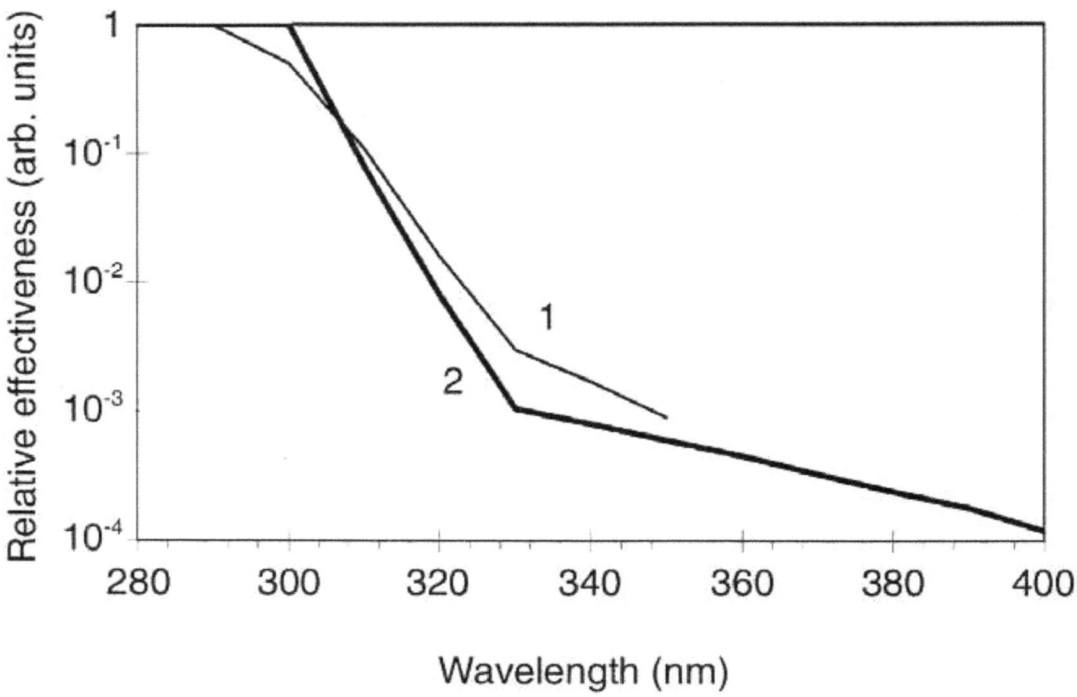

Fig. 16. Comparison of OSL excitation spectrum of AlN ceramics (1) with action spectrum of human erythema (2) (CIE Research Note, 1987)

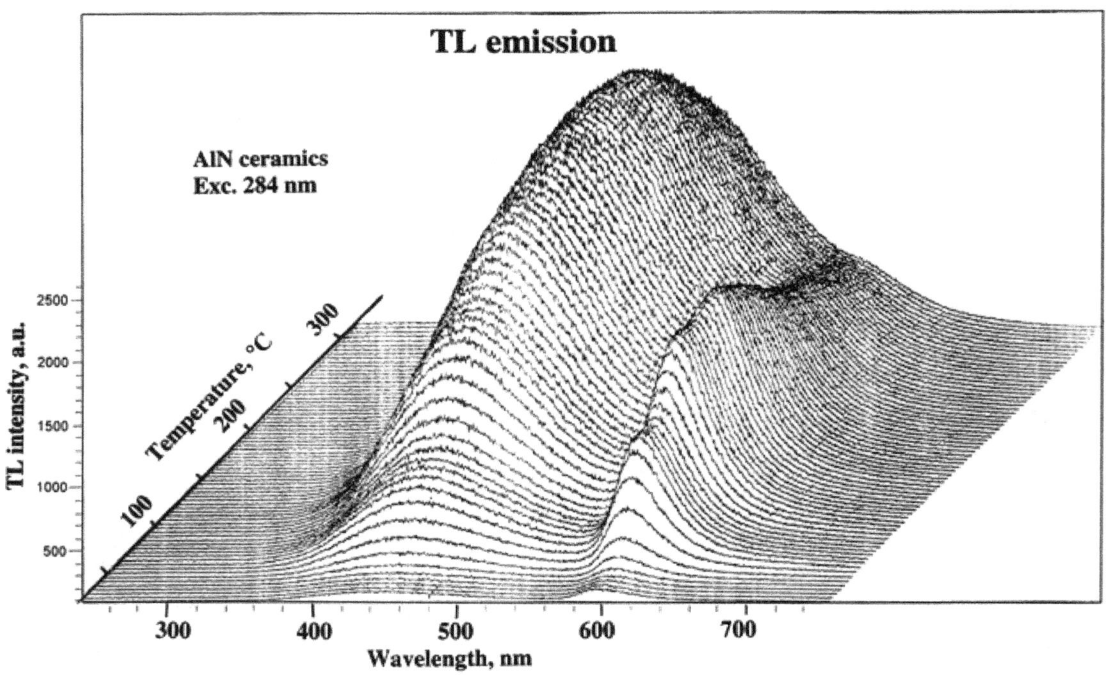

Fig. 17. 3D image of UV light induced (excitation 284 nm) TL emission from AlN ceramics

OSL is characterized also by stimulation spectrum, determined by energy levels of trapping centres, emptied during optical stimulation. For AlN ceramics it extends from 400 up to 1000 nm with an expressed maximum at 500 nm, see Fig.15, curve 3. Large spectral interval and complex structure of the OSL emission spectrum means that numerous different trap levels participate in the stimulated luminescence process. Most of halogen and incandescent lamps with continuous distribution of emission in combination with glass filters as well as monochromatic LEDs and lasers are suitable for use as stimulation sources of OSL from AlN ceramics.

4.2.3 Use of 480 nm emission band for UV induced TL and OSL

It has been shown already that the oxygen-related UV-blue PL emission band contains two subbands around 400 and 480 nm, each subband having its own excitation spectrum, as shown in Fig.18, curves 1 and 2. The same emission bands at 400 and 480 nm can be distinguished also in TL and OSL emission spectra. Similarly to PL case, each of these bands is characterised with a distinct shape of TL or OSL excitation spectra. The case of TL excitation spectra recording emission at 400 and 480 nm is shown in Fig. 18 (curves 3 and 4). The results on TL and OSL experiments mentioned in previous sections were obtained selecting either the 400 nm emission band or the whole UV-blue band. However, it appeared that selection of the 480 nm for detection of stimulated emission results in better parameters of the TL and OSL processes (Trinkler et al, 2007a, 2007b).

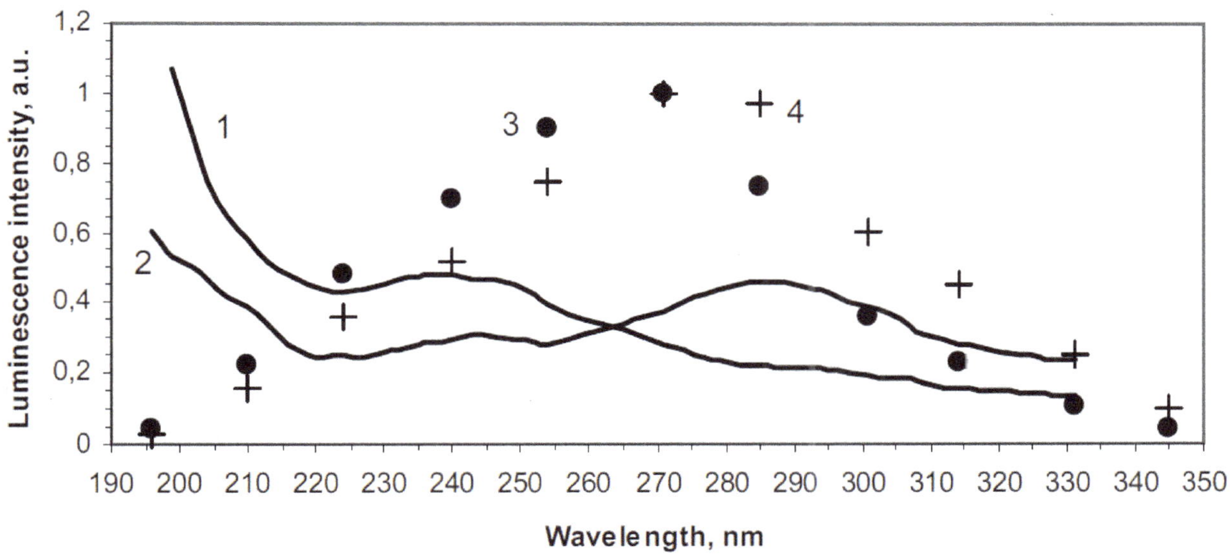

Fig. 18. Excitation spectrum of PL (1, 2) and TL (3, 4) of AlN ceramics recording emission at 400 nm (1, 3) and 480 nm (2, 4)

It was found that intensity and shape of TL curves measured in the same conditions depend on the selected excitation and emission wavelength. Two peaks are distinguished in TL curves: those at 80 and 220 °C (see Fig.19). The low temperature peak is better pronounced selecting 400 nm emission under irradiation in its characteristic excitation band (curve 1), while the high temperature peak appears for selection of 480 nm emission after excitation in the long wavelength region (curve 4). Other curves (2 and 3) contain both peaks due to overlapping of the emission bands. The higher temperature of the TL glow peak implies the lower fading rate of the stored signal. Indeed, measurement of TL signal after one hour of storage had shown 50 % of the initial yield of the TL response at 400 nm emission excited at 245 nm contrary to 80 % of that for 480 nm emission excited at 315 nm.

Correlation of emission bands with the definite TL glow peaks speaks in favor of pair distribution of trap centers and recombination centers, as it was approved already for $(v_{Al}\text{-}O_N)^+$ and $(O_N)^-$ centers, responsible for the 400 nm band. The trap centers responsible for the 480 nm emission evidently are more stable, than those participating in the 400 nm emission, and the corresponding TL and OSL signal fades slower. Similar effect was observed for the OSL process (Trinkler et al., 2007b): after one hour storage selection of the 480 nm band provided 40% of the initial value of the OSL yield versus 18 % for selection of 400 nm band.

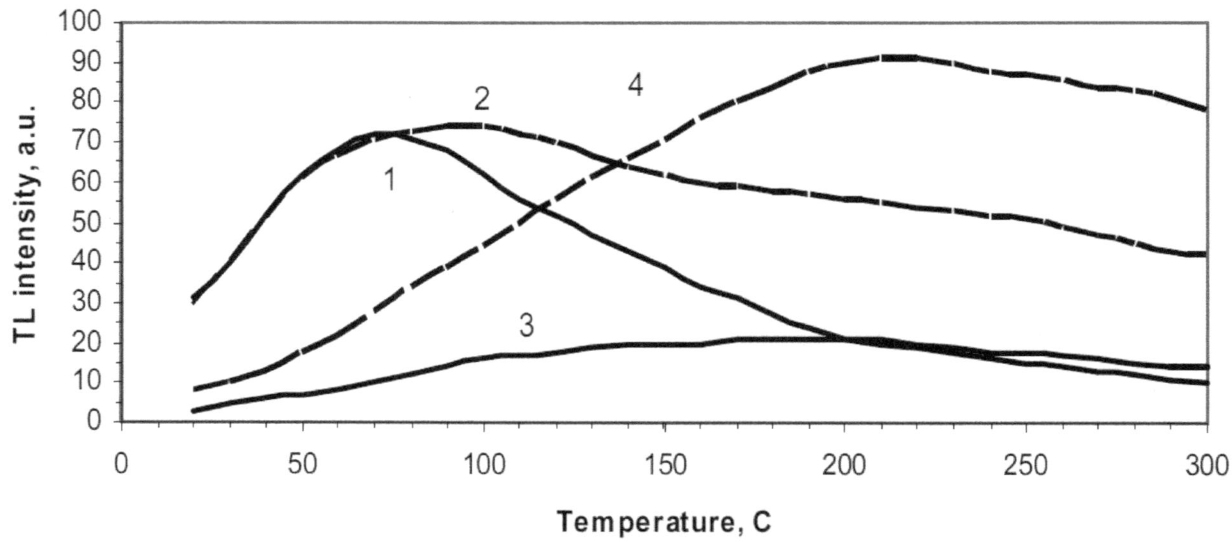

Fig. 19. TL curves of AlN ceramics after UV light irradiation at 245 nm (1, 2) and 315 nm (3, 4) detecting emission at 400 nm (1, 3) and 480 nm (2, 4)

We have done a special study of stimulated processes in AlN ceramics subjected to ion implantation with oxygen ions. It was found that after such treatment relative contribution of the 480 nm band into the total emission spectrum increases. In the same time this procedure provides very little improvement in the intensity and stability of the OSL signal, but gives some improvements of the dosimetric characteristics for TL.

From the results obtained it follows that for UV dosimetry applications it is better to use the 480 nm emission band, since the response signal is more stable than that for the 400 nm band and its excitation spectrum falls in the UV-B region, which is the detection range of practical interest. Comparing different types of dosimetric applications of AlN ceramics it should be recognized that OSL method is less suitable because of a higher rate of decrease of the signal at room temperature.

4.2.4 TL in the low temperature region

Recently we have done experiments on TL measurements of AlN ceramics at low temperature. The sample was irradiated with UV light of different wavelengths at 8 K. TL signal was recorded during heating of the sample up to 300 K. Two glow curves corresponding to the main oxygen-related bands are shown in Fig. 20.

The common feature of all glow peaks is a peak at 60 K, at further rise of temperature behavior of glow curves is different until 270 K, when both curves begin to rise. This rise is the beginning of the main TL peaks, whose maxima were observed at 80 °C (353 K) for 400 nm emission band and 220 °C (493 K) for 480 nm emission band. It means that room temperature 20 °C (293 K) is sufficiently high to start the spontaneous thermoluminescence process without any external heating. This process will continue until complete depletion of the trapped charge carriers. Storage of the dosed samples at room temperature will lead to the fading of the stored TL and OSL signal.

Thermal region of the main peaks is determined by the set of trapping centrers present in AlN ceramics. The fading problem could be solved if the dosed AlN samples are stored at reduced temperature (around 250-260 K). Another way to eliminate fading is to introduce impurities providing deep trapping levels in the energy band diagram of AlN, which are stable at room temperatures. Further practical studies of this phenomenon are planned.

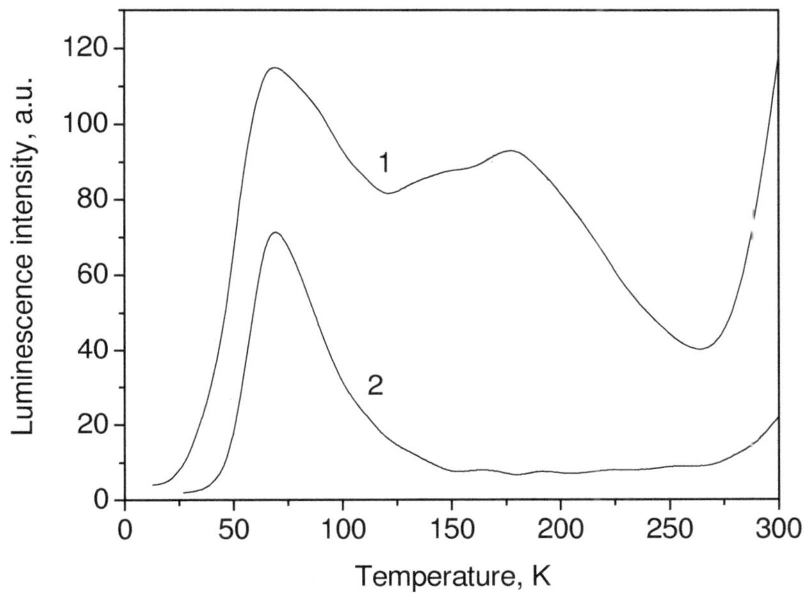

Fig. 20. TL glow curves of AlN irradiated at 8K: (1) excitation 243 nm, emission 400 nm, (2) excitation 298 nm, emission 500 nm

5. Conclusions

AlN ceramics is a wide-gap material whose spectral properties are determined by presence of uncontrolled impurities and intrinsic defects, the main role played by oxygen-related defects. The fulfilled investigations show that AlN ceramics is a material, highly sensitive to irradiation with both ionizing radiation and UV light. The dose of the obtained irradiation can be retrieved using TL and OSL method. In many aspects AlN ceramics outperforms other actually used TL and OSL dosimeters. However, this material has an important drawback, hampering its practical application in dosimetry area – it is high fading rate of the signal during storage at room temperature.

Different measures have been tried to diminish the effect of TL or OSL signal fading, such as optimization of ceramics sintering conditions, use of preheat procedure, detection of the definite emission band (480 nm) and oxygen ion implantation. These measures helped to slow down, but not to eliminate the fading process, which is determined by intrinsic properties of AlN ceramics. That is why at present AlN is not offered for practical application in the area of solid state dosimetry. The further studies of this potentially perspective material are planned.

6. Acknowledgement

This work was supported by ERAF Project 2010/0204/2DP/2.1.1.2.0/10/APIA/VIAA/010.

7. References

Akselrod, M.S. ; Kortov, V.S. ; Kravetsky, D.J & Gotlib, V.I. (1990). High sensitive thermoluminescence anion-defctive Al_2O_3:C single crystal detector. *Radiation Protection dosimetry,* Vol 32, Nos. 1-4, (April 1990), pp. 15-29, ISSN: 0144-8420

Baur, J. ; Maier, K. ; Kunzer, M. ; Kaufman, U. & Schneider, J. (1995). Determination of the GaN/AlN band discontinuities via (-/0) acceptor level of iron. *Materials Science and Engineering B,* Vol. 29, No. 1-3, (January 1995), pp. 61-64 ; ISSN: 0921-5107

Benabdesselam, M.; Iacconi, P.; Lapraz, D.; Grosseau, P. & Guilhot, B. (1995). Thermoluminescence of AlN – Influence of synthesis processes. *J. Phys. Chem.,* Vol. 99, No. 25, (June 1995), pp. 10319-10323, ISSN: 1089-5639

Berzina, B. ; Trinkler, L. ; Sils, J. & Palcevskis, E. (2001). Oxygen-related defects and energy accumulation in aluminum nitride ceramis. *Rad. Eff. & Def. In Solids,* Vol. 156, No.1, (January 2001), pp. 241-247, ISSN: 1029-4953

Berzina, B. ; Trinkler, L. ; Sils, J. & Atobe, K. (2002). Luminescence mechanisms of oxygen-related defects in AlN. *Rad. Eff. & Def. In Solids,* Vol. 157, No. 6-12, pp. 1089-1092, (January 2001), ISSN: 1029-4953

Berzina, B. ; Trinkler, L. ; Jakimovica, D. ; Korsaks, V. ; Grabis, J. ; Steins, I. ; Palcevskis, E. ; Belluci, S. ; Chen, L.—Ch. ; Chattopadyay, S. & Chen K.-H. (2009). Spectral charcterisation of bulk and nanostructured aluminum nitride. *Journal of Nanophotonics,* Vol. 3, 031950, (December 2009), pp. 1-16, ISSN: 1934-2608

CIE (International Commission on Illumination) Research Note (1987). A reference action spectrum for ultraviolet induced erythema in human skin. *CIE Journal* Vo. 6, No. 1, (January 1987), pp. 17-22 ; ISSN: 0252-9246

Harris, R.A.; Youngman, R.A. & Teller, R.G. (1990). On the nature of the oxygen-related defects in in aluminum nitride. *J. Mater. Res.,* Vol. 5, No. 8, (August 1990), pp. 1763-1773, ISSN: 2044-5326

Karel, F.; Pastrnak, J.; Hejduk, J. & Losik, V. (1966). Fine structure of emission spectra of the red AlN:Mn luminescence. *phys. stat. sol. (b),* Vol. 15, No. 2, (February 1966), pp. 693-699, ISSN: 0370-1972

Miyajima, T.; Kudo, Y.; Uruga, T. & Hara, K. (2006). Analysis of the local structure of AlN:Mn using X-ray absorption fine structure measurements. *phys. stat. sol. (c),* Vol.3, No. 6, (May 2006), pp. 1742-1745, ISSN: 1862-6351

Nappe, J.C.; Grosseau, Ph.; Benabdesselam, M., Beauvy, M. & Guilhot, B. (2009). Characterization of aluminum nitride material under swift heavy ion irradiations. Proceedings of the 11 th ECERS conference, 2009. *11th International Conference and Exhibition of the European Ceramic Society,* pp. 1105-1108, ISBN:9788360958544, Cracow, Poland, May 21-25, 2009

Nappe, J.C.; Benabdesselam, M.; Grosseau, Ph. & Guilhot, B. (2011). Effect of swift heavy ion irradiations in polycrystalline aluminum nitride. *Nuclear Instruments and Methods in*

Physics Research Section B: Beam Interactions with Materials and Atoms, Vol. 269, No. 2, (February 2011), pp. 100-104; ISSN: 0168-583X

Palcevskis, E. ; Berzina, B. ; Trinkler, L. ; Ulmanis, U. & Mironova-Ulmane, N. (1999). Ceramics from fine plasma processed AlN powder, sintering and properties. *Latvian Journal of Physics and Technical Sciences,* No.1, (February 1999), pp. 34-51, ISSN: 0868-8257

Palcevskis, E. ; Jakobsen, L. ; Trinklere, L. & Ulmanis, U. (1997) . Ceramics from fine plasma-processed AlN powder. *Key Engineering Materials,* Vols. 132-136, (1997), pp. 185-188, ISSN: 1662-9795

Sarua, A.; Rajasingam, S.; Kuball, M.; Garro, N.; Sancho, O.; Cros, A.; Cantarero, A.; Olguin, D.; Liu, B.; Zhuang, D. & Edgar, J. (2003). Effect of impurities on Raman and photoluminescence spectra of AlN bulk crystals. *Materials Research Society Symposium – Proceedings,* 798, (May 2003), pp. 297-302, ISSN: 0272-9172

Schweizer, S.; Rogulis, U.; Spaeth, J.-M.; Trinkler, L. & Berzina, B. (2000). Investigation of oxygen-related luminescence centres in AlN ceramics. *phys. stat. sol. (b),* Vol. 219, No. 1, (January 2000), pp. 171-180, ISSN: 0370-1972

Slack, G.A. (1973). Nonmetallic crystals with high thermal conductivity. *J. Chem. Phys. Solids,* Vol. 34, No. 2, (February 1973), pp. 321-335, ISSN : 0022-3697

Tale, I. & Rosa, J. (1984). Fractional glow technique spectroscopy of traps in heavily doped AlN:O. *phys. stat. sol.(a),* Vol. 86, No.1, (January 1984), pp. 319-326, ISSN: 0031-8965

Trinkler, L.; Christensen, P.; Larsen, N.A. & Berzina, B. (1998). Thermoluminescence properties of AlN ceramics. *Rad. Meas.,* Vol. 29, (May 1998), No. 3-4, ISSN: 1350-4487

Trinkler, L. ; Bos, A.J.J. ; Winkelman, A.J.M. ; Christensen, P. ; Agersnap Larsen, N. & Berzina, B. (1999). Thermally and optically stimulated luminescence of AlN-Y_2O_3 ceramics after ionising irradiation. *Radiation Protection Dosimetry,* Vol. 84, Nos. 1-4, (August 1999), pp. 207-210, ISSN: 0144-8420

Trinkler, L.; Botter-Jensen, L.; Christensen, P. & Berzina, B. (2000). Studies of aluminium nitride ceramics for application in UV dosimetry. *Radiation Protection Dosimetry,* Vol. 92, No. 4, (March 2000), pp. 299-306, ISSN: 0144-8420

Trinkler, L. & Berzina, B. (2001a). Radiation induced recombination processes in AlN ceramics. *J. Phys. : Condens. Matter,* Vol. 13, (September 2001), pp. 8931-8938 ; ISSN: 0953-8984

Trinkler, L ; Botter-Jensen, L. ; Christensen, P. & Berzina, B. (2001b). Stimulated luminescence of AlN ceramics induced by ultraviolet radiation. *Rad. Meas.,* Vol. 33, No. 5, (October 2001), pp.731-735, ISSN: 1350-4487

Trinkler, L. ; Botter-Jensen, L. & Berzina, B. (2002). Aluminium nitride ceramics: a potential dosemeter material. *Radiation Protection Dosimetry,* Vol. 100, Nos. 1-4, (April 2002), pp. 313-316, ISSN: 0144-8420

Trinkler, L. ; Berzina, B. & Benadbesselam, M. (2003). Use of AlN ceramics in ultraviolet radiation dosimetry. *Proceedings of SPIE* Vol. 5123, (November 2003), pp. 49-54, ISSN: 0277-786X

Trinkler, L. ; Berzina, B. ; Shi, S. C. ; Chen, L.-Ch. ; Benabdesselam, M. & Iacconi, P. (2005). UV ligth induced luminescence processes in AlN nanotips and ceramics. *phys. stat. sol. (c),* Vol. 2, No. 1, (January 2005), pp. 334-338, ISSN: 1862-6351

Trinkler, L.; Berzina, B.; Auzina, A.; Benabdesselam, M. & Iacconi, P. (2007a). UV light energy storage and thermoluminescence in AlN ceramics. *phys. stat. sol. (c),* Vol. 4, No. 3, (March 2007), pp. 1032-1035, ISSN: 1862-6351

Trinkler, L.; Berzina, B.; Auzina, A.; Benabdesselam, M. & Iacconi, P. (2007b). Use of aluminum nitride for UV radiation dosimetry. *Nuclear Instr. And Methods in Physics Research A,* Vol. 580, (May 2007), pp. 354-357, ISSN: 0168-9002

Youngman, R.A. & Harris, J.H. (1990). Luminescence studies of oxygen-related defects in aluminum nitride. *J. Am. Ceram. Soc.,* Vol. 73, No. 11, (November 1990), ISSN: 0002-7820

11

Characterization of Clay Ceramics Based on the Recycling of Industrial Residues – On the Use of Photothermal Techniques to Determine Ceramic Thermal Properties and Gas Emissions during the Clay Firing Process

Faria Jr. R. T.[1], Souza V. P.[1], Vieira C. M. F.[2], Toledo R.[1], Monteiro S. N.[2], Holanda J. N. F.[2] and Vargas H.[1]
Universidade Estadual do Norte Fluminense Darcy Ribeiro – UENF
[1]Laboratório de Ciências Físicas,
[2]Laboratório de Materiais Avançados,
Parque Califórnia, Campos dos Goytacazes, RJ
Brasil

1. Introduction

Since Sumerians about 4000 - 3000 BC had created ceramics (Heydorn *et al.*, 1989), the ceramic production has worldwide increased, but the methodology has not suffered substantial evolution in relation to the ancient process. Research work continues to a better understanding of the processes involved and tailoring the desired properties of the final product. The most of ceramists has just worried up with the morphological, physical and structural quality of their products (Okada *et al.*, 1986; Dubois *et al.*, 1995; Nastro *et al.*, 2006). However, one of the main ceramics features was not yet well concerned by the community, that is, its thermal behaviour. Urban buildings of equatorial and tropical countries are often subjected to constant solar radiation. Thus, thermal properties play a significant role in the quality of end-products. A better indoor ambient of the houses should be ever pursued. While technological properties of those materials are frequently found in the literature, information about the thermophysical characteristics of red ceramics is very scarce (Alexandre *et al.*, 1999).

In the early eighties, a new physical device to generate and detect thermal waves was proposed (Vargas and Miranda, 1988). It has been widely used in last few years as a tool to measure the thermal properties of a large range of materials (Marín *et al.*, 2002; Faria *et al.*, 2005). The photoacoustic (PA) technique via OPC (an open photoacoustic cell) is used to evaluate the thermal diffusivity, an important parameter that determines the light-to-heat conversion efficiency, and furthermore, it allows us to study how heat diffuses through the sample (Perondi and Miranda, 1987). The thermal diffusivity is an intrinsic property of matter. Moreover, it is dependent upon the effects of composition and microstructural variables, as well as processing conditions. Thermal properties are very sensitive to amorphous–crystalline phase transitions of clays heated above 900 °C, typical temperatures applied in the manufacturing process of ceramic materials.

Clay deposit areas have a high economic potential, and soil mineralogical composition, plasticity and porosity are fundamental properties for industrial applications. The resulting ceramic products present a network of capillary pores that vary according to the firing cycle. Important ceramic properties such as hardness, water absorption, gas permeability, shrinkage, density, thermal properties, gas emission as well as the vitrification temperature range are intrinsically correlated to the porous system (Rice and Dekker, 1998).

In addition, the whole world presently draws increasing attention to the environment, requiring the adjustment of industrial activities according to laws and norms that rule the sector. Thus, increasing environmental problems related to climate changes and pollution are going attention and raising special concern regarding to the dispersal of industrial wastes. As a consequence, the society is demanding not only actions towards waste reduction but also measures for redirecting or post-using wastes mainly those of industrial origin. Most of the residues generating in the production process is normally handled by land disposal in certain areas, thus recycling of these residues may produce the value-added products and prevent environmental deterioration. As solid wastes are becoming a matter of increase world concern due to their amount as well as difficulty and cost of final disposal, ceramic products can well supply as matrices for industrial residues incorporation such as giving an option for the solid waste disposal with a possible increase in the properties of the clay ceramic products. The incorporation of wastes from several industrial activities is a technological alternative to reduce the environmental impact due to their indiscriminate disposal. One such a measure is the waste incorporation into intensive-produced clay ceramics like bricks and tiles (Parsons *et al.*, 1997). This incorporation is now a common practice and benefits both, the industry that generates the waste and the ceramic industry that saves on the clay, which is its basic raw material. In some cases the incorporation also improves the properties of the ceramic product (Oliveira and Holanda, 2004).

Research works on the structural and physical changes that wastes, mainly industrially produced, cause to the final ceramic product has been rapidly expanding in these last two decades. See, for instance references in (Vieira and Monteiro, 2009). In spite of these examples of a continuous effort to incorporate wastes into clay ceramic motivated by technical and economical advantages, environmental issues are still a matter of concern. The firing stage during the clay ceramic process can promote the elimination of potentially toxic constituents present in the solid residues through their volatilization. However, the gas emission due to the clay firing process and its related atmospheric pollution may be enhanced as a consequence of an incorporated waste (Souza *et al.*, 2008). It is known that the firing of clay ceramics in conventional furnaces using fuels such as wood, charcoal, heavy oil and natural gas generates appreciable amounts of gaseous components, mainly carbon dioxide (CO_2), carbon monoxide (CO) and methane (CH_4). Additionally to the atmospheric pollution, these gas sets could be harmful to the human health and can corrode equipments (Morgan, 1993).

Works on the effect of gas emission caused by the firing of clay ceramics incorporated with industrial wastes are practically inexistent. Therefore one of the mean objectives of this chapter was to investigate the gas emission resulting from the firing of clay ceramics incorporated with different industrial residues. Here, slag type of waste from steel refining operation and sanitary ware mass wastes were the selected residues.

In order to show methodologies to characterize ceramics here depicted, the samples under study were clay from Rio de Janeiro state and from São Paulo state. Firstly, the raw materials without residues is presented and throughout characterized. In a second stage, the aggregation of the residues is pointed out. Characterizations are performed and compared with the clays without residues.

2. Ceramic samples

Clayey soils from two sites of the Southeast Brazilian region were studied for the production of red ceramics. The samples preparation was pointed out in the following.

2.1 Kaolinite clay from Rio de Janeiro state (CG clay)

The raw material was collected from a quaternary sedimentary basin situated in Campos dos Goytacazes, from a layer localized between 0.30 m and 1.80 m below the surface, where large clay deposits are quarried as raw material for red bricks and tiles. About 20 kg of material collected in situ were dried, grounded, and passed through a sequence of sieves yielding a homogeneous powder with particle size smaller than 75 μm, corresponding to 95 % of the original mass.

For technological measurements, a quantity of the powder was mixed with water and then extruded at 36 MPa into a rectangular prismatic mould with a 20x10 mm cross-section, cut into bars 100 mm long which were left drying in air for two weeks at room temperature. Before the firing process, the samples were subjected to an additional drying cycle in an oven at (110 ± 5 °C) for 24 hours. Then, the sample weight and dimensions were measured just before starting the firing process. X-ray diffraction measurements were done on the same samples used for technological measurements.

To simulate the continuous sintering process of the ceramic material up to 1200 °C, and to accompany the changes that occur, 13 thermal treatment steps were chosen, each step being denoted T_F henceforth. For firing temperatures up to 600 °C the heating rate was set to 2 °C/min. Once the desired T_F was reached, the sample was left at constant temperature for 3 hours, and then cooled down to room temperature at a rate of 1.5 °C/min. For T_F above 600 °C, the heating cycle was such that the sample was heated at a rate of 2 °C/min up to 600 °C, left at this temperature for one hour, and then heated at a rate of 10 °C/min up to the selected firing temperature, where it was left for 3 hours before cooling down to room temperature at a rate of 1.5 °C/min.

To perform the thermal characterization the raw material was passed through a sieve with nominal aperture of 75 μm (mesh 200 ASTM). In order to prepare the samples for analysis, a press (SHIMADZU) was used. 90 mg of each sample was weighed and submitted to a 9 tonne pressure for 10 minutes, using disc samples with 10 mm diameter and a thickness between 200 μm and 500 μm. After that, the samples were fired in a furnace (MAITEC 12 kW/32 A) at six different temperature steps, ranging from 900 to 1100 °C.

Finally, to analyse the gas emissions from the CG clay and its residue, rectangular 114.5 x 25.4 x 10 mm samples with 8 % of humidity were 20 MPa moldpress by mixing the clay with different amounts of 5 and 10 wt% of steel slage. Samples of pure clay were also considered for comparison. All samples were initially dried at 110 °C in a laboratory stove and then fired in a continuous way, from room temperature to a maximum of 1100 °C, inside a tubular model FT-1200 BI, MAITEC electrical furnace. The firing was conducted at a heating rate of 2 °C/min from RT to 500 °C and 4 °C/min between 500 and 1100 °C.

2.2 Illitic clay from São Paulo state (J clay)

The raw material was collected from a neocenozoic sedimentary soil placed in Jundiaí city, 60 km far from São Paulo city in São Paulo state. The preparation process was similar to the sample of Campos dos Goytacazes for both applications. However, the J clays do not have such plasticity as CG clays. According to their chemical properties (sub-section 3.1), they can be considered fluxes clays mainly by the alkaline elements (K_2O, CaO and MgO), which are responsible for the ceramic sintering. The value of K_2O comes from the mineral illite, while CaO + MgO are linked with calcite and dolomite minerals. Al_2O_3 is related mainly to the kaolinite mineral.

3. Characterization by X-Ray

The chemical composition and crystalline phases of the raw material with and without residues and only of the residues were presented. The microstructure of the samples as in function of the firing temperature was represented by X-ray diffractions.

3.1 Energy dispersive X-Ray fluorescence

Using X-ray fluorescence spectrometry methods described in previous work Mota *et al.*, 2008), the percentual contribution of the main oxides present in our material was determined.
The chemical characterizations were obtained by energy dispersive X-ray spectroscopy (EDS), using a SHIMADZU equipment. The resulting values of this quantitative analysis from CG clay and its steel slag residue and from J clay and its sanitary ware mass residue are shown in Table 1 and Table 2, respectively.
Observing Table 1 and Table 2 we can classify CG clay as kaolinite-based clay (higher Al_2O_3) and J clay as illitic-based clay (higher fluxes oxides), as explained in the previous section. In addition to SiO_2 and Al_2O_3, there are considerable amounts of Fe_2O_3, K_2O and TiO_2,, which is favourable for red ceramics production. Iron oxide accounts for the bodies´ red pigmentation after firing.

Oxides in wt%	CG Clay	Steel slag residue
SiO_2	46.42	10.29
Al_2O_3	27.90	21.5
MnO	0.11	5.40
MgO	0.71	10.33
CaO	0.22	45.10
Na_2O	0.36	-
K_2O	1.67	-
TiO_2	1.32	0.34
Fe_2O_3	9.10	*
P_2O_5	-	1.85
SO_3	-	0.54
SrO		0.18
LoI	11.96	12.96
* Total Fe from the residue (%) - 23.62		

Table 1. Chemical composition of the CG clay and steel slag (% in weight). The slag predominantly is formed by composites of Ca, Al, Fe, Mg and Si

Oxides in wt%	J Clay	Sanitary ware mass residue
SiO_2	59.61	64.8
Al_2O_3	21.19	21.5
MnO	0.05	0.06
MgO	1.07	0.78
CaO	0.33	0.32
Na_2O	0.12	0.51
K_2O	2.33	2.53
TiO_2	0.94	0.89
Cr_2O_3	-	0.05
P_2O_5	0.13	0.13
SO_3	-	0.06
Fe_2O_3	5.45	5.96
ZnO	-	0.03
ZrO_2	-	0.13
LoI	9.28	2.25

Table 2. Chemical composition of the J clay and the sanitary ware mass residue (wt.%). The residue is a silt-clay with very low plasticity which should enhance the structural quality of the ceramic

3.2 X-Ray diffraction

The X-ray diffraction experiments were performed at room temperature Using the Co-Kα or Cu-Kα radiation of a Seifert URD65 diffractometer, equipped with a diffracted beam graphite monochromator. The CG clay diffractograms were obtained from 3 to 50° using the Co-Kα, while the slag steel residue from 3 to 70°.The J clay diffractograms were obtained from 3 to 75° with step sizes of 0.03° and accumulation time of 3 s. From the diffractograms the sample crystallinities were determined as the ratio between the integrated intensity of the sharp diffraction peaks to the total area of the diffraction pattern, which includes the non-coherent intensity. The CG clay sample was thoroughly analysed as in function of the firing temperature, whereas the J clay was just analysed the 110 °C sample.

Fig. 1 shows the XRD patterns as a function of the firing temperature TF. A quantitative phase analysis was performed for the natural powdered soils of Campos dos Goytacazes using the Rietveld refinement method (Manhães *et al.*, 2002). The Rietveld analysis was performed considering the angular range $10° < 2\theta < 70°$, thus not taking into account the illite phase due to its peak shape. Kaolinite is the main crystalline phase with 86% mass fraction, followed by quartz (5%), anatase (5%) and gibbsite (4%).

The diffractograms of the samples treated up to 400 °C show a similar mineral composition with kaolinite as the major phase. Between 400 and 500 °C the kaolinite dehydroxylation occurs, with the transformation to a non-crystalline phase, metakaolin: $Al_2Si_2O_5(OH)_4 \rightarrow Al_2Si_2O_7 + 2H_2O\uparrow$. The other crystalline phases (illite and quartz) remain unchanged up to 900 °C. For this temperature range no more organic material is present in the amorphous component. . The main crystalline phases present in our non-treated samples were kaolinite ($Al_2Si_2O_5(OH)_4$), quartz (SiO_2), anatase (TiO_2), gibbsite ($Al(OH)_3$), goethite ($FeO(OH)$) and illite ($(K,H_3O)Al_2Si_3AlO_{10}(OH_2)$), while the sinterized samples showed the presence of mullite ($Al_6Si_2O_{13}$), cristobalite (SiO_2) and hematite (Fe_2O_3). Quantitative phase analysis using the Rietveld refinement method showed that kaolinite is the main crystalline phase responding for approximately 86 wt. %.

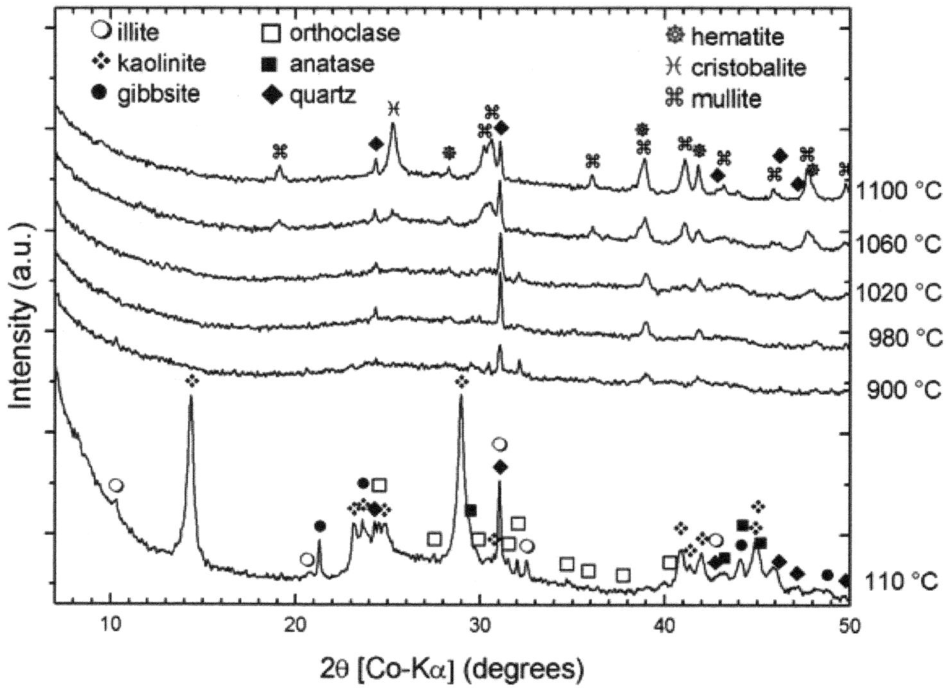

Fig. 1. X-Ray diffractograms of CG clay as in function of firing temperature

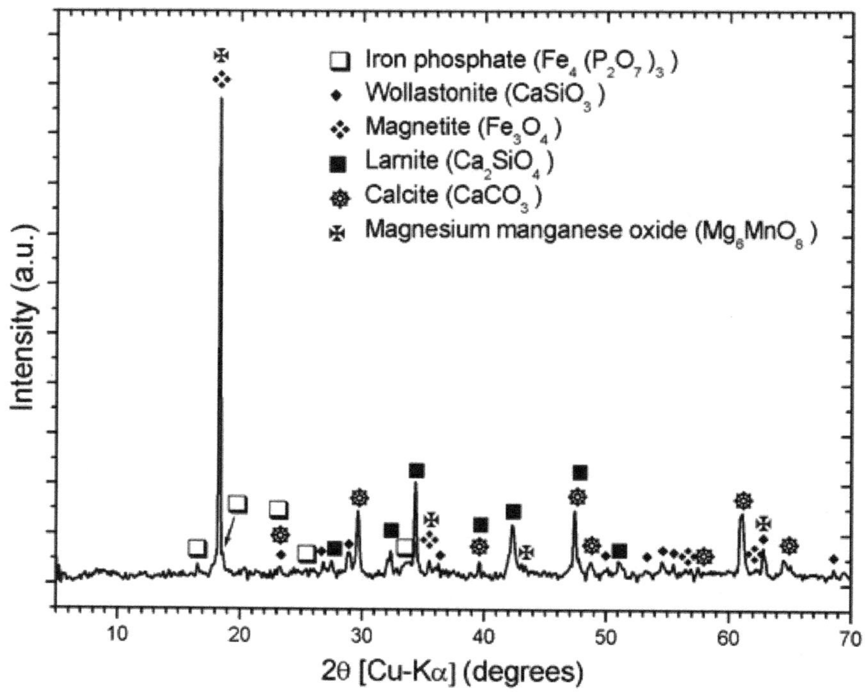

Fig. 2. Steel slag residue as processed diffractogram

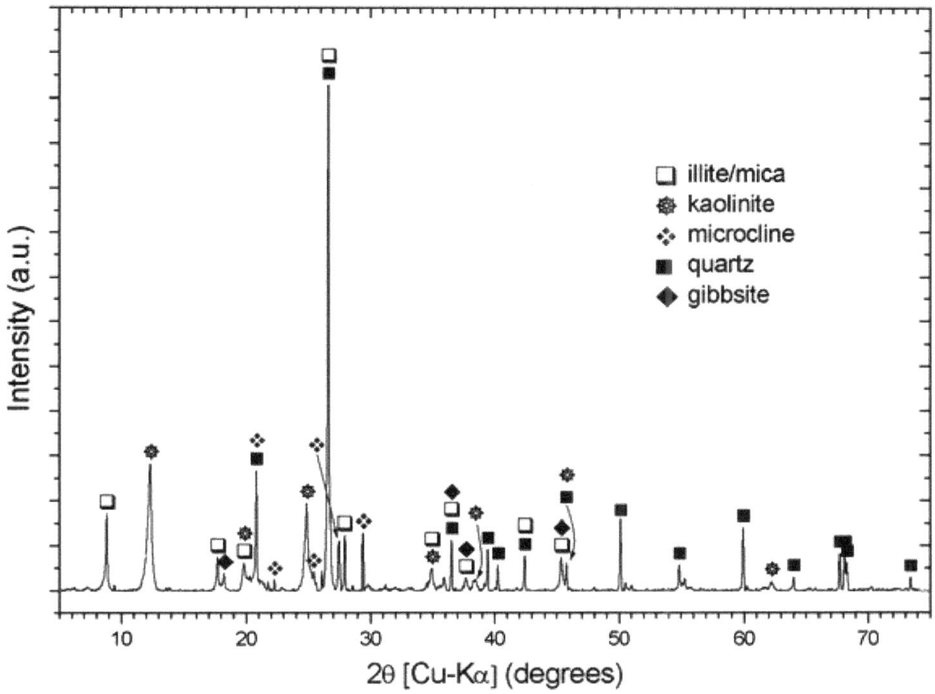

Fig. 3. X-ray diffractogram of J clay at room temperature

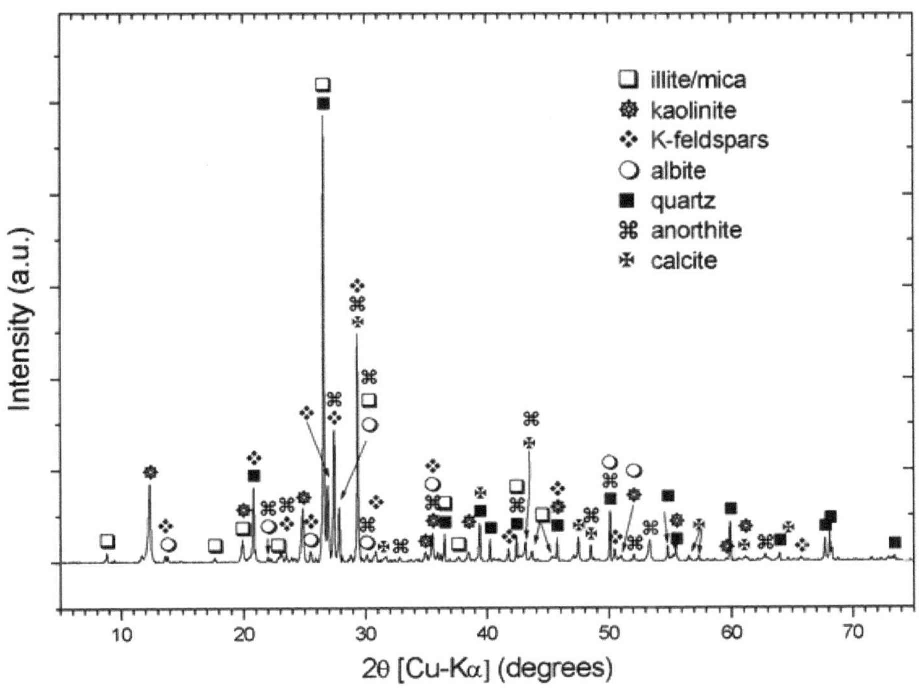

Fig. 4. X-ray diffractogram of the sanitary ware mass residue

Fig. 2 shows the X-rays diffraction pattern of the steel residues. In the respective figure the predominantly formed crystalline phases of Ca and Fe had been identified to associate diffraction peaks. The rich phases in calcium (Ca) are silicates and the calcite. On the other hand, iron (Fe) is present in the form of magnetite and iron phosphate. Diffraction peaks of the magnesium manganese oxide were identified. In accordance to Fig. 2, Ca is present in the wollastonite ($CaSiO_3$) and larnite (Ca_2SiO_4) form and also as carbonate (calcite – $CaCO_3$). Mg and Mn are presented as magnesium manganese oxide (Mg_6MnO_8), while Fe is given in form of magnetite (Fe_3O_4).

Fig. 3 shows the XRD patterns of the J clay raw material. This diffractogram indicates the presence of a micaceous mineral, quartz, microcline, gibbsite and kaolinite.

Fig. 4 points out the sanitary ware mass diffractogram, which shows the presence of micaceous material, albite, anorthite, k-feldspars, clay minerals, quartz, and calcite. We observed a variety of flux materials. Feldspars and microcline are sources of alkaline flux materials, such as K_2O and Na_2O, which make possible the formation of a liquid phase above 700 °C (Reed, 1976). The fluxing capacity of this residue, which is associated with lower porosity after firing, is also confirm by the presence of K_2O and Na_2O containing minerals. Then, the sanitary ware mass is the source of K_2O and Na_2O, which act as fluxes to improve the sintering process.

4. Technological essays

4.1 Density and porosity measurements

The samples dimension and mass were measured before and after each firing cycle so that the volumetric shrinkage and bulk densities could be straightforwardly determined.

Between 500 and 1200 °C, the porosity measurement consisted in measuring first the mass of the dry sample after firing at temperature T_F, denoted henceforth as m_{TF}, then two water-saturated sample masses: one, denoted as m_i, inside water after 2 h boiling, the other, denoted as m_o, outside taking care to remove the excess water from the sample surfaces. The sample open porosity was obtained as the ratio $(m_O - m_{TF}) / (m_O - m_i)$.

For samples that undergone heat treatments at 110, 300 and 400 °C, we have adopted the conventional pycnometer method, the reason for this being that these samples usually exhibit larger porosity and lower particle cohesion than those treated at higher temperatures. In this way, we have avoided eventual sample disintegration due to prolonged water immersion.

4.2 Mechanical testing

The flexural rupture tension of our samples was determined by a universal testing machine using the three-point loading test with a crosshead speed of 0.1 mm/min, designed following the ABNT standard procedure[17], method equivalent to ASTM C67. The bending strength was calculated from the breaking load using the relation $BS = 3PL/2bd^2$, where P is maximum load, L is the distance between the supports of the beam, b is the brick average width and d is the brick average depth.

Figure 5 indicates that after 1000 °C the open porosity decreases abruptly from 39.6 % to 5.6 %, while the volumetric shrinkage and the bending strength both exhibit a large increase from 16.8 % to 40.0 % and from 9.0 MPa to 25.2 MPa, respectively. These results together with the X-ray diffraction data indicate that these changes are associated with recrystallization and sintering processes setting in between 1000 and 1200°C with the formation of mullite, cristobalite, and hematite phases, with great improvement of the ceramic properties. To further strength this aspect we present in Fig. 6 the correlations of the sample density, bending strength and volumetric shrinkage with the sample crystallinity, in the temperature range above 1000 °C. The solid lines in this figure correspond to the data linear regression. These results indicate that, indeed, these physical properties are all well correlated with the crystallinity changes taking place in our sample in this temperature interval.

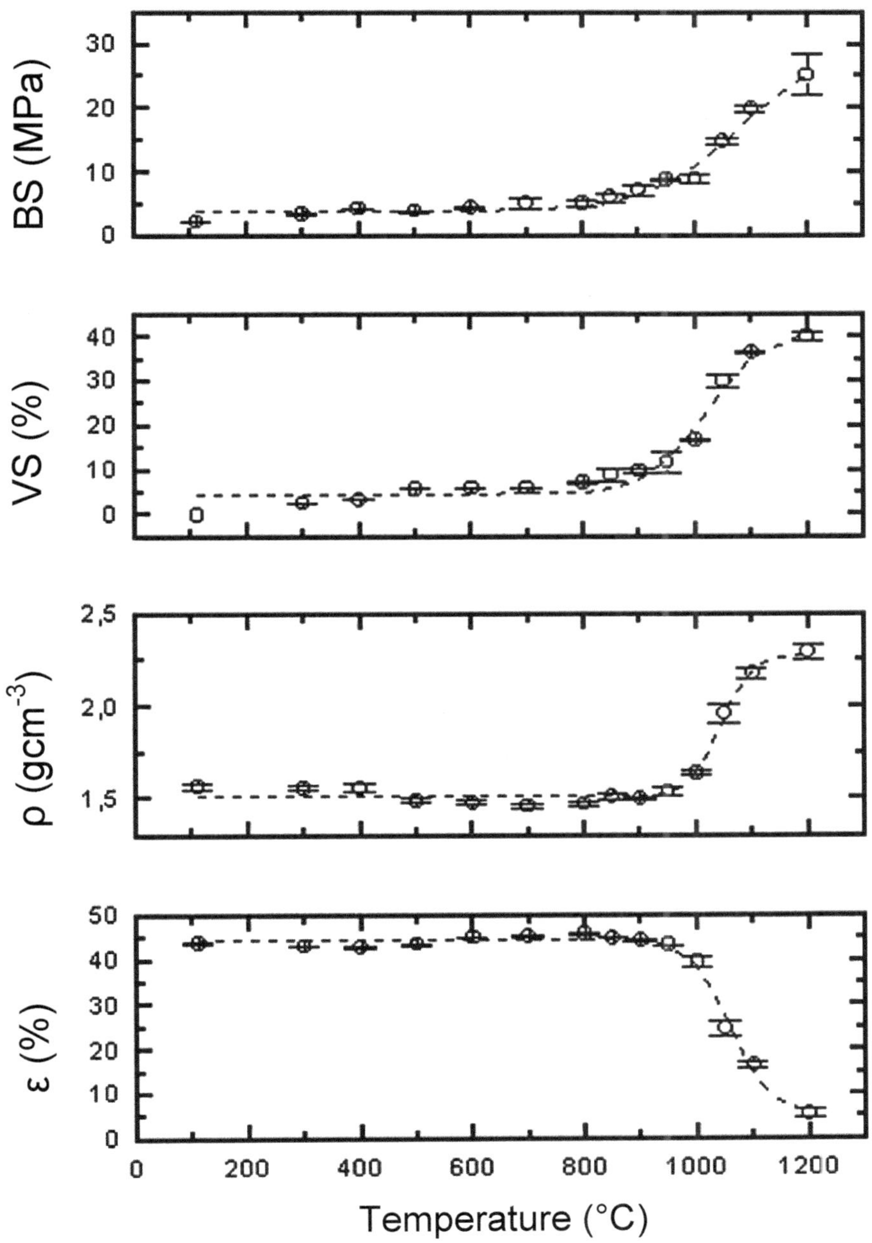

Fig. 5. The dashed lines in the bending strength, volumetric shrinkage, density and porosity plots correspond to the data fitting using a logistic function (Eq. 1)

In Fig. 5 we present the evolution of the technological properties (bending strength (BS), volumetric shrinkage (VS), bulk density (ρ) and open porosity (ε)) as a function of the firing temperature. In order to single out the temperature point at which the phase transitions taking place in our sample affect most the corresponding physical properties we have performed a fitting of our data to a logistic curve of the following form,

$$f(T) = A + B\frac{\exp((T - T_0) / \Delta T)}{1 + \exp((T - T_0) / \Delta T)}, \quad (1)$$

where A, B, T_0 and ΔT are all adjustable parameters. The dashed lines Fig. 5 correspond to this logistic data fitting. In Eq. 1, T_0 indicates the point at which the transition occurs whereas ΔT gives us an idea of how wide this transition is. The resulting values obtained from this data fitting procedure for the parameters T_0 and ΔT are summarized in Table 3.

Data	T_0 (°C)	ΔT (°C)
Open Porosity (ε)	1057.03	35.62
Bulk Density (ρ)	1042.57	28.00
Volumetric Shrinkage (VS)	1021.61	52.68
Bending Strength (BS)	1079.97	80.39

Table 3. Values of the data fitting parameters characterizing the temperature and its excursion where the main changes of the physical properties of our samples occur

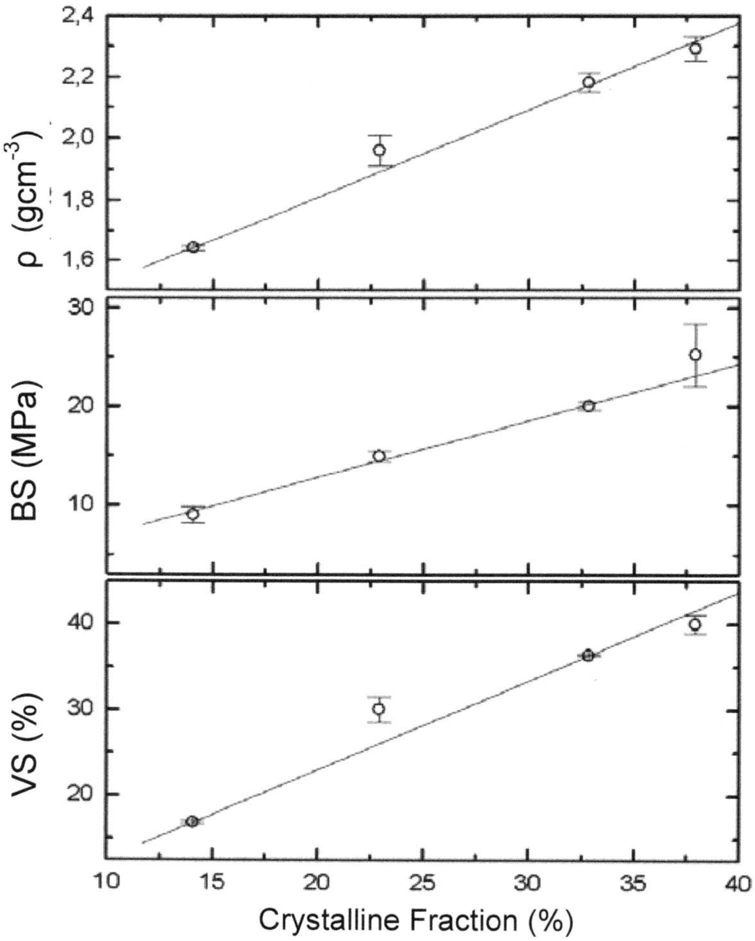

Fig. 6. Correlations of the sample density, bending strength and volumetric shrinkage with the sample crystallinity, in the temperature range above 1000 °C. The solid lines correspond to the data linear regression

The data shown in Fig. 6 also suggest some other interesting correlations among the different measured physical properties. For instance, the changes in bending strength and the volumetric shrinkage, as well as those of the porosity and density, seem to be well correlated.

We present in Figs. 7 and 8 the correlations between these physical properties. The straight lines in these figures correspond to the data linear regression. It follows from Fig. 6 that the bending strength and the volumetric shrinkage are well correlated up to 1050 °C, at which point a sharp discontinuity in the correlation takes place.

As to the correlation between the sample porosity and density, shown in Fig. 5, the same behaviour is observed, that is, a good correlation among these properties up to 1050 °C followed by a sharp discontinuity at this temperature. These critical transitions taking place around 1050 °C, are in good agreement with the values found from the logistic data fitting for T_0, as shown in Table 3, as well as with the argument presented above indicating that a sintering process is set in around this temperature.

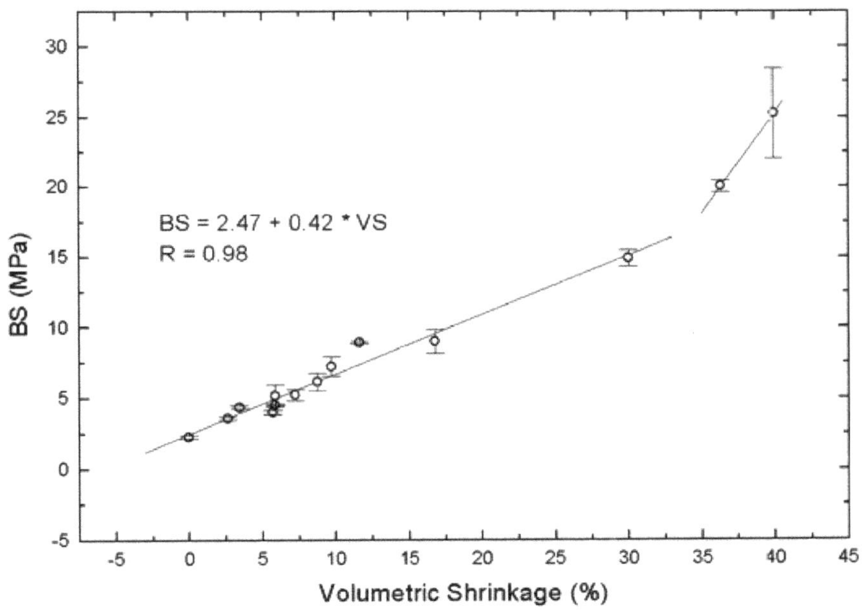

Fig. 7. Correlation between the bending strength and the volumetric shrinkage. The straight lines correspond to the data linear regression

The phase transformations from X-ray diffraction data are well known. Up to 400 °C the dominant phase transformations are due to intense dehydroxylation processes of components, such as gibbsite and goethite, with kaolinite remaining as the major mineral phase. Between 400 and 500 °C, we observed a reduction in crystaline fraction, due the transformation of kaolinite to a noncrystalline phase, metakaolinite ($Al_2Si_2O_5(OH)_4 \rightarrow Al_2Si_2O_7 + 2H_2O$) took place, while the other crystalline phases (quartz, anatase and illite) remained unchanged up to 900 °C. Heat treatments above 1000 °C cause new structural changes with the formation of mullite, cristobalite, and hematite phases, among others.

Considering the technological measurements of the J clay and the sanitary ware mass residue, Table 4 presents the technological properties of volumetric shrinkage, bending strength and water absorption for the product fired at 980 °C, considered the best temperature for commercial purposes.

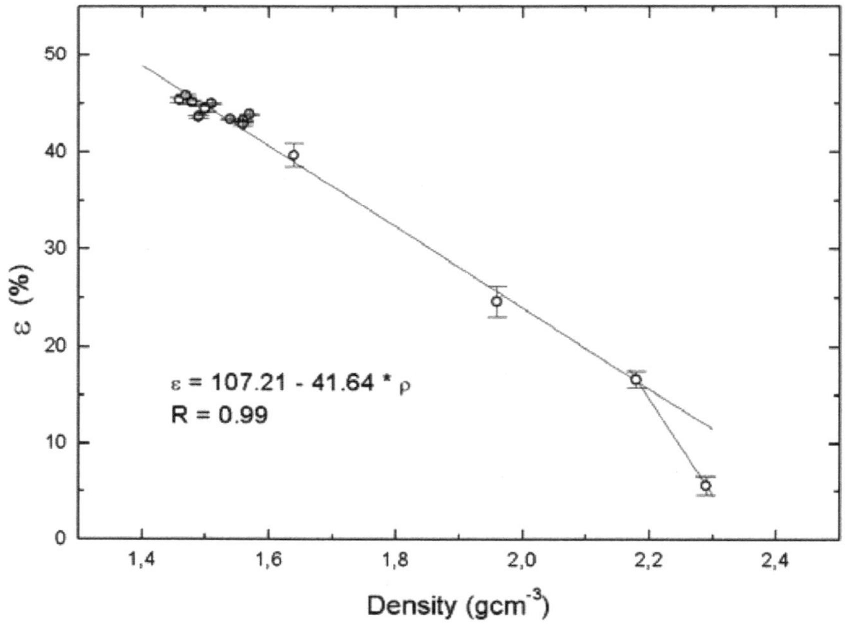

Fig. 8. Correlation between the samples porosity and density. The straight lines correspond to the data linear regression

Sample	Volumetric Schrinkage (%)	Bending Strength (MPa)	Water Absorption (%)
J Clay	2.1	3.3	20.3
J Clay 30	1.5	2.6	20.9

Table 4. Technological properties of end-product ceramic (~980 °C) without residue (J clay) and with 30 wt.% of sanitary ware residue (J Clay 30)

Comparing the Kaolinitic CG clay and the illitic J clay, we noted that the volumetric shrinkage value of J clay is half of the value of CG clay. In this case, J clay has advantages for the ceramic serial production in relation to dimensions uniformity. On the other side, CG clay presents a double value of bending strength, in relation to the J clay, i. e., more resistant pieces are shaped.

5. Thermal characterisation of CG and J clays using photothermal techniques, in relation to structural changes on firing

By analysing the thermal properties, some information can be collected from the material. In the following, we define the thermal properties and point out the values and their correlations with the sample microstructure.

Thermal diffusivity, a particular property that evaluates the heat propagation behaviour within the sample, is quite sensitive to the material structure as well as preparation and sintering conditions.

When values of thermal diffusivity (α) are evaluated from the amplitude data of the photoacoustic signal, we should pay attention to the microphone non-linear frequency response in relation to acoustic vibrations. Practically, all microphones present this irregularity. In our case, our microphone had a good linear frequency response above 20 Hz. In order to certify our set-up, a calibration measurement was performed. Figure 9 shows the dependence of the photoacoustical (PA) signal on the modulation frequency for the aluminium (Al) sample.

For frequencies used in this work, the signal exhibited a frequency dependency close to $f^{-1.5}$. This is the typical behaviour we would expect from the thermal diffusion model for a thermally thin sample. In fact, for a 25 μm thick Al sample and a thermal diffusivity of 93.28 x 10^{-6} m^2/s (Almond and Patel, 1996) the characteristic frequency f_c for the transition between the thermally thin and thick regime is about 47.5 kHz.

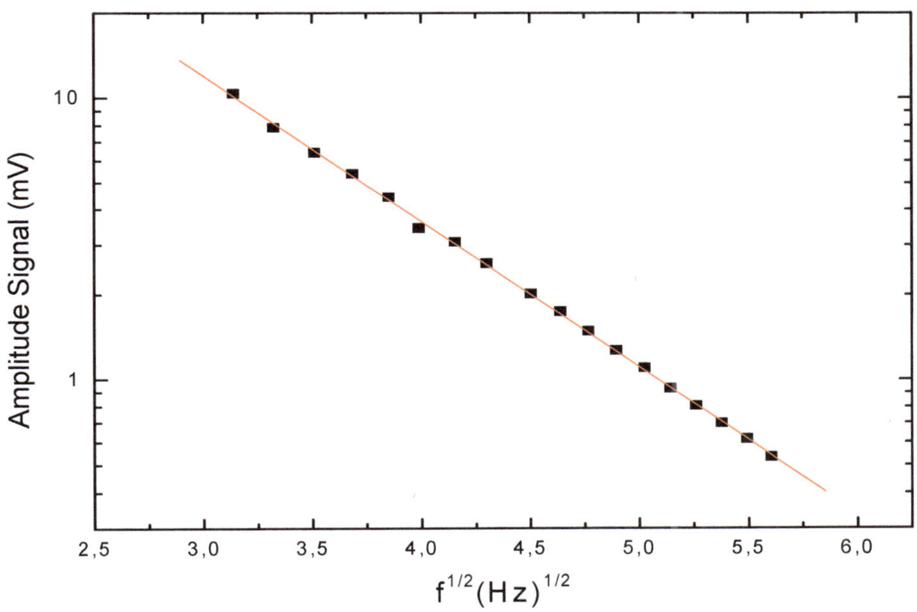

Fig. 9. Microphone output voltage as a function of the chopping frequency for the 25 μm thick Al sample. The signal behaves roughly as $f^{-1.5}$

The thermal conductivity (κ) is associated with thermal conduction, which is the phenomenon where the heat is carried from regions of high temperature to regions of low temperature in a material.

The thermal effusivity (ε) denotes the material thermal impedance and is a measurement of the heat energy stored in a solid per degree of temperature rise from the start of a surface heating process (Marín *et al.*, 2002).

Here, thermal conductivity and thermal effusivity are reached applying the following expressions:

$$k = \alpha \rho c \tag{2}$$

$$\varepsilon = \sqrt{k\rho c} \tag{3}$$

Experiments concerning with thermal diffusivity, samples thickness and specific heat capacity measurements were performed five, ten and three times to produce the deviations, respectively.

5.1 Thermal diffusivity measurement

The thermal diffusivity experimental set-up is shown in Fig. 10. The sample is mounted directly onto a commercial electret microphone (Omnidirectional back electret condenser microphone cartridge, model WM-61A - Panasonic), fixed with a silicone grease, illuminated by the light beam from a He-Ne laser (Unilaser mod. 025) and modulated with a mechanical chopper (EG&G Instruments mod. 651), before it reaches the sample's surface. It consists of an OPC configuration in the sense that the sample is placed on top of the detection set-up itself (Vargas and Miranda, 1988). As a result of the periodic sample heating by modulated light absorption, the pressure inside the cell oscillates at the chopping frequency and can be detected by the microphone. The resulting PA signal is then subsequently fed into a field-effect-transistor (FET) pre-amplifier and leads directly to a "Lock-in" amplifier (Perkin Elmer Instruments mod. 5210), where it is possible to obtain the photoacoustic amplitude and the phase signal, which are recorded as a function of the modulation frequency in an appropriate software program.
According to the model proposed by (Rosencwaig and Gersho, 1976) for thermal diffusion, the equation that leads us to the pressure fluctuation (δP) in the air chamber is

$$\delta P = \frac{\gamma P_0 I_0 (\alpha_g \alpha_s)^{1/2}}{2\pi l_g T_0 k_s f} \frac{\exp j(\omega t - \pi/2)}{\sinh(\sigma_s l_s)}, \tag{4}$$

where γ is the air specific heat ratio, P_0 the ambient pressure, T_0 ambient temperature, I_0 is the absorbed light intensity, f is the modulation frequency, and l_i, k_i and α_i are the length, thermal conductivity and the thermal diffusivity of the sample, respectively. Here, the subscripts denote the absorbing samples *(s)* and the gas *(g)* media, respectively, and $\sigma_i = (1+j)a_i$, is the complex thermal diffusion coefficient of *i* medium. It is assumed in the Eq. 4 that the sample is optically opaque $l_\beta << l_s$, where l_β is the optical penetration depth. For thermally thin sample $\mu_s > l_s; f < f_c$, where μ_s is the thermal diffusion length and $f_c = \alpha / \pi l_s^2$ is the cut-off frequency, Eq. 4 reduces to

$$\delta P = \frac{\gamma P_0 I_0 (\alpha_g)^{1/2} \alpha_s}{2\pi^{3/2} T_0 l_g l_s k_s} \frac{\exp j(\omega t - 3\pi/4)}{f^{3/2}} \tag{5}$$

Here, the amplitude of the PA signal decreases as $f^{-1.5}$. On the other hand, at modulation frequencies above f_c, the sample is regarded as thermally thick and the corresponding equation is

$$\delta P = \frac{\gamma P_0 I_0 (\alpha_g \alpha_s)^{1/2}}{\pi l_g T_0 k_s} \frac{\exp - l_s \sqrt{\pi f / \alpha_s}}{f} \exp j(\omega t - \pi/2 - l_s \alpha_s) \tag{6}$$

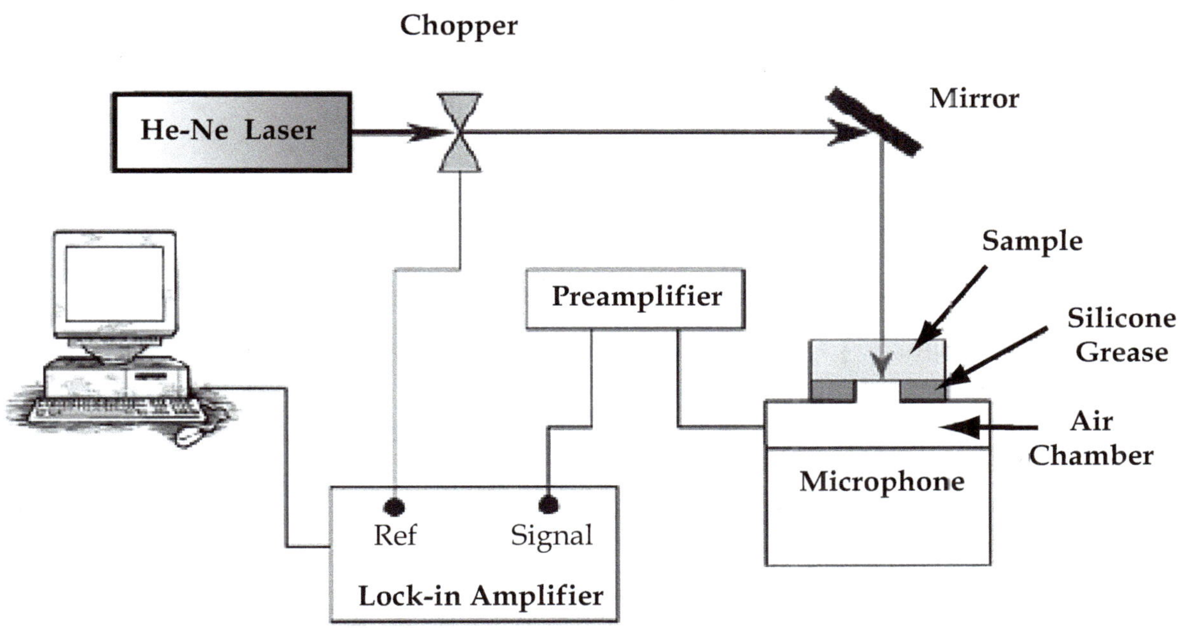

Fig. 10. Experimental arrangement for the thermal diffusivity measurements following (Yunus *et al.*, 2002)

The amplitude of the PA signal for a thermally thick sample decreases exponentially with the modulation frequency as $S_{PA} \propto 1/f \exp(-b\sqrt{f})$, where $b = l_s\sqrt{\pi/\alpha_s}$. In this case, α_s is obtained from the experimental data fitting from the parameter *b*.

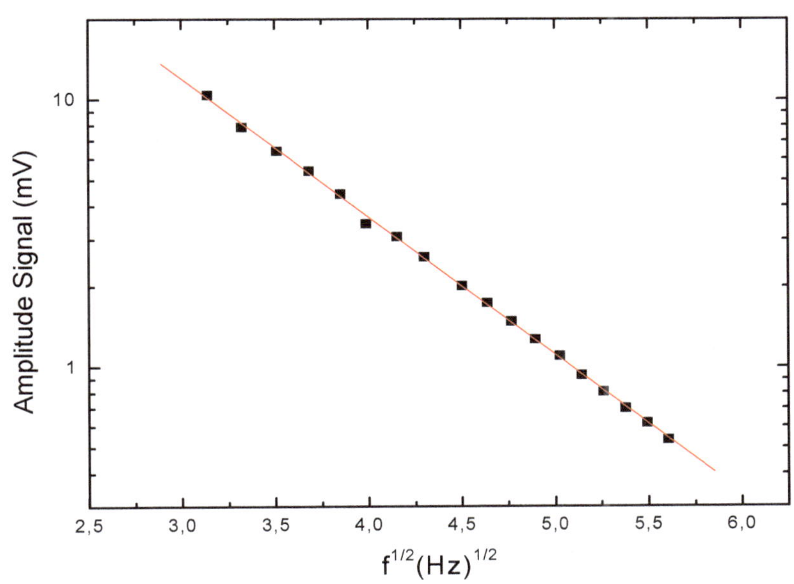

Fig. 11. Photoacoustic signal amplitude measurements as a function of the frequency square root for the 900 °C sample, together with the theoretical line

Taking values of thermal diffusivity from the literature, the cut-off frequency f_c is about 5.87 Hz, i.e. the frequency domain is regarded as thermally thick. As an example, Fig. 11 shows one of our sets of experimental points and the theoretical line for the 900 °C sample.

5.2 Specific heat capacity measurement

Specific heat capacity illustrates how a material can store large amount of heat, without suffering drastic temperature changes. The specific heat capacity (ρc) was measured using the photothermal technique of temperature evolution induced by continuous illumination of the sample in vacuum. The sample surface was painted black to make the parameter characterising the radiative properties of the surface equals to unit $(\varepsilon = 1)$ and placed inside a Dewar, which is subsequently vacuum-sealed, As shown in Fig. 12. The front surface was illuminated by the He-Ne laser focused on the sample through an optical glass window on the Dewar. On the sample backside was connected a thin-wire T-type thermocouple. The thermocouple output was measured as a function of time by using a thermocouple monitor (Model SR630, Stanford Research Systems) attached to a computer. Care has been taken to prevent the heating light from reaching the thermocouple. Since the sample thicknesses, typically of the order of micrometers was much smaller than their widths (e.g., 10 mm), the simple one-dimensional heat diffusion equation with radiation losses can be applied to our experiment (Mansanares *et al.*, 1990).

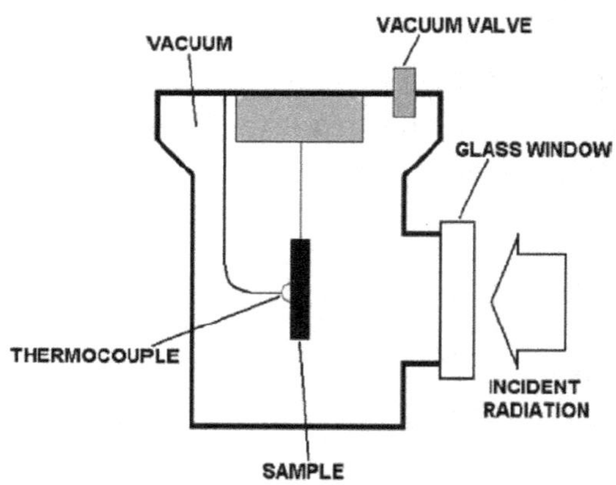

Fig. 12. Experimental arrangement for the specific heat capacity measurements

The temperature evolution was monitored from the ambient temperature up to the sample temperature saturation, obeying Eq. 7 and then the light was turned off and the temperature variation recorded until it reached the ambient temperature again (Eq. 8).

$$\Delta T \uparrow = \frac{I_0}{H}(1 - \exp(-t / \tau)) \tag{7}$$

$$\Delta T \downarrow = \frac{I_0}{H}\exp(-t / \tau) \tag{8}$$

Here, I_0 is the incident light intensity, $\tau = l\rho c/2H$ is the thermal relaxation time, $H = 4\sigma T_0^3$ is the heat transfer coefficient, where σ is the Stefan-Boltzmann constant, T_0 is the ambient temperature and t is the time variable (Mota *et al.*, 2008).

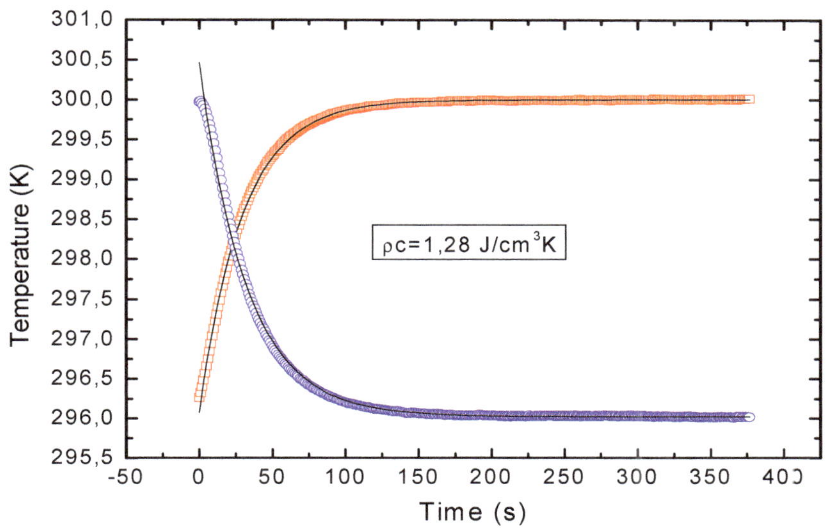

Fig. 13. Experimental data and theoretical lines of rising (square) and decreasing (circle) temperatures as a function of time for the 1060 °C sample

As an example, Fig. 13 shows the experimental data (open symbols) and theoretical lines for the 1060 °C sample (CG clay).

5.3 Results and discussion

Results of measurements of the thermal properties from CG clay are shown in Table 5 as a function of firing temperature.

Temperature (°C)	Thickness (μm)	Thermal diffusivity (cm^2/s)	Specific heat capacity (J/cm^3 K)	Thermal conductivity (W/cmK)	Thermal effusivity ($Ws^{1/2}/cm^2$ K)
900	275±1.56%	0.012±2.58%	1.367±4.13%	0.016	0.148
950	273±3.58%	0.025±4.96%	1.064±3.08%	0.027	0.168
980	419±0.61%	0.012±2.99%	1.178±3.64%	0.014	0.129
1020	323±1.41%	0.021±2.16%	1.201±1.34%	0.025	0.174
1060	330±0.63%	0.031±0.22%	1.141±2.14%	0.035	0.200
1100	323±1.56%	0.013±2.58%	1.067±1.37%	0.014	0.122

Table 5. Thermal properties values and samples thickness as a function of firing temperature with its standard deviation

Thermal diffusivity values ranged from 0.0016 to 0.0063 cm^2/s and are in close agreement with the literature (Alexandre *et al.*, 1999; García *et al.*, 2002). Table 5 presents values for specific heat capacity that were around 1.0 (J/cm^3K). Thus, we can infer that thermal conductivity and effusivity tend to follow the thermal diffusivity profile in relation to the firing temperature. Moreover, for this kind of material, thermal properties are strongly dependent on the material microstructure.

In relation to the structure, Fig. 1 shows the evolution of the X-ray diffraction patterns as a function of firing temperature. For natural clay the diffractogram reveals predominant phases of kaolinite, quartz and small amounts of gibbsite, illite, orthoclase and anatase. Following the treatment at 900 °C, illite, quartz, anatase and orthoclase remain unchanged and a hematite phase is revealed. Treatments at 950 °C or higher cause sample recrystallization in the Al-Si spinel (Okada *et al.*, 1986) and a liquid phase formation. Data clearly show the evolution of mullite and cristobalite in samples over the temperatures 1060 °C and 1100 °C.

Analysing the thermal diffusivity and the diffractograms, we can explain that the lower diffusivity value in 980 °C and 1020 °C is possibly due to the Al-Si rearrangements, owing to a crystallization process followed by a lattice formation (Okada *et al.*, 1986), according to

$$2Al_2(OH)_4Si_2O_5 \xrightarrow{500^\circ C} 2Al_2Si_2O_7 + 4H_2O \xrightarrow{980^\circ C} Al_4Si_3O_{12} + SiO_2. \tag{9}$$

Equation 9 shows kaolinite $2Al_2(OH)_4Si_2O_5$ evolution, which reaches the spinel phase $Al_4Si_3O_{12} + SiO_2$ formation, with an intermediate phase, namely metakaolin $2Al_2Si_2O_7 + 4H_2O$. The high diffusivity value at 1060 °C could possible be connected to the liquid phase formation within the beginning of the vitreous phase. The heat can be transported easier in this kind of phases than through the clay mineral grains. In support of this explanation for the highest thermal diffusivity, we note the clear modification of structure shown to occur above 1060 °C.

At 1100 °C, the thermal diffusivity decreases as explained by the reaction.

$$Al_4Si_3O_{12} \xrightarrow{\approx 1100^\circ C} \frac{2}{3}[3Al_2O_3 2Si_2] + \frac{5}{3}SiO_2 . \tag{10}$$

Through sintering at this temperature, formation of mullite $2/3[3Al_2O_3 2Si_2]$ and cristobalite $5/3\,SiO_2$ starts from the spinel phase $Al_4Si_3O_{12}$, as reported by references in (Okada *et al.*, 1986). We suggest that there is a reduction of liquid phase and pores, with the formation of cristobalite, which probably is responsible for the thermal diffusivity decay.

Summarizing, the red clay from Campos dos Goytacazes-RJ region are used mainly for bricks and roofing tiles. In general, ceramic plants fire their products at temperatures below 900 °C. In the raw state the red clay, as shown in Fig. 1, is composed of a high proportion of kaolinite clay mineral besides accessory minerals including oxy-hydroxide, illite, quartz, and anatase. In this case sintering is dominated by particle-to-particle contact mainly of metakaolinite platelets, resulting in a more open structure and also ceramic products of low quality. High porosity results in a lower thermal diffusivity and thermal conductivity. On the other hand, the sintering accelerates above 950 °C with drastic microstructural changes due to the formation of a vitreous phase, resulting in better densification behaviour of the red clay. As a consequence, the thermal diffusivity increases abruptly. However, a decrease in the thermal diffusivity occurred between 1060 °C - 1100 °C. This is due mainly to formation of mullite, which is characterized by a high proportion of oxygen vacancies that lead to low thermal diffusivity. It will be reasonable then to consider that the optimum firing temperature range for the studied red clay is above 950 °C.

Temperature (°C)	Thermal diffusivity (cm^2/s)	Specific heat capacity (J/cm^3 K)	Thermal conductivity (W/cm K)	Thermal effusivity ($Ws^{1/2}/cm^2$ k)
900	0.003 ± 0.0004	0.61 ± 0.11	**0.002 ± 0.00**	0.034 ± 0.0013
950	0.005 ± 0.0000	0.83 ± 0.02	**0.004 ± 0.00**	0.061 ± 0.0005
1000	0.007 ± 0.0000	0.82 ± 0.13	**0.006 ± 0.00**	0.072 ± 0.0008
1050	0.004 ± 0.0000	0.92 ± 0.09	**0.004 ± 0.00**	0.061 ± 0.0001
1100	0.004 ± 0.0002	0.67 ± 0.11	**0.003 ± 0.01**	0.045 ± 0.0010

Table 6. Thermal properties of the J clay as in function of some possible sintering temperatures

Temperature (°C)	Thermal diffusivity (cm^2/s)	Specific heat capacity (J/cm^3 K)	Thermal conductivity (W/cm K)	Thermal effusivity ($Ws^{1/2}/cm^2$ k)
900	0.008 ± 0.0001	0.98 ±0.00	**0.007 ± 0.00**	0.088 ± 0.0011
950	0.004 ± 0.0005	1.12 ± 0.15	**0.005 ± 0.11**	0.072 ± 0.0008
1000	0.009 ± 0.0008	1.57 ± 0.04	**0.014 ± 0.08**	0.148 ± 0.0005
1050	0.006 ± 0.0004	1.43 ± 0.17	**0.008 ± 0.06**	0.111 ± 0.0010
1100	0.004 ± 0.0009	1.25 ± 0.02	**0.005 ±0.09**	0.082 ± 0.0009

Table 7. Thermal properties of the J clay with 30% sanitary ware mass residues as in function of some possible sintering temperatures

Tables 6 and 7 show the thermal properties of the J clay and J clay with 30% residue, respectively.
Regarding to the ceramic thermal properties, the CG clay ceramic sample presented much higher values than the J clay ceramic. Thus, depending on the application, considering thermal properties we can use a kaolinite raw material if we need ceramics with fast thermalization or illitic clays for a better thermal isolation. We noted that the aggregation of 30% of sanitary ware mass residue in the J clay matrix generates higher values of the ceramic thermal properties.

6. Pollutant gases from the firing of clay ceramic added with residues

The gas emission concentrations during the CG and J clays firing process were measured. The figures were displayed in such way that it is possible to compare the whole samples for each gas, obeying the CG clay samples and the J clay samples.

6.1 Experimental procedure

The following materials were studied in this present research: **a**) the CG clay and steel slag considered as a waste by the National Steel Plant (CSN) located at the south of the state. This slag was generated during the conventional oxygen blowing, LD (Linz Donawitz) process, of steel refining; **b**) The J clay and the sanitary ware mass rejects from a ceramic industry of São Paulo state.
The emitted gases from the firing process were quantitatively measured by means of a photothermal technique.

The gas released from the furnace was directly connected to an infrared model URAS 14, ABB gas analyser under a suction flux of 0.3 L/min This gas analyser detected and quantified simultaneously CO, CO_2, CH_4, NO, N_2O, NH_3 e SO_2. Gas samples were collected 20 min. after settled temperature stages of respectively, 150, 300, 450, 550, 650, 800, 950, 1050 and 1100 °C. The samples were brought to room temperature following the inertial cooling of the resistive furnace.
The gas analyser measurement process is based on resonance absorption at the characteristic vibrational rotation spectrum bands of non-elemental gases in the middle infrared range between 2 μm and 12 μm. Because of their bipolar moment, the gas molecules interact with infrared emissions. For selectivity, the receiver is filled with the applicable sample components to establish reference and sensitivity to these components (~1 ppm). It consists of the cell divided into two identical compartments: one in the measuring cell, through which the sampled gas is flowed, and the other acts as reference, filled with nitrogen. The light emitted from a hot filament is modulated by a mechanical chopper and divided by a beam splitter. Each beam goes simultaneously through the measuring cell and the reference cell. The detector consists of two sealed chambers separated by a diaphragm capacitor. Both chambers are filled with pure gas of the chemical species under study. The light beams emerged from the sample and reference cells reach independently the two detector chambers, causing a differential pressure that is proportional to the light absorption by the sample. The pressure difference is converted by the diaphragm capacitor into an electrical signal. A detailed description of the Uras performance can be found in the literature (11). Before each sample analyses the cells were calibrated using pure standard N_2.

6.2 Results and discussion

Figures 14 and 15 show the released CO_2 and CO gases as in function of the temperature from the CG clay with (5%, 10%) and without (0%) steel slag residue in different concentrations in weight. The other gases had not presented significant values.

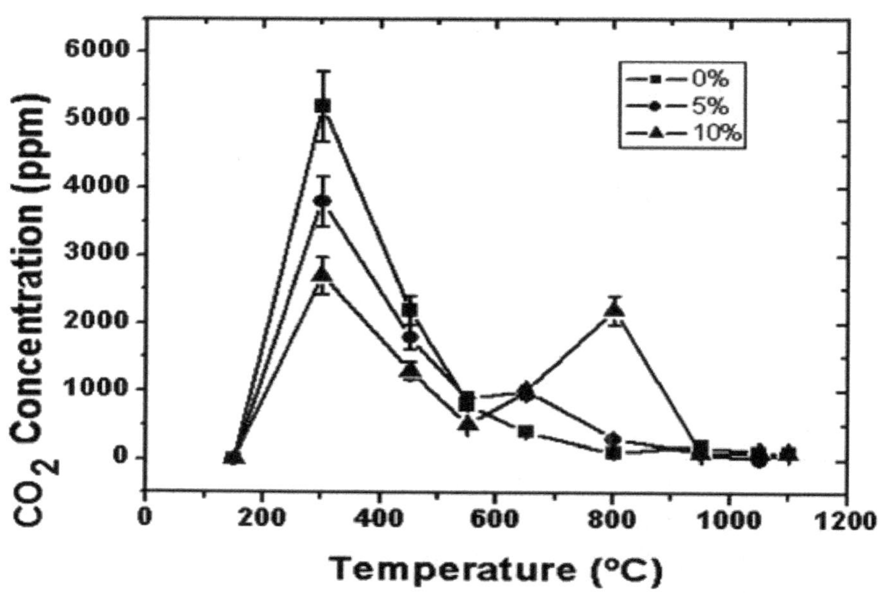

Fig. 14. CO_2 profiles as in function of the firing temperature for the samples with 0 %, 5 %, and 10 %

CO_2 concentration values were emitted more significantly in the presence of the residue. Generally, carbon composites, as CO_2, in the range of 300 °C to 500 °C are emitted due to the organic matter oxidation. Moreover, the reaction of the kaolinite and goethite dehydroxilation deals to a metakaolinite, that is, an amorphous phase.

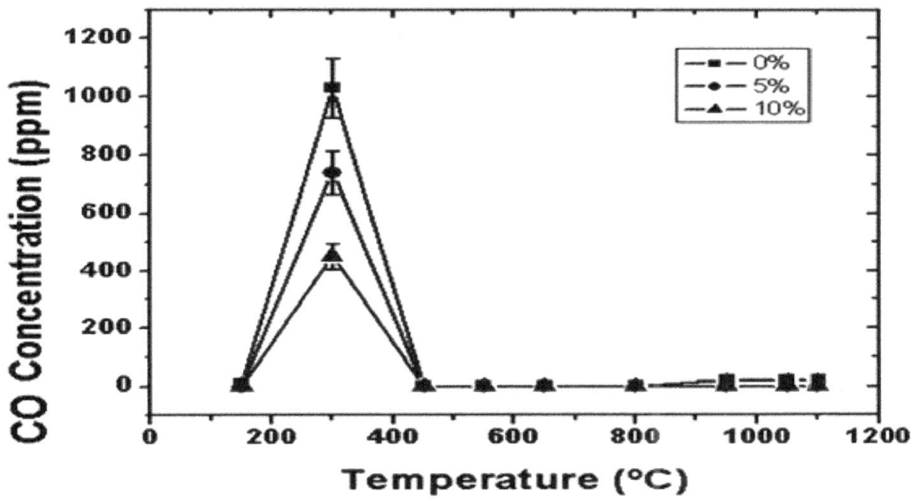

Fig. 15. CO profiles as in function of the firing temperature for the samples with 0 %, 5 %, and 10 %

The CO_2 emission in temperatures higher than 850 °C, is possibly due to: the hydroxil (OH) release, proceeding from the gases that were locked up in the pores, which are set free later when the pores begin processes of volumetric reduction, due to the high temperature; the chemical dehydration of minerals that contains mica (muscovite); and the calcite decomposition ($CaCO_3 = CaO+CO_2$). Values reached in this section were just proportional to a laboratory furnace.

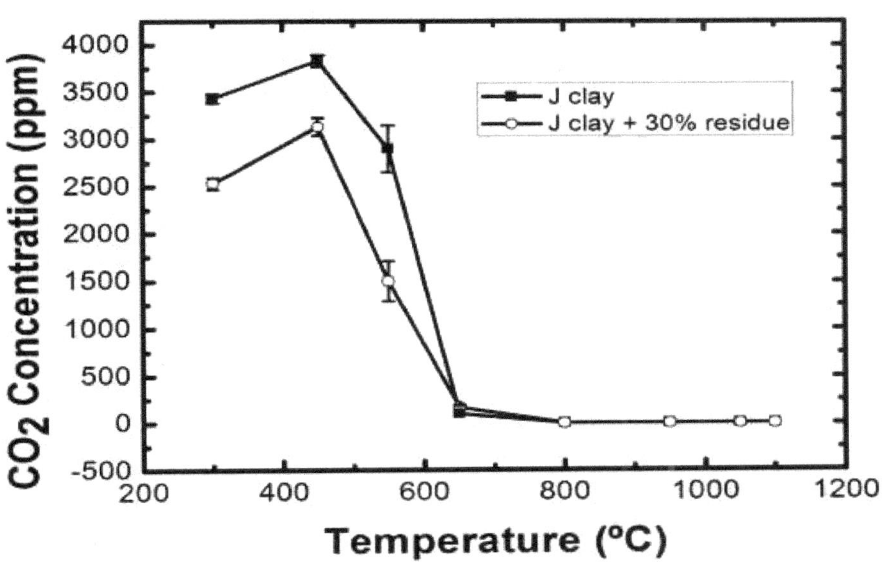

Fig. 16. CO_2 emmited during the ceramic firing process as in function of the firing temperature

Figures 16 and 17 show the CO_2 and CO emissions as in function of firing temperature from the J clay without and with 30% in weight of sanitary ware mass residue. Firstly, we observed that the addition of the residue diminishes the amount of CO_2 emission. The amount of CO_2 emitted between 300 and 550 °C has the same explanation previously cited, that is, due to organic matter oxidation and the dehydroxilation of clay minerals.

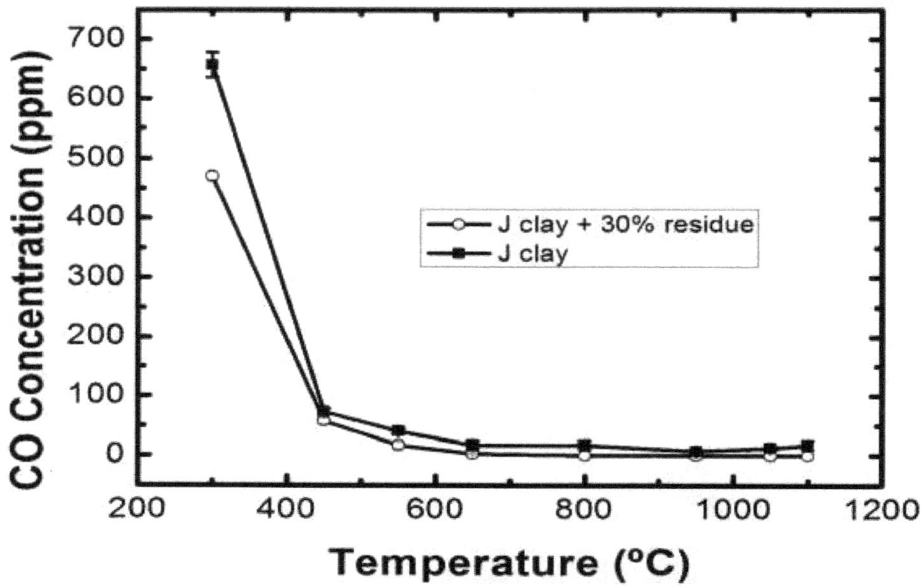

Fig. 17. CO emmited during the ceramic firing process as in function of the firing temperature

7. Conclusions

Here, red ceramic samples from Rio de Janeiro state and from São Paulo state were analysed as in function of firing temperature. In order to study the ceramics quality, technological essays were prepared. Comparisons between the Kaolinite CG clay and the illitic J clay were made, indicating the advantages and disadvantages of each specimen.

The thermal property of red ceramic samples as well as their crystalline phases were analysed as a function of sintering temperature. Photothermal techniques were used to measure the thermal properties and the microstructures were identified by x-ray diffraction, in order to compare structural and thermal features of the samples. It was shown that there is an intrinsic relationship between the structure and the thermal diffusivity. When the crystalline structure undergoes strong perturbations the thermal diffusivity values decrease and vice-versa. When liquid phases achieve the highest concentration at around 1060 °C (CG clay) and 1000 °C (J clay), the thermal diffusivity shows its maximum peak. Afterwards, there is a decrease, although the thermal diffusivity maintains high values, due to mullite enhancement and vitreous phase consolidation in the red ceramics. The J clay sintering temperature is lower than to the CG clay, due to a larger amount of flux material.

Thermal property values are consistent with the literature. As the specific heat capacity has values of approximately unity, thermal conductivity and effusivity follow that of thermal diffusivity. Considering thermal properties, it was shown that red ceramic sintering temperatures should be greater than 950 °C, but below 1060 °C from both clays.

The steel slag investigated, considered as a steelmaking waste, has an elevated amount of Ca, Fe, Mg and Si. The main crystalline phases identified were magnetite, Fe_3O_4, calcite, $CaCO_3$, magnesium manganese oxide, Mg_6MnO_8 and calcium silicate, Ca_2SiO_4, $CaSiO_3$. The incorporation of steel slag into clayey ceramics, must be done in amounts of up to 10 wt.%. In relation to the steel slag incorporation, it had been mainly detected CO and CO_2. Comparing the diverse percentages of residues into the ceramic mass, it observed that the CO_2 concentrations decrease in the range of 300 °C to 500 °C, when increased the percentage of residue to the mass. On the other side, between 500 °C and 950 °C, there was a reasonable increase of the concentration (in relation to the pure clay), when the residues percentage was enlarged. The calcite starts to decompose around 675 °C and its complete decomposition occurs up to 950 °C. When mixed to another minerals, the calcite presents lower temperature of decomposition (880 °C, approximately).

It is possible that sanitary ware residues can attenuate the pollutant gas release from the ceramic firing. We concentrate in this chapter the CO and CO_2 gases, in order to facilitate the point of view of our study. It will appear elsewhere the complete set of measured gases from this research.

8. Acknowledgment

The authors thank the Brazilian agencies FAPERJ, CNPq and CAPES for the financial support.

9. References

Alexandre, J.; Saboya, F.; Marques, B. C.; Ribeiro, M. L. P.; Salles, C., da Silva, M. G.; Sthel, M. S., Auler, L. T. & Vargas, H. (1999). Photoacoustic thermal characterization of kaolinite clays. The *Analyst*, vol. 124, pp. 1209 –1214.

Almond, D.P., & Patel, P.M. (1996). *Photothermal Science and Techniques*. Chapman & Hall, London.

Dubois, J.; Murat, M.; Amroune, A.; Carbonneau, X. & Gardon, R. (1995). High-temperature transformation in kaolinite: the role of the crystallinity and of the firing atmosphere. *Appl. Clay Science,* vol. 10, pp. 187–198.

Faria Jr., R.T.; Filgueira, M.; Esquef, I.A.; Machado, F.A.; Rodrigues, M.F.; Bobrovnitchii, G..S. & Vargas, H. (2005). Thermal characterization of sintered hardmetal. *Journal de Physique. IV* , vol. 125,pp. 237–239.

García, E., Osendi, M.I., Miranzo, P., 2002. Thermal diffusivity of porous cordierite ceramic burners. Journal of Applied Physics 92, 2346–2349.

Heydorn, K. & Thuesen, I. (1989). Classification of ancient Mesopotamian ceramics and clay using SIMCA for supervised pattern recognition. *Chemometrics and Intelligent Laboratory Systems,* vol.7, pp. 181-188.

Manhães, R. S. T.; Auler, L. T.; Sthel, M. S.; Alexandre, J.; Massunaga, M. S. O.; Carrió, J. G.; dos Santos, D. R.; da Silva, E. C.; Garcia-Quiroz, A. & Vargas, H. (2002). Soil characterisation using X-ray diffraction, photoacoustic spectroscopy and electron paramagnetic resonance. *Appl. Clay Science*, vol. 21, pp. 303.

Mansanares, A.M.; Bento, A.C.; Vargas, H., Leite, N.F. & Miranda, L.C.M. (1990). Photoacoustic measurement of the thermal properties of two-layer systems. *Physical Review B*, vol. 42, pp. 4477–4486.

Marín, E.; Marín-Antuna, J. & Díaz-Arencibia, P. (2002). On the wave treatment of the conduction of heat in photothermal experiments with solids. *European Journal of Physics*, vol. 23, pp. 523–532.

Morgan, D. J. (1993). Thermal analysis including evolved gas analysis of clay raw materials, *Applied Clay Science*, vol. 8, pp. 81.

Mota, L.; Toledo, R. ; Machado, F.A.L.; Holanda, J.N.F.; Vargas, H. & Faria Jr., R.T. (2008). Thermal characterization of red clay from the Northern Region of Rio de Janeiro State, Brazil using an open photoacoustic cell, in relation to structural changes on firing. *Appl. Clay Science, vol.* 42, pp. 168–174.

Nastro, V.; Vuono, D.; Guzzo, M.; Niceforo, G.; Bruno, I. & de Luca, P. (2006). Characterisation of raw materials for production of ceramics. Journal of Thermal Analysis Calorimetry, vol. 84, pp. 181–184.

Okada, K.; Otsuka, N. & Ossaka, J. (1986). Characterization of spinel phase formed in the kaolin-mullite thermal sequence. *J. Am. Ceram. Soc.*, vol. 69, pp. C251-C253.

Oliveira, G.E., & Holanda, J.N.F., (2004). Utilization of solid waste from siderurgical industry in red ceramic, *Cerâmica*, vol. 50, pp. 75-80.

Parsons, A. J.; Inglethorpe, S. D. J.; Morgan, D. J. & Dunham, A. C. (1997). Evolved gas analysis (EGA) of bricks clays. *J. Therm. Analysis*, vol. 48, pp.49.

Perondi, L.F. & Miranda, L.C.M. (1987). Minimal volume photoacoustic cell measurement of thermal diffusivity: effect of the thermoelastic sample bending. *Journal of Applied Physics* , vol. 62, pp. 2955–2959.

Reed J.S. (1976). *Principles of the Ceramics Processing*. John Wiley and Sons, New York.

Rice, R.W. & Dekker, M. (1998). *Porosity of Ceramics,* Elsevier, New York

Rosencwaig, A. & Gersho, A. (1976). Theory of the photoacoustic effect with solids. *Journal of Applied Physics, vol.* 47, pp. 64–69

Souza, V.P.; Toledo, R.; Holanda, J.N.F.; Vargas, H. & Faria Jr, R.T. (2008). Pollutant gas analysis evolved during firing of red ceramic incorporated with water treatment plant sludge. *Cerâmica*, vol. 54, pp. 351-355.

Vargas, H. & Miranda, L.C.M. (1988). Photoacoustic and related photothermal techniques. *Physics Reports, vol.* 161, pp. 43–101.

Vieira, C.M.F. & Monteiro, S.N. (2009). Incorporation of solid wastes in red ceramics: an updated review. *Matéria (Rio J.)*, vol.14, nº .3, pp. 881-905. ISSN 1517-7076.

Yunus, W.M.M.; Fanny, C.Y.J.; Phing, T.E.; Mohamed, S.B.; Halim, S.A. & Moskin, M.M. (2002). Thermal diffusivity measurement of Zn, Ba, V, Yand Sn doped Bi–Pb–Sr–Ca–Cu–O ceramics superconductors by photoacoustic technique. *Journal of Material Science*, vol. 37, pp. 1055–1060.

Permissions

All chapters in this book were first published in ACCRMPPDH, by InTech Open; hereby published with permission under the Creative Commons Attribution License or equivalent. Every chapter published in this book has been scrutinized by our experts. Their significance has been extensively debated. The topics covered herein carry significant findings which will fuel the growth of the discipline. They may even be implemented as practical applications or may be referred to as a beginning point for another development.

The contributors of this book come from diverse backgrounds, making this book a truly international effort. This book will bring forth new frontiers with its revolutionizing research information and detailed analysis of the nascent developments around the world.

We would like to thank all the contributing authors for lending their expertise to make the book truly unique. They have played a crucial role in the development of this book. Without their invaluable contributions this book wouldn't have been possible. They have made vital efforts to compile up to date information on the varied aspects of this subject to make this book a valuable addition to the collection of many professionals and students.

This book was conceptualized with the vision of imparting up-to-date information and advanced data in this field. To ensure the same, a matchless editorial board was set up. Every individual on the board went through rigorous rounds of assessment to prove their worth. After which they invested a large part of their time researching and compiling the most relevant data for our readers.

The editorial board has been involved in producing this book since its inception. They have spent rigorous hours researching and exploring the diverse topics which have resulted in the successful publishing of this book. They have passed on their knowledge of decades through this book. To expedite this challenging task, the publisher supported the team at every step. A small team of assistant editors was also appointed to further simplify the editing procedure and attain best results for the readers.

Apart from the editorial board, the designing team has also invested a significant amount of their time in understanding the subject and creating the most relevant covers. They scrutinized every image to scout for the most suitable representation of the subject and create an appropriate cover for the book.

The publishing team has been an ardent support to the editorial, designing and production team. Their endless efforts to recruit the best for this project, has resulted in the accomplishment of this book. They are a veteran in the field of academics and their pool of knowledge is as vast as their experience in printing. Their expertise and guidance has proved useful at every step. Their uncompromising quality standards have made this book an exceptional effort. Their encouragement from time to time has been an inspiration for everyone.

The publisher and the editorial board hope that this book will prove to be a valuable piece of knowledge for researchers, students, practitioners and scholars across the globe.

List of Contributors

G.M. Revel, E.P. Tomasini, G. Pandarese and A. Cavuto
Università Politecnica delle Marche, Ancona, Italia
Dipartimento di Meccanica, Ancona, Italia

Bandar M. A. AL-Makramani
Department of Conservative Dentistry, Faculty of Dentistry, Sana'a University, Yemen

Abdul A. A. Razak
Department of Conservative Dentistry, Malaysia
Faculty of Dentistry, Malaysia
University of Malaya, Kuala Lumpur, Malaysia

Mohamed I. Abu-Hassan
Department of Restorative Dentistry, Malaysia
Faculty of Dentistry, Malaysia
University Teknology MARA, Selangor, Malaysia

Jaime Vite-Torres, María del Carmen Carreño de León, Manuel Vite-Torres and Juan Rodrigo Laguna-Camacho
Instituto Nacional de Investigaciones Nucleares (ININ), México
SEPI-ESIME-Campus Zacatenco-IPN, México
Universidad Veracruzana, Campus Poza Rica, Veracruz, México

Pavel Ctibor
Institute of Plasma Physics, ASCR, v.v.i., Prague,

Josef Sedlacek
Department of Electrotechnology, Prague
Faculty of Electrical Engineering, Prague
Czech Technical University, Prague , Czech Republic

Anatoly Fishman
Institute of Metallurgy, Ural Division, Russian Academy of Sciences, Russia

Tatyana Kurennykh and Vladimir Vykhodets
Institute of Metal Physics, Ural Division, Russian Academy of Sciences, Russia

Evgeniya Vykhodets
Ural Federal University, Yekaterinburg, Russia

Xavier Gustavo de Castro, Saboya Fernando, Maia Paulo Cesar de Almeida and Alexandre Jonas
State University of Norte Fluminense Darcy Ribeiro - UENF, Department of Civil Engineering, Campos, Rio de Janeiro, Brazil

Zdeněk Jonšta, Evelyn A. Bolaňos C., Monika Hrabalová and Petr Jonšta
VŠB-Technical University of Ostrava, Czech Republic

Igor Štubňa and Libor Vozár
Department of Physics, Constantine the Philosopher University, Nitra, Slovakia

Anton Trník
Department of Physics, Constantine the Philosopher University, Nitra, Czech Republic
Department of Materials Engineering and Chemistry, Czech Republic
Czech Technical University, Prague, Czech Republic

František Chmelík
Department of Physics of Materials, Charles University Prague, Prague, Czech Republic

Gang Deng
Faculty of Engineering, University of Miyazaki, Japan
State Key Lab of Mechanical Transmission, Chongqing University, China

Tsutomu Nakanishi
Faculty of Engineering, University of Miyazaki, Japan

Laima Trinkler and Baiba Berzina
Institute of Solid State Physics, University of Latvia, Latvia

Faria Jr. R. T., Souza V. P., Toledo R. and Vargas H.
Laboratório de Ciências Físicas, Parque Califórnia, Campos dos Goytacazes, RJ Brasil

Monteiro S. N., Holanda J. N. F. and Vieira C. M. F.
Laboratório de Materiais Avançados, Parque Califórnia, Campos dos Goytacazes, RJ Brasil

Index

U

W

X

www.ingramcontent.com/pod-product-compliance
Lightning Source LLC
LaVergne TN
LVHW081034210926
848485LV00004B/1023